CU00766604

Methodological Advances in Experimental Philosophy

Advances in Experimental Philosophy

Series Editor:

James Beebe, Professor of Philosophy, University at Buffalo, USA

Editorial Board:

Joshua Knobe, Yale University, USA

Edouard Machery, University of Pittsburgh, USA

Thomas Nadelhoffer, College of Charleston, UK

Eddy Nahmias, Neuroscience Institute at Georgia State University, USA

Jennifer Cole Wright, College of Charleston, USA

Joshua Alexander, Siena College, USA

Empirical and experimental philosophy is generating tremendous excitement, producing unexpected results that are challenging traditional philosophical methods. *Advances in Experimental Philosophy* responds to this trend, bringing together some of the most exciting voices in the field to understand the approach and measure its impact in contemporary philosophy. The result is a series that captures past and present developments and anticipates future research directions.

To provide in-depth examinations, each volume links experimental philosophy to a key philosophical area. They provide historical overviews alongside case studies, reviews of current problems and discussions of new directions. For upperlevel undergraduates, postgraduates and professionals actively pursuing research in experimental philosophy, these are essential resources.

Titles in the series include:

Advances in Experimental Epistemology, edited by Professor James R. Beebe

Advances in Experimental Moral Psychology, edited by Hagop Sarkissian and Jennifer Cole Wright

Advances in Experimental Philosophy of Language, edited by Jussi Haukioja

Advances in Experimental Philosophy of Mind, edited by Justin Sytsma

Advances in Religion, Cognitive Science, and Experimental Philosophy, edited by Helen De Cruz and Ryan Nichols

Methodological Advances in Experimental Philosophy

Edited by
Eugen Fischer and Mark Curtis

BLOOMSBURY ACADEMIC
LONDON • NEW YORK • OXFORD • NEW DELHI • SYDNEY

BLOOMSBURY ACADEMIC
Bloomsbury Publishing Plc
50 Bedford Square, London, WC1B 3DP, UK
1385 Broadway, New York, NY 10018, USA

BLOOMSBURY, BLOOMSBURY ACADEMIC and the Diana logo are
trademarks of Bloomsbury Publishing Plc

First published in Great Britain 2019
Paperback edition published 2021

Copyright © Eugen Fischer and Mark Curtis, 2019

Eugen Fischer and Mark Curtis have asserted their right under the Copyright,
Designs and Patents Act, 1988, to be identified as Editors of this work.

Series design by Catherine Wood
Cover image © Dieter Leistner / Gallerystock

All rights reserved. No part of this publication may be reproduced or transmitted in
any form or by any means, electronic or mechanical, including photocopying,
recording, or any information storage or retrieval system, without prior
permission in writing from the publishers.

Bloomsbury Publishing Plc does not have any control over, or responsibility for, any
third-party websites referred to or in this book. All internet addresses given
in this book were correct at the time of going to press. The author and publisher
regret any inconvenience caused if addresses have changed or sites have ceased
to exist, but can accept no responsibility for any such changes.

A catalogue record for this book is available from the British Library.

A catalog record for this book is available from the Library of Congress.

ISBN: HB: 978-1-3500-6899-5
        PB: 978-1-3501-9039-9
        ePDF: 978-1-3500-6900-8
        eBook: 978-1-3500-6901-5

Series: Advances in Experimental Philosophy

Typeset by Integra Software Services Pvt. Ltd.

To find out more about our authors and books visit www.bloomsbury.com
and sign up for our newsletters.

Contents

List of Illustrations

Figures

Tables

List of Contributors

Lara Alcock – Mathematics, Loughborough University, UK

Mark Alfano – Philosophy, Delft University of Technology, Netherlands/ Australian Catholic University, Brisbane, Australia

Adrian J. T. Alsmith – Psychology, Institut Jean Nicod, Paris, France

Hein van den Berg – Philosophy, Institute of Logic, Language and Computation, University of Amsterdam, Netherlands

Arianna Betti – Philosophy, Institute of Logic, Language and Computation, University of Amsterdam, Netherlands

Roland Bluhm – Philosophy and linguistics, Independent Scholar, Berlin, Germany

Justin Bruner – Philosophy, University of California, Irvine, CA, USA

Mark Curtis – Philosophy, University of East Anglia/Anglia Ruskin University, Cambridge, UK

Rodrigo Díaz – Philosophy, University of Bern, Switzerland

Michelle DiBartolo – Psychology, John Hopkins University, Baltimore, MD, USA

Paul E. Engelhardt – Psychology, University of East Anglia, Norwich, UK

Eugen Fischer – Philosophy, University of East Anglia, Norwich, UK

Annie Galizio – Psychology, Utah State University, Logan, UT, USA

Andrew Higgins – Philosophy, Illinois State University, Normal, IL, USA

Matthew Inglis – Mathematical Cognition, Loughborough University, UK

Kristen Lew – Mathematics, Texas State University, San Marcos, TX, USA

Matthew R. Longo – Cognitive Neuroscience, Birkbeck, University of London, UK

Shaun Nichols – Philosophy, University of Arizona, Tucson, AZ, USA

Cailin O'Connor – Logic and Philosophy of Science, University of California, Irvine, CA, USA

Yvette Oortwijn – Philosophy, Institute of Logic, Language and Computation, University of Amsterdam, Netherlands

Paolo Rago – Independent data manipulation consultant, São Paulo, Brazil

Juan Pablo Mejía-Ramos – Mathematics, Rutgers University, New Brunswick, NJ, USA

Evan Reinhold – Psychology, College of Charleston, SC, USA

Kevin Reuter – Philosophy, University of Bern, Switzerland

Hannah Rubin – Philosophy, University of California, Irvine, CA, USA

Chris Sangwin – Science Education, University of Edinburgh, UK

Justin Sytsma – Philosophy, Victoria University of Wellington, New Zealand

Caspar Treijtel – Librarian, University of Amsterdam, Netherlands

Pascale Willemsen – Philosophy, Ruhr-University Bochum, Germany

Jennifer Cole Wright – Psychology and Philosophy, College of Charleston, SC, USA

Introduction

Eugen Fischer and Mark Curtis

Experimental philosophy is undergoing exciting transformations. Until recently, the discipline has prominently been associated with the questionnaire-based study of philosophically relevant intuitions (Alexander 2012; Fischer and Collins 2015; Mallon 2016). However, experimental philosophy – conceived, quite broadly, as the practice of systematically collecting and analysing empirical data to address philosophical questions or problems (Sytsma and Livengood 2016; *cf.* Knobe and Nichols 2017) – adapts a wide range of empirical methods for a variety of new uses in philosophy, in the study of intuitions and beyond (*cf.* Nado 2016). To some extent, the recruitment of methods 'beyond the questionnaire' has been part of experimental philosophy since the inception of the movement (e.g. Greene et al. 2001). Currently, however, the uptake of further methods is gathering momentum: this might mark a step change in the larger enterprise of making philosophical questions and problems more empirically and scientifically tractable.

Experimental philosophy's new methods include sophisticated paradigms for behavioural experiments from across the social sciences – ranging from psycholinguistics to experimental economics – as well as computational methods from the digital humanities that can process large bodies of text and data. This volume offers a conspectus of these cutting-edge empirical methods and explores how they can complement questionnaire-based approaches in a variety of philosophical pursuits, across several key areas. The studies reported or discussed herein demonstrate – in some cases pioneer – the use of their chosen methods in experimental philosophy. The areas of philosophy covered include epistemology and metaphysics, the philosophies of mind and language, the philosophies of science and mathematics, moral philosophy and psychology, as well as the history of ideas.

The questionnaire-based surveys and experiments that have so far been the standard tools of experimental philosophy are convenient, can demonstrably be used to address a very wide range of philosophically relevant questions (see Alexander 2012 and Knobe & Nichols 2017 for reviews) and have generally led to replicable findings (Cova et al. 2018). This volume will make the case for complementing these familiar and useful approaches not by belabouring their limitations and shortcomings (which the following chapters will broach, where relevant), but by positively demonstrating what the alternative methods allow us to do, and the philosophical uses to which their findings can profitably be put.

The first part of the volume focuses on experimental methods drawn from different social sciences. The second part examines computational methods from the digital humanities. In each part, the first chapters discuss how further empirical methods can be put to the service of (natural extensions of) philosophical projects that have been prominently associated with experimental philosophy and rely on or respond to intuition-driven methodologies. In line with a broad conception of experimental philosophy, subsequent chapters then explain how further methods can be recruited for philosophical projects that have not seen much reliance on intuitions. All the methods discussed can be profitably put to more philosophical uses than are presented here, including uses in other areas of philosophy. We therefore encourage readers to also consider chapters which address philosophical topics outside their own immediate areas of interest.

Each chapter combines presentation, demonstration and discussion of a chosen method or set of methods with an explanation and assessment of how the methods' application can contribute to traditional philosophical concerns, or to ongoing philosophical debates, or can open up new philosophical research questions. Usually, the demonstration and discussion of the philosophical application will include the report of a new (previously unpublished) empirical study. Contributions provide accessible explanations of the methods used. They do not assume prior familiarity with the methods presented, but convey the methods' potential, workings, strengths and limitations. The chapters thus address experimental philosophers who would like to expand their methodological repertoire as well as further philosophers at different levels of the profession – including advanced undergraduates – who are curious about how a new range of empirical methods can be put to the service of philosophical projects. Each chapter is appended with some suggested readings on the introduced methodology (which precede the bibliographies).

Chapter outlines

The empirically informed conceptions of intuitions that are prevalent in experimental philosophy conceptualize them, in slightly different ways, as judgements generated by largely automatic cognitive processes (Gopnik and Schwitzgebel 1998; Fischer and Collins 2015; *cf.* De Cruz 2015). An early 'Experimental Philosophy Manifesto' already advocated examining the evidentiary value of philosophically relevant intuitions by studying their internal sources (Knobe and Nichols 2008, 8), i.e. the automatic cognitive processes that generate them. A proper understanding of their workings might help us develop 'epistemological profiles' for such cognitive processes, which tell us when and why we may (or may not) trust their outputs (Weinberg 2015). It is then natural to extend the scope of the inquiry: to ask how automatic cognition shapes philosophical thought, more generally, for better or worse, and to investigate through what representations and processes automatic cognition shapes inferences, judgements, decisions and arguments in thought that is relevant either as the medium or the topic of philosophical reflection. The first five chapters explore how new empirical methods can be recruited for these purposes.

In *Chapter 1*, Shaun Nichols uses a *statistical learning* approach employed in different branches of cognitive science, to examine how children acquire concepts and distinctions which continue to shape automatic cognition – as manifest in judgement and reasoning under cognitive load – in adults. The studies discussed provide evidence for a surprisingly demanding, infallibilist, concept of knowledge and for remarkably complex rules which conform to the principle of double effect and inform adults' distinctions between moral and immoral actions. This approach allows us to understand, for example, why an infallibilist concept of knowledge might inform some sceptical arguments. However, unlike familiar debunking accounts, the statistical learning paradigm also allows us to understand why it is perfectly rational for learners to initially acquire the philosophically troublesome concepts and rules.

In *Chapter 2*, Eugen Fischer and Paul Engelhardt introduce several *experimental methods from psycholinguistics* and explain how they can be used to examine automatic inferences and assess intuitions and arguments. One recent and one new eye-tracking study pioneer the use of *pupillometry* and *reading-time measurements*, respectively, in experimental philosophy. They examine automatic inferences that constantly occur in language comprehension – such as when we hear or read verbal case descriptions or premises of arguments. They serve to study how stereotypes (implicit knowledge structures built up subconsciously through sustained interaction with the physical and discourse environment) are deployed in understanding and facilitate automatic inferences from verbal case descriptions. The eye-tracking studies contribute to an epistemological profile of the key comprehension process of stereotypical enrichment, provide evidence of a cognitive bias besetting it and deploy results to expose seductive fallacies in a philosophical paradox about perception.

In *Chapter 3*, Jennifer Cole Wright and colleagues use different *eye-tracking methods*, which work with visual scenes (rather than text), to examine how people form and deploy their moral judgements. Two new studies elicit moral judgements through a text-based task and then present visual scenes in which people get helped or harmed by the actions previously judged right or wrong. By measuring gaze-fixation frequency, gaze duration and scan path on these visual scenes, the studies examine whether participants attend more to 'beneficiaries' or 'victims' of their judgements. Findings support the hypothesis that people reinforce their moral judgements through subsequent patterns of sensory processing and suggest that we subconsciously employ approach-and-avoidance strategies that have developed to maintain optimal mental health and well-being – but which need not help us correct errors or make good judgements.

In *Chapter 4*, Rodrigo Díaz discusses the use of *fMRI neuroimaging* techniques in two philosophical projects: (i) to explain and assess moral intuitions and (ii) to assess the empirical adequacy of extant mental (viz. emotion) categories. The first project seeks to show that non-utilitarian judgements derive from emotional distortions in rational thought. It uses fMRI to identify mental processes (emotions) involved in a moral judgement task. This move relies on 'reverse inference' from observed brain activation during an experimental task to the conclusion that the task involves a mental process previously found associated with such brain activity. The second project involves ontology testing which deploys the logic 'same pattern of brain activity equals same mental processes' in meta-analysis of fMRI studies. The chapter assesses these two inference strategies, finds the first problematic and promotes the second.

In *Chapter 5*, Adrian Alsmith and Matthew Longo review the use of *Virtual Reality (VR) technology* to study people's use of the self-concept. Their topic is the flexibility of people's self-conception: to what extent people can be induced to apply the self-concept to different kinds of things ('I am a physical body' vs 'I am a non-physical soul') and to credit themselves with properties attributable to distinct entities in similar contexts ('I will die one day' vs 'I am immortal') – a topic of conflicting philosophical intuitions. The chapter reviews studies that have employed VR technology to study participants' sense of self-location, ownership and agency over virtual entities (avatars) which are increasingly less similar to them, and discuss methodological challenges involved in building on such studies to examine the flexibility of our self-conception.

In *Chapter 6*, Hannah Rubin, Cailin O'Connor and Justin Bruner use paradigms from *experimental economics* that involve *induced valuation*: They have participants make choices in strategic situations where choices have real (financial, for example) consequences that the participants care about. The chapter presents extant and new studies which employ these methods to study the emergence of communication and language – e.g. whether similarity structures facilitate communication, and categories of what width people form to describe a complex world with a limited stock of signals. The experimental paradigms demonstrated allow them to test predictions from game theory and evolutionary game theory. They can be usefully employed wherever philosophers have turned to these formal frameworks to study strategic interactions – communication, coordination, altruism, cooperation, social dilemmas, social norms, resource distribution, etc. – in areas ranging from political philosophy to social epistemology.

These latter chapters may also help experimental philosophers address another important challenge: People don't always do as they say. For example, the decisions about moral dilemmas that people make when immersed in VR simulations (Francis et al. 2017) or when real-life consequences attach to decisions (Gold et al. 2015) are out of line with their moral judgements about hypothetical cases. The moment we are interested in people's decisions and actions, we need to go beyond eliciting, explaining and assessing judgements or decisions about hypothetical cases, and methods involving VR techniques or induced valuation become instrumental.

In another, currently central and potentially transformative, methodological development, experimental philosophers are beginning to recruit computational methods. Computational modelling has always been an integral part of cognitive science – to which many experimental philosophers take themselves to belong (*cf.* Knobe 2016). But experimental philosophers are now deriving inspiration also from the digital humanities and are beginning to recruit several methods of corpus analysis, data mining and text mining, to take advantage of the ready availability of large amounts of text and other data through the internet and of software to process and analyse them. The rapidly increasing importance of these methods is aptly illustrated by the fact that half of the studies in the first part of the volume, though focusing on other methods, build on corpus analyses (Chapters 1 and 2) or employ an automated database using text-mining techniques (the Neurosynth database of fMRI studies, Chapter 4). The second part of the volume is devoted to computational methods and seeks to build a bridge between experimental philosophy and the digital humanities.

In *Chapter 7*, Justin Sytsma and colleagues introduce qualitative and quantitative methods of *corpus analysis* and deploy them in a new study of causal attributions ('A causes B'). The authors hypothesize that these attributions are not purely descriptive but are similar to attributions of responsibility in that they express normative evaluations – which are more frequently made when outcomes are negative. Using a general-purpose corpus, they investigate which nouns most frequently follow 'caused the', its thesaurus-recognized synonyms, and 'responsible for the', and which verbs most frequently accompany the nouns identified ('death', 'accident', etc.). The computational methods of *Latent Semantic Analysis* and *Distributional Semantic Analysis* help follow up first findings. To extract information about semantic similarities, these methods look at the distribution of words across a corpus and consider, for example, the extent to which words co-occur with the same other words, in the same proportion. Comparisons between a general purpose and a philosophy corpus reveal that philosophers use causal language differently from ordinary folk.

In *Chapter 8*, Pablo Mejia-Ramos and colleagues explain further aspects of *corpus linguistics* and examine occurrence frequencies in mathematics, physics and general-purpose corpora, in a study that addresses a central issue from the philosophy of mathematics, considered as applied philosophy of science: How central to mathematical practice is explanation – even though standard (causal or statistical) accounts of scientific explanation do not seem to apply there? They examine absolute and relative occurrence frequencies of 'explain' words and related expressions identified through concordance searches across the three corpora, and replicate findings across different parts of the corpora. Results suggest that explanation occurs in mathematical practice but does so (even) less frequently than in physics. Where mathematicians do give explanations, they explain more often how to do something in mathematics rather than why certain mathematical facts obtain.

In *Chapter 9*, Mark Alfano and Andrew Higgins offer a crash course in *semantic network analysis* for experimental philosophers, with illustrations drawn from research that applies data-mining techniques to an obituary corpus, to extract and visualize information about the values, virtues and constituents of well-being recognized by a community. The chapter explains how these methods can be put to the service of a familiar philosophical project whose implementation has not been feasible previously. David Lewis developed the idea of devising implicit definitions of folk-theoretical terms (such as virtue or mental terms) by constructing Ramsay sentences that collect all the accepted platitudes in which the terms ('belief', 'desire', etc.) occur, replace these terms by variables and quantify over them. The chapter explains how a modified form of this method of Ramsification can be implemented with semantic networks.

In *Chapter 10*, Arianna Betti and colleagues introduce the nascent field of *computational history of ideas*. Their latest study examines the origin, development and spread of the notion of 'conceptual scheme', much used in the philosophy of science and in several social sciences. Betti and colleagues apply a mixed-methods approach to a large interdisciplinary corpus of research articles from the 1880s to the 1950s. Automated search in a digital corpus facilitates expert annotations of occurrences of the term of interest, in their immediate textual context, guided by a simple model of the concept of interest that represents it as a relational structure with stable/determinate and variable/determinable elements. Quantitative analysis then also takes into account institutional affiliations and author career information. The approach permits researchers to advance beyond analysis of a small number of canonical texts and to develop and assess hypotheses about how key concepts were used and disseminated in larger research communities and across disciplines.

Acknowledgements

While most papers were commissioned independently, this volume has been developed in parallel with the eighth annual conference of the Experimental Philosophy Group UK, 'Alternative Methods in Experimental Philosophy – Beyond the Questionnaire': Early drafts of four of the ten papers in the volume were presented – mostly as keynotes – at this international conference, held at the University of East Anglia (UEA), Norwich, UK, in July 2017. For financial support of this event we thank the Mind Association, the Analysis Trust and the UEA. For excellent advice and enthusiastic support throughout the book project, we would like to thank the series editor, James Beebe. We are also greatly indebted to the anonymous referees, all experts in the relevant methodologies, who reviewed the chapters for this volume. Our greatest thanks are due to the authors who provided chapters that combine accessible introductions to promising methodologies with original research which the pressures of career development could easily have steered towards prestigious journals, instead. We are deeply grateful they contributed their work to the present volume, to facilitate a comprehensive and up-to-date demonstration of the use and usefulness of cutting-edge empirical methods in experimental philosophy.

References

Alexander, J. (2012). *Experimental Philosophy*. Cambridge: Polity.

Cova, F., Strickland, B., Abatista, A. et al. (2018). Estimating the reproducibility of experimental philosophy. *Review of Philosophy and Psychology*. https://doi. org/10.1007/s13164-018-0400-9.

De Cruz, H. (2015). Where philosophical intuitions come from. *Australasian Journal of Philosophy*, *93*, 233–249.

Fischer, E., and Collins, J. (2015). Rationalism and naturalism in the age of experimental philosophy. In their (eds.), *Experimental Philosophy, Rationalism and Naturalism* (pp. 3–33). London: Routledge.

Francis, K. B., Terbeck, S., Briazu, R. A., Haines, A., Gummerum, M., Ganis, G., and Howard, I. S. (2017). Simulating moral actions: An investigation of personal force in virtual moral dilemmas. *Scientific Reports*, *7*, Article no. 13954.

Greene, J. D., Sommerville R. B., Nystrom L. E., Darley J. M., and Cohen, J. D. (2001). An fMRI investigation of emotional engagement in moral judgment. *Science*, *293*, 2105–2108.

Gold, N., Pulford, B., and Colman, A. (2015). Do as I say, don't do as I do: Differences in moral judgments do not translate into differences in decisions in real-life trolley problems. *Journal of Economic Psychology*, *47*, 50–61.

Gopnik, A., and Schwitzgebel, E. (1998). Whose concepts are they anyway? In M. R. DePaul and W. Ramsay (eds.), *Rethinking Intuition* (pp. 75–91). New York: Rowman & Littlefield.

Knobe, J. (2016). Experimental philosophy is cognitive science. In J. Sytsma and W. Buckwalter (eds.), *A Companion to Experimental Philosophy*. Malden, MA: Blackwell.

Knobe, J., and Nichols, S. (2008). An experimental philosophy manifesto. In their (eds.), *Experimental Philosophy* (pp. 3–14). Oxford: OUP.

Knobe, J., and Nichols, S. (2017). Experimental philosophy. In E. N. Zalta (ed.), *The Stanford Encyclopedia of Philosophy* (Winter 2017). https://plato.stanford.edu/ archives/win2017/entries/experimental-philosophy/.

Mallon, R. (2016). Experimental philosophy. In H. Cappelen, T. Szabo Gendler, and J. Hawthorne (eds.), *Oxford Handbook of Philosophical Methodology* (pp. 410–433). Oxford: OUP.

Nado, J. (ed.) (2016). *Advances in Experimental Philosophy and Philosophical Methodology*. London: Bloomsbury.

Sytsma, J., and Livengood, J. (2016). *The Theory and Practice of Experimental Philosophy*. Boulder: Broadview.

Weinberg. J. (2015). Humans as instruments, on the inevitability of experimental philosophy. In E. Fischer and J. Collins (eds.), *Experimental Philosophy, Rationalism, and Naturalism* (pp. 171–187). London: Routledge.

Part One

Behavioural Experiments beyond the Questionnaire

1

Experimental Philosophy and Statistical Learning

Shaun Nichols

Introduction

Much experimental philosophy aims to uncover the processes and representations that guide judgements about philosophically relevant issues. This is the agenda in discussions about whether people are incompatibilists about free will (e.g. Murray and Nahmias 2014), whether moral judgement is driven by distorting emotions (e.g. Greene 2008) and whether judgements about knowledge are sensitive to irrelevant details (e.g. Swain et al. 2008). Much less attention has been paid to *historical* questions about how we ended up with the processes and representations implicated in philosophical thought. There are different kinds of answers to these historical questions. One might offer distal answers that appeal to the more remote history of the concept. For instance, an evolutionary psychologist might argue that some of our concepts are there because they are adaptations. Or a cultural theorist might argue that some of our concepts are there because they played an important role in facilitating social cohesion. On the proximal end of things, we can attempt to say how the concepts might have been acquired by a learner. Those proximal issues regarding acquisition will be the focus in this paper.[1] Recent cognitive science has seen the ascendance of statistical learning accounts which draw on statistical learning to explain how

Acknowledgements

I'd like to thank Eugen Fischer and an anonymous referee for very helpful comments on an earlier draft of this chapter. This material is based upon research supported in part by the U. S. Office of Naval Research under award number 11492159.

[1] Of course proximal and distal issues are not unrelated. For an evolutionary psychologist, the proposal that a concept is an adaptation will typically be accompanied by the expectation that the characteristic development of the concept is not explicable in terms of domain-general learning mechanisms.

we end up with the representations we have (Perfors et al. 2011). In this chapter, I'll describe how this approach might be extended to explain the acquisition of philosophically relevant concepts and distinctions.

1. Background on statistical learning

Accounts of acquisition in terms of statistical learning include not only very sophisticated statistical techniques from machine learning, but also humble and familiar forms of statistical inference. Imagine you're on a road trip with a friend and you have been sleeping while he drives. You wake up wondering what state you're in. You notice that most of the licence plates are Kansas plates. You will likely use this information to conclude that you are in Kansas. This is a simple form of statistical learning. You are consulting samples of licence plates (the ones you see) and using a principle on which samples (in this case, of licence plates) reflect populations. This, together with the belief that Kansas is the only state with a preponderance of Kansas plates, warrants your new belief that you are in Kansas.

Early work on statistical reasoning in adults indicated that people are generally bad at statistical inference (e.g. Kahneman and Tversky 1973). But over the last decade, work in developmental and cognitive psychology suggests that children actually have an impressive early facility with statistical reasoning (Girotto and Gonzalez 2008; Xu and Garcia 2008; Xu and Denison 2009; Fontanari et al. 2014). In tandem with this, computational psychologists have offered statistical learning accounts for a wide range of representations and processes including categorization (Smith et al 2002; Kemp et al. 2007), word learning (Xu and Tenenbaum 2007) and causal judgement (Bramley et al. 2017). I will now review some key characteristics of a statistical learning account of acquisition, and then present philosophical applications that draw on several statistical learning principles from this growing body of literature.

2. Theories of acquisition

A theory of acquisition involves many components, and it's helpful to pull them apart. We will be considering accounts of acquisition that are, broadly speaking, empiricist. In this section, I'll say a little bit about the empiricist/nativist debate give a sketch of the structure of statistical learning accounts.

2.1. Empiricism vs nativism

In the contemporary idiom of cognitive science, the debate between empiricists and nativists is all about learning (see e.g. Laurence and Margolis 2001; Cowie 2008). Take some capacity like the knowledge of grammar. How is that knowledge acquired? Empiricists about language learning typically maintain that grammatical knowledge is acquired from general purpose learning mechanisms (e.g. statistical learning) operating over the available evidence.[2] Nativists about language learning maintain instead that there is some domain-specific learning mechanism (e.g. a mechanism specialized for learning grammar) that plays an essential role in the acquisition of language. In the case of grammatical knowledge, debate rages on – e.g. Perfors et al. (2011) against Berwick et al. (2011). But it's critical to appreciate that there is some consensus that for certain capacities, an empiricist account is most plausible, while for other capacities, a nativist account is most plausible.

On the empiricist end, research shows that infants can use statistical evidence to segment sequences of sounds into words. The speech stream is largely continuous, as is apparent when you hear a foreign language as spoken by native speakers. So how can a continuous stream of sounds be broken up into the relevant units? In theory, one way that this might be done is by detecting 'transitional probabilities': how likely it is for one sound (e.g. a syllable) to follow another. In general, the transitional probabilities between words will be lower than the transitional probabilities within words. Take a sequence like this:

> happy robin

As an English speaker, you will have heard 'PEE' following 'HAP' more frequently than ROB following PEE. This is because 'HAPPEE' is a word in English but 'PEEROB' isn't. This sort of frequency information is ubiquitous in language. And it could, in principle, be used to help segment a stream into words. When the transitional probability between one sound and the next is very low, this is evidence that there is a word boundary between the sounds. In a ground-breaking study, Jenny Saffran and colleagues (1996) used an artificial language experiment to see whether babies could use transitional probabilities to segment a stream. They created four nonsense 'words':

> pabiku tibudo golatu daropi

---

[2] Note that empiricists allow that these general purpose learning mechanisms themselves might be innate; after all, we are much better at learning than rocks, trees and dust mites.

These artificial words were strung together into a single sound stream, varying the order between the words (the three orders are depicted on separate lines but are seamlessly strung together in the audio):

pabikutibudogolatudaropi
golatutibudodaropipabiku
daropigolatupabikutibudo

By varying the order of the words, the transitional probabilities are also varied. Transitional probabilities between syllabus pairs within a word (e.g. bi-ku) were higher than between words (e.g. pi-go) ($p=1.0$ vs $p=.33$). After hearing 2 continuous minutes of this sound stream, infants were played either a word (e.g. pabiku) or part word (e.g. pigola). Infants listened longer (i.e. showed more interest) when hearing the part word, which indicates that they were tracking the transitional probabilities. This ability to use statistical learning to segment sequences isn't specific to the linguistic domain. It extends to segmenting non-linguistic tones (Saffran et al. 1999) and to the visual domain (Kirkham et al. 2002). Perhaps humans have additional ways to segment words, but at a minimum, there is a compelling empiricist account of one way that we can segment streams of continuous information into parts using statistical learning.

Nativists can claim victories too, however. Birdsong provides a compelling case. For many song birds, like the song sparrow and the swamp sparrow, the song they sing is species-specific. It's not that the bird is born with the exact song it will produce as an adult, but birds are born with a 'template song' which has important elements of what will emerge as the adult song. One line of evidence for this comes from studies in which birds are reared in isolation from other birds. When the song sparrow is raised in isolation, it produces a song that shares elements with the song the normally reared adult song sparrow; similarly, an isolated swamp sparrow produces a song that shares elements with normally reared swamp sparrows. Critically, the song produced by the isolated song sparrow *differs* from the song produced by the isolated swamp sparrow. This provides a nice illustration of a nativist capacity. It's not that experience plays no role whatsoever – the specific song that the bird produces does depend on the experience. But there is also an innate contribution that is revealed by the song produced by isolate birds. The template gives the bird a head start in learning the appropriate song (see, e.g. Marler 2004). We can also cast the point in terms of a poverty-of-the-stimulus argument. The evidence in the stimulus is not adequate to explain the isolate songs of the two kinds of sparrows. If the evidence in the stimulus were adequate, then we should find that song and swamp sparrows produce the same song if they are provided with the same evidence. The fact that they produce different songs shows that there is something contributed by the organism that is not present in the stimulus.

The examples of bird song and segmentation of acoustic strings show that it's misguided to think that there is a general answer to the nativist/empiricist debate. The debate needs always to be focused on particular capacities. The examples I'll give in Sections 3 and 4 are all empiricist learning stories for specific pieces of knowledge, based on principles of statistical inference. Let's now turn to the characteristics of such learning accounts.

2.2. A schema for statistical learning accounts of acquisition

Suppose we want to argue that some concept was acquired via statistical inference over the available evidence. Several things are needed.

1. The first task is to describe the concept (belief, distinction, etc.) the acquisition of which is to be explained. We can call this target the *acquirendum* (Pullum and Scholz 2002). Part of the work here will be to argue that we do in fact have the concept or distinction or belief that is proposed as the acquirendum (A).
2. Insofar as statistical learning is a form of hypothesis selection, one needs to specify the set of hypotheses (S) that the learner considers in acquiring A. This set of hypotheses will presumably include A as well as competing hypotheses.
3. One will also need an empirical assessment of the evidence (E) that is available to the learner. For this, one might consult, inter alia, corpora on child-directed speech.
4. One needs to articulate the statistical principles (P) that are supposed to be implicated in acquiring the concept. These principles should make it appropriate for a learner with the evidence E and the set of hypotheses S to infer A. This step will thus provide an *analysis* of how the principle would yield the concept given certain evidence.
5. Finally, a complete theory of acquisition would tie this all together by showing that the learner in fact does use the postulated statistical principle P and the evidence E to select A among hypotheses S.

Few theories of acquisition manage to provide convincing evidence for all of these components. Furthermore, the last item is well beyond what most learning theorists hope to achieve. In place of this daunting demand, a learning theorist might aim for a weaker goal – to show that learners are capable of using the relevant kind of evidence to make the inferences that would be appropriate given the postulated statistical principles. That is, instead of trying to capture the learner's actual acquisition of the concept, one might settle for something a good deal weaker:

> 5*. Show that the learner is appropriately sensitive to the evidence; that is, given evidence like *E,* she makes inferences that would be appropriate if she were deploying the postulated statistical principles *P.*

3. A first application: 'Knowledge'

The foregoing is a recipe in a sense, but the critical ingredient is coming up with a specific idea for how principles of statistical learning might explain some aspect of our philosophical outlook. In this section, I'll explain how the method has been applied to a core issue in epistemology and in the next section, I'll review four different aspects of moral cognition that we have attempted to explain using statistical principles. The reviews will be rather brief, but the idea is to give a sense of the range of possibilities, not a detailed exposition of any one.

Acquirendum

It's an old idea in epistemology, stretching back to Plato, that knowledge demands infallibility. Infallibilism is characterized in different ways, but one intuitive characterization is modal: if S knows that p on the basis of his evidence only if, given the evidence, S's belief that p *could not have been wrong.*[3] In contemporary discussions, several philosophers have suggested that infallibilism is supported by the fact that it's odd to say things like 'He knows that it's raining, but he could be wrong about that' (see e.g. Rysiew 2001; Dodd 2011). This forms an acquirendum: how do we acquire an infallibilist notion of knowledge. Ángel Pinillos and I have recently offered a statistical learning explanation for the acquisition of an infallibilist notion of knowledge (Nichols and Pinillos forthcoming).

---

[3]   More broadly, infallibilism holds that if S knows that p at t, then S's epistemic state (e.g. his evidence, reason, justification…) guarantees the belief that p, where 'guarantee' can be cashed out in different ways, for example, in terms of modality (as above), entailment or probability.

Hypotheses

We suggest that when a learner is trying to acquire the concept *knowledge*, the competing hypotheses will include the following: (1) the possibility that knowledge is true belief, (2) the possibility that knowledge is true belief with justification of some sort but without the requirement for infallibility and (3) the possibility that knowledge is infallible. We can think of these hypotheses are forming a subset structure as in Figure 1.1.

Thus, the hypotheses here are as follows:

($H_{TrueBelief}$) *Knowledge* corresponds to true belief.
($H_{Fallibilist}$) *Knowledge* corresponds to fallible knowledge.
($H_{Infallibilist}$) *Knowledge* corresponds to infallible knowledge.

Statistical principle: The size principle

The statistical principle we invoke to explain why children select the infallibilist hypothesis is the 'size principle' (Xu and Tenenbaum 2007). To get an intuitive

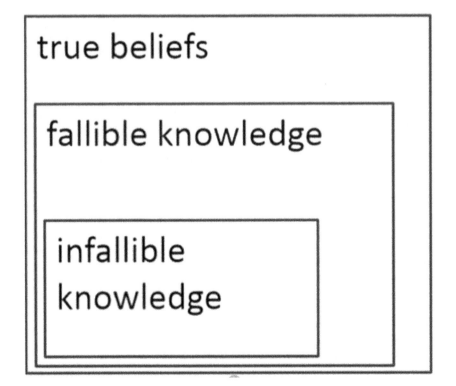

Figure 1.1 Epistemic categories represented as a subset structure.

sense of the principle, imagine your friend has four dice, each with a different denomination: four-sided, six-sided, eight-sided and ten-sided. He picks one at random, hides it from your view and rolls it 10 times. He reports the results: 3,2,2,3,1,1,1,4,3,2. Which die do you think it is? Intuitively, it's probably the four-sided die. But all of the evidence is consistent with it being any of the other dice, so why is it more probable that it's the four-sided one? Because otherwise it's a suspicious coincidence that all of rolls were 4 or under. One way to think about this is that the four-sided die hypothesis generates a proper subset of the predictions generated by each of the other hypotheses. We can illustrate this in terms of a nested structure (Figure 1.2).

The size principle states that when all of the evidence is consistent with the 'smaller hypothesis' (in this case the hypothesis that it's the four-sided die), that hypothesis should be preferred.

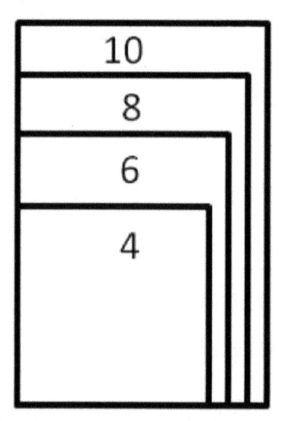

Figure 1.2 The numbers represent the highest denomination of the die; the rectangles represent the relative sizes of the hypotheses.

Xu and Tenenbaum use the size principle to explain how the absence of evidence might play a role in word learning. When learning the word 'dog', children need only a few positive examples in which different dogs are called 'dog' to infer that the extension of the term is –dog– rather than –animal–. Pointing to a Poodle, a Labrador and a Chihuahua suffices. You don't also need to point to a bluejay or a halibut and say 'that's not a dog.' Xu and Tenenbaum explain this in terms of the size principle: the likelihood of getting those particular examples (a Poodle, a Labrador and a Chihuahua) is higher if the extension of the word is –dog– as compared with –animal–. Xu and Tenenbaum conducted word learning experiments to confirm that children and adults use the absence of evidence to infer word meanings. For example, participants were told 'these are <u>feps</u>' while being shown three Dalmatians and no other dogs. In that case, people tended to think that <u>fep</u> refers to Dalmatians rather than dogs. The absence of other dogs in the sample provides evidence for the more restricted hypothesis that *fep* refers to Dalmatians.[4]

Now let's return to the learner trying to decide between the various hypotheses for meaning of 'knowledge'. We already have reason to think that the learner is adept at using something like the size principle to infer the extension of words (Xu andTenenbaum 2007). If the learner hears the word 'know' often, and if all of the evidence she gets is consistent with $H_{\text{Infallibilism}}$, then the size principle dictates that this is the most likely hypothesis. Otherwise it would be a suspicious coincidence that all of the evidence is consistent with that hypothesis.

Evidence available to the learner

The foregoing suggest that <u>if</u> the evidence that children receive regarding the term 'knowledge' is consistent with an infallibilist interpretation, then children should prefer such an interpretation. What is their evidence like? To explore this, we had research assistants code the standard US corpus of child-directed speech, the CHILDES database. Assistants first identified all the cases in which an adult used cognates of the words 'know' and 'think' to attribute an attitude. We then had assistants go through those identified items to determine whether the attribution was consistent or inconsistent with an infallibilist interpretation,

---

[4] Of course, the details of the hypothesis space matter here in various ways. For one thing, there is an assumption that the prior probability for each of the hypotheses is the same. In addition, it is assumed that certain hypotheses are excluded from the space. For instance, when shown three Dalmatians, one logically possible hypothesis is that 'fep' refers to all and only those three dogs. These kinds of hypotheses are excluded, on the plausible grounds that word learning will involve generalization to further cases (cf. Nichols et al. 2016, 541; Xu and Tenenbaum 2007, 362).

as reflected by hedges such as 'I φ that p, but I'm not sure' or 'He φs that p but he could be wrong'. For *think*, there were many such cases, for example, 'I think that was right Peter. I'm not sure', 'I think that'd be alright we should ask Ninny first though', 'I think it's behind you. Is it?' For *know*, however, there were no such instances. Parents don't say 'I know that P but I could be wrong'. Of course this is hardly a surprise. But the important point from the perspective of the learning theory is that this is the evidence that kids get. Given the absence of such qualifications on attributions of knowledge, it would be appropriate for a rational learner to select the infallibilist hypothesis.

Sensitivity to the evidence

When learning an epistemic term, are people sensitive to evidence regarding the presence or absence of fallibilist hedging? There are no direct tests of this yet. But there is compelling if indirect reason to think that children are sensitive to the presence of fallibilist hedging. The corpus evidence on 'thinking' shows that children are exposed to a doxastic verb – thinks – that is hedged as fallible, and we know that children acquire a fallibilist conception of thinking. This makes it plausible that if 'knowing' were hedged as fallible, children would acquire a fallibilist conception of knowledge. It's a further question how sensitive children are to the absence of hedging as evidence. Hopefully future studies will examine this.

4. Applications in moral psychology

Now I'd like to turn to a broader set of studies in moral psychology that draws on statistical learning models to explain several different aspects of moral psychology. This will allow us to review a range of different learning principles invoked by such models.

4.1. The size principle revisited: Act/outcome

Acquirendum

The rules that guide moral judgement have an apparent complexity that demands explanation. For instance, children judge that actions that produce harms are worse than omissions that produce equal harm; children also distinguish intentional harms from unintended but foreseen harms (see e.g. Pellizzoni et al. 2010; Mikhail 2011). These are subtle distinctions to acquire, and children are presumably never given any explicit instruction on the matter. Few parents tell children 'It is wrong to intend to do X, but it is permissible to allow X to occur.' So if we are to explain people's facility with moral distinctions in terms of complex rules, we need some account of how they arrive at such complex rules despite scant explicit instruction. Although there are several distinctions that might be explored, for the acquirendum we will focus on the fact that rules tend to be framed over actions rather than outcomes. Children arrive at rules that apply to what an agent *does*, but not to what that agent *allows*. Given that children don't receive any explicit instructions on the matter (Dwyer et al. 2010, 492), how do they acquire rules that specify a focus on actions rather than outcomes?

Hypotheses

The question is; why does the child infer that a given rule applies to actions rather than outcomes?[5] Thus, for a prohibition, we might characterize the competing hypotheses as follows:

(H_{Act}) Act-based: the rule being taught prohibits agents from producing a certain outcome.

($H_{Outcome}$) Outcome-based: the rule being taught prohibits agents from producing a certain outcome and also from allowing such an outcome to come about or persist.

Statistical principle: The size principle

The same principles appealed to above (Section 3) can explain how children might infer that a rule prohibits *actions* (Nichols et al. 2016). Even though children don't get explicit instruction on the act/outcome distinction, it's possible that they can infer that a rule is act-based from the kinds of violations they learn about. If all of the violations they learn about are cases in which the agent intentionally produced the negative outcome, this might count as evidence that the operative rule applies narrowly to what an agent *does* rather than more widely to include what people *allow*. This is just the basic insight of the size principle. In the case of the dice, the fact that none of the rolls were over 4 would

---

5   As is familiar from discussions in moral philosophy, characterizing the precise boundaries of these distinctions is a delicate issue (see e.g. McNaughton and Rawlings 1991; Nagel 1986; Parfit 1984). Here we are not trying to solve this issue of precise boundaries, but we hope that the core distinction is clear enough that it can be used by people to make inferences in acquiring rules.

be a suspicious coincidence if the die was eight-sided. Similarly, when none of the violations children have observed are 'allowings' this would be a suspicious coincidence if the rule explicitly prohibited allowings. The *absence* of evidence is itself evidence.

Evidence available to the learner

The foregoing gives us an analysis of the problem. A rational learner trying to decide between the hypotheses would infer that a rule is act-based when all the sample violations of the rule are actions. The next question is, what does the child's evidence look like? To answer this, we coded a portion of the standard database for child-directed speech (CHILDES) and found that the vast majority (over 99%) of observed violations concerned intentional actions. Children heard lots of cases of 'don't', 'stop it', 'don't sit on that' and 'don't put that there', but almost no cases of statements like 'don't let that happen'. Thus, for the vast majority of rules the child learns, there is a conspicuous lack of evidence in favour of $H_{Outcome}$. And this lack of evidence for $H_{Outcome}$ counts as evidence in favour of H_{Act}.

Sensitivity to the evidence

It is a further question whether people are sensitive to evidence about whether a rule applies to agents producing an outcome or to both agents producing and agents allowing an outcome. We investigated this using novel and unemotional rules. Participants were asked to infer the content of a novel rule based on sample violations. In one condition, they were given examples like the following: (i) John violated the rule by putting a truck on the shelf, (ii) Jill violated the same rule by putting a ball on the shelf and (iii) Mike violated the rule by putting a doll on the shelf. Then they were asked to determine whether the rule applied to other cases. For instance, is Mary violating the rule when she sees a puzzle on the shelf and doesn't remove it? Note that sample violations (i)–(iii) above are all examples of a person *intentionally* producing the outcome, and when the rule is taught through such examples, participants tend to infer that Mary (who allowed the outcome) did *not* break the rule. However, when two of the sample violations involve a person allowing the outcome to persist (e.g. leaving a jump rope on the shelf), participants overwhelming infer that Mary is breaking the rule by leaving a puzzle on the shelf. This suggests that people are sensitive to evidence that bears on whether the rule applies only to what an agent does, or also to what an agent allows.

4.2. Overhypothesis formation: Act bias

Acquirendum

In the previous section, we provided a statistical learning account of how children might acquire an act-based rule based on the available evidence. But there is a further phenomenon here that requires explanation. Studies on rule learning show a pronounced bias in favour of act-based rules. When learning a new rule, participants tend to expect that the rule will be act-based rather than outcome-based (Nichols et al. 2016; Millhouse et al. forthcoming). One explanation is that people have an innate bias to think rules are act-based. We suggest instead that people learn a second-order generalization about rules, an 'overhypothesis', according to which rules tend to be act-based (Ayars and Nichols 2017). Such an overhypothesis would then lead people to expect that a new rule will also be act-based. Thus, we want to explain the acquisition of the overhypothesis that most rules are act-based.

Hypotheses

The candidate hypotheses are as follows:

$(H_{Act-Overhyp})$ Overhypothesis that rules tend to be act-based.

$(H_{Outcome-Overhyp})$ Overhypothesis that rules tend to be outcome-based.

$(H_{No-Overhyp})$ No overhypothesis about whether rules tend to be act- or outcome-based.

Statistical principle: Overhypothesis formation

Imagine you are presented with a bag of marbles and you are trying to predict the colour of the next marble you are to draw randomly from a bag. You first draw a green marble. What will the colour of the next marble be? Without any additional information about the colour distribution of the marbles in the bag, there is little way to predict the colour of the next marble. But now suppose that before drawing from the bag, you draw marbles from three other bags first. In the first bag, all the marbles are blue. In the second, all the marbles are pink. In the third, all the marbles are yellow. Finally, you draw a single marble from the fourth bag and find that it is green. What will the colour of the next marble from that fourth bag be? In this case you can reasonably infer that the next marble will also be green. The fact that all bags encountered contained marbles of a uniform colour suggests a higher-order hypothesis about the contents of the bags, viz., within bags, marbles are uniform in colour. The example here comes from Nelson Goodman, and he calls these kinds of higher-level hypotheses 'overhypotheses' (1955).

There is evidence that children learn overhypotheses. Early work demonstrated a 'shape-bias' in word learning – children tend to think nouns will refer to classes of similarly shaped (rather than similarly coloured) objects (Heibeck and Markman, 1987). One explanation for this is that there is an innate bias that leads children to expect nouns to refer to objects of the same shape. Another possibility, though, is that children form an overhypothesis. Since for most of the nouns they learn the category includes objects of the same shape, the child might *infer* that nouns tend to refer to objects that have the same shape (Smith et al. 2002). In a lovely study, Linda Smith and colleagues brought 17-month-old children into the lab weekly for 7 weeks; during these sessions they taught the children several novel names, each of which was associated with objects that had the same shape (but differed in size, shape and colour). After this exposure, the children tended to expect a new noun to refer to objects with the same shape, and this was not the case for a control group of children (Smith et al. 2002, 16).[6]

We can extend the idea of overhypotheses to the normative domain (Ayars and Nichols 2017). One can think of overhypotheses about rules as analogous to the overhypothesis about the uniformity of marble colour within bags. In the case of the marbles, the overhypothesis is that for any arbitrary bag (in the relevant population of bags), the colour of marbles tends to be uniform. In the case of rules, the overhypothesis might be for any arbitrary prohibitory rule (in the relevant population of rules); rules tend to be act-based. This overhypothesis about rule uniformity could constrain inferences about new rules just as the overhypothesis about marble-colour uniformity constrains inferences about the colour of marbles in a new bag. Such an overhypothesis could be acquired if the majority of rules the learner is exposed to are act-based, i.e. the rules that prohibit bringing about some outcome but do not prohibit allowing such an outcome to occur.

Evidence available to the learner

There hasn't been a direct measure of the relative proportion of act-based rules in the population of rules to which children are exposed. Nonetheless, it seems likely, just from reflecting on our lives, that most of the rules we have are act-

---

[6] More recently, Dewar and Xu (2010) adapted Goodman's marbles example for an infancy experiment and found that babies had expectations that suggest that they formed overhypotheses that the colour of objects was uniform within containers.

based. Our rules against, say, littering, prohibit littering but do not require that we pick up others' litter or prevent others from littering. In addition to this armchair reflection, the evidence mentioned in Section 4.1 provides some reason to think that most of the rules that children are exposed to are act-based. For, as noted above, in the CHILDES corpus there was almost no evidence that would lead children to infer an outcome-based rule.

Sensitivity to the evidence

Our analysis is that learners have an overhypothesis that rules tend to be act-based, formed from registering the fact that most extant rules are act-based. But do people really form these kinds overhypotheses about rules, based on evidence of the available rules? To investigate this, we developed learning tasks that varied whether participants were exposed to several act-based rules or to several outcome-based rules. We found that after learning three outcome-based rules, participants were more likely to think that a new rule prohibited allowing the outcome to occur. This suggests that people can use evidence to learn overhypotheses about rules – the participants seemed to have moved to the overhypothesis that rules (in the relevant population set up in the experiment) tend to be outcome-based. In addition, the results suggest that the bias for act-based rules can be overturned in light of evidence.

4.3. Pedagogical sampling: The principle of natural liberty

Acquirendum

When we learn normative systems, we do so based on limited evidence. Many of the possible actions that are available to an agent have never been explicitly permitted or prohibited. But people tend to think that most actions that have never been explicitly permitted or prohibited are, in fact, permissible. For instance, I've never heard a rule about whether it's okay to look out my window or to open the cupboard or to turn on the stereo. But I think each of these is permissible. Why do I think that? One possible explanation is that I've learned a 'closure principle' regarding the moral system which states that anything that isn't expressly forbidden is permitted (Gaus and Nichols 2017; Nichols and Gaus forthcoming). This closure principle is one of *Natural Liberty* (Mikhail 2011). We take as the acquirendum the knowledge that moral systems are characterized by a principle of Natural Liberty.

Hypotheses

The principle of Natural Liberty is one closure principle, but there is an obvious alternative: the principle of *Residual Prohibition*. On this principle, an action-type is *not expressly permitted*, then acts of that type are prohibited (Mikhail 2011, 132–3). We take the relevant hypotheses then to be as follows:

(H_{NatLib}) The closure principle is one of *Natural Liberty*: if an action-type is not expressly forbidden, then acts of that type are permitted.

$(H_{ResProh})$ The closure principle is one of *Residual Prohibition*: if an action-type is not expressly permitted, then acts of that type are prohibited.

$(H_{NoClosure})$ There is no closure principle.

Statistical principle: Pedagogical sampling

Closure principles might be learned, we argue, through *pedagogical sampling*. The core idea of pedagogical sampling is quite intuitive, and it's reflected in the following normative claim:

> If you know that a teacher is trying to teach you a set of rules (say), and you know that the teacher assumes that you are rational, then you should expect the teacher to provide examples that are maximally useful for a rational learner trying to identify the set.

There is empirical evidence that learners make inferences that conform to this normative claim (Shafto et al. 2014). That is, learners make inferences that suggest that they expect teachers to provide helpful examples. One study that illustrates this has participants play the *rectangle game*. In this game, a teacher is supposed to help a learner select a rectangle with certain dimensions. The teacher picks two points on a board, one that falls within the rectangle, and one that falls outside the rectangle. The inside point is indicated by a green circle and the outside point is indicated by a red X. These points constitute the evidence for the learner, and the learner is to infer the dimensions of the rectangle from the two points provided by the teacher (Shafto et al. 2014, 64–65). This is all common knowledge between teacher and learner. As a teacher, knowing that the learner will use these points to infer the dimension, the best strategy is to pick an inside point right next to one of the inside corners of the rectangle (say, the bottom left) and pick an outside point that falls just outside of the diagonal corner (the top right). Learners should be able to figure out this pedagogical strategy and thus expect teachers to provide these kinds of examples, and, in fact the experiments show learners tend to select rectangles that would be appropriate given the assumption that the teacher picked such examples.

Now consider the acquisition of closure principles. When a norm system has more prohibitions than permissions, an efficient teacher would provide examples of permissions and expect that learner to infer the principle of *Residual Prohibition*; and when a norm system has more than permissions than prohibitions, an efficient teacher would provide examples of prohibitions with the expectation that the learner will infer *Natural Liberty*. Now, imagine that I'm the learner, and I assume that the teacher (i) aims for efficiency and (ii) knows that I am a rational learner capable of learning closure principles then. In that case, I should operate via the following rules:

> If for domain X, I'm taught via prohibitions, that is evidence that the domain is characterized by Natural Liberty;

> If for domain Y, I'm taught via permissions, that's evidence that the domain is characterized by Residual Prohibition.

Evidence available to the learner

The above analysis suggests that if learners rely on pedagogical sampling and for the moral domain, they receive mostly prohibitions, then the learners should infer that morality is characterized by the closure principle of Natural Liberty. Does the evidence available to learners mostly consistent in prohibitions (rather than permissions)? Again, there is little direct evidence on this. As far as we know, no quantitative measures of frequency have been conducted, and this would be a valuable area to have real data. Nonetheless, intuitively it certainly is easier to call to mind moral prohibitions. For present purposes we are going to tentatively suggest that intuition is right here – that common-sense morality is primarily composed of prohibitions.

Sensitivity to the evidence

If the *pedagogical sampling* analysis is right, then when people learn a new norm system, they should be sensitive to whether the rules they learn are prohibition rules or permission rules. Thus, we predicted that participants would infer Natural Liberty when trained on prohibitions, and they would infer Residual Prohibition when trained on permissions.[7] This is exactly what we found across several experiments.

---

[7] Our experiments only used prohibitions of the form 'Ss are not allowed to do A'. One might classify obligations (e.g. 'Ss must do B') as a kind of prohibition too, though (see e.g. Mikhail 2011). It will be important to see whether people also treat obligations the way they do prohibitions.

4.4. Fit/flexibility trade-off: Explaining folk relativism

Acquirendum

Many philosophers maintain that ordinary people reject moral relativism (Mackie 1977; Joyce 2001). Recent empirical work indicates that, at least for some moral claims, people do indeed reject relativism in favour of the view that the moral claims are universally true (Nichols 2004; Goodwin and Darley 2008; Wright et al. 2014). Psychologists have investigated the issue by drawing on the idea that if a claim is universally true, then if two people make different judgements about the truth of the claim, at least one of them must be wrong. This stands in contrast to claims that are only relatively true, like claims about the time of day – if Paul in Minneapolis says it's 3 PM and Shane in Dublin says it's not 3 PM they can both be right. Goodwin and Darley found that people were more likely to give the relativist response – that both people could be right – for claims of taste and social convention than for moral claims (1352–1353). Why is this? Why do people believe of some moral claims that they are universally true? We take that as the acquirendum (Ayars and Nichols forthcoming). More broadly, we want to explain what leads to the belief that a claim is relatively or universally true. To anticipate, the core idea is that the amount of consensus regarding the claim counts as evidence when deciding whether a claim is universally or only relatively true.

Hypotheses

The exact characterization of universalism is delicate.[8] To ease exposition, we will rely on a simple dichotomy to characterize the set of hypotheses. For a given claim that is true in at least some contexts, there will be two possible hypotheses:

($H_{Universalism}$) The claim is universally true.
($H_{Relativism}$) The claim is only relatively true.

Statistical principle: Tradeoff between fit and flexibility

The statistical principle that we appeal to here is more global than in the preceding cases, and it will take a little more exposition. In deciding between two hypotheses, there is a key trade off. Obviously the extent to which a hypothesis *fits the data* counts in favour of the hypothesis; but the extent to which a hypothesis

---

[8]   For present purposes, let's assume a modest notion of 'universal' that doesn't have extra-terrestrial ambitions. For a claim to be universally true will mean that it's not specific to one's culture, geographic location, etc.

is *flexible in its ability to fit data* counts against the hypothesis. To illustrate how these different factors play into hypothesis selection, imagine you're a physician trying to determine which disease is present in your community. You're deciding between two hypotheses, each with a prior probability of 5.

H_M: The disease present in the community is M, which typically produces high fever and doesn't produce any other symptoms.

H_D: The disease present in the community is D, which typically produces either high fever or a sore throat, but never both. There is no prior reason to think that a person with D will be more likely to have the fever or the sore throat.

Now imagine that 20 patients have come to see you. Patients 1, 2, 4, 6, 8, 11, 12, 14, 16, 17 and 20 have a high fever and patients 3, 5, 7, 9, 10, 13, 15, 18 and 19 have a sore throat. H_M does a poor job of fitting the data – it has to dismiss almost half of the data points as noise. By contrast, H_D can fit the data perfectly since each patient has one of the possible symptoms produced by disease D. In this case, it seems quite plausible that the disease in the community is D. Now imagine instead that among the 20 patients come to see you, patients 1–3 and 5–20 have a high fever and patient 4 has a sore throat. H_D can again fit the data perfectly, for one can simply say that patient 4 has the sore throat symptom of D and all the rest of have the fever symptom of D. But at this point, it should seem like things are a bit too easy for H_D. No matter what the distribution of fever/throat symptoms, H_D can perfectly fit the data. H_D would fit the data if patient 5 had the sore throat instead of patient 4. H_D can fit the data if patients 2 and 6 have sore throats, or if 3 and 4 have sore throats. *Any* pattern of symptoms with fevers or sore throats can be fitted by H_D. By contrast, H_M makes a much less-flexible prediction: it predicts lots of high fevers and nothing else. The great flexibility of H_D can make it a worse explanation of the 19:1 data than H_M, and this holds despite the fact that H_D fits the data better.[9] H_M does have to attribute some of the data, namely patients with sore throats, to noise (e.g. perhaps the sore throats were misreported or were caused by some environmental irritant). But a highly flexible hypothesis that can fit the data perfectly can be less probable than an inflexible hypothesis that has to attribute some data to noise. In hypothesis selection, we want a hypothesis that fits the data well without being overly flexible (see e.g. Perfors et al. 2011, 310–2).

---

[9] Of course, this conclusion will be affected by the likelihood of finding a sore throat among individuals without disease D. So, if it turns out that sore throats are extremely rare except among patients with disease D, then this will offset the penalty to some extent.

Just as the distribution of symptoms provide evidence on whether the population has disease D or M, distributions of opinion might provide evidence on whether a claim is universally or relatively true. If we suppose that people are reasonably good at knowing the truth about matters with which they are familiar, then a good explanation for widespread divergence regarding a claim is that the claim is only relatively true. To see this, imagine a child learning about months and seasons. She thinks that the claim 'July is a summer month' is true, and she's trying to figure out whether it's universally true or only relatively true. She learns that while 55% of the global population thinks, as she does, that July is a summer month, 45% thinks that July is not a summer month. Given this broad diversity, it would be reasonable for her to conclude that the claim 'July is a summer month' is only relatively true. The hypothesis that it is a universal truth fits the consensus data too poorly. On the flipside, if there is widespread consensus regarding some claim, that might count as evidence that the claim is universally true. Imagine the same child learning that 95% of people around the world think that summer is the hottest season. The consensus surrounding this claim provides reason to think that the claim 'summer is that hottest season' is a universal truth. A relativist account can, of course, fit the responses – one could say that the claim is true relative to one group and false relative to another, but just as with the case of the diseases, the massive flexibility of the relativist hypothesis counts against it. It will often be more plausible to count a small minority as mistaken about a universal truth rather than correct about a relative one.

The same principles might apply to normative claims. If people think that there is widespread divergence on some claim in aesthetics, that can be regarded as evidence that the claim is only relatively true. And if people think that almost everyone thinks that some moral claim is true, that can be evidence that the claim is universally true.

Evidence available to the learner

There is near universal consensus about the wrongness of hitting innocent people, stealing from others and cheating. As in other cases, there is no direct measure of whether the child gets evidence of this consensus, but it seems very likely that the evidence is available to them.

Sensitivity to the evidence

The foregoing suggestion is that people believe that moral claims are objective because that provides the best explanation of the evidence of consensus regarding the claims. Are people actually sensitive to consensus evidence when making judgements about whether a claim is universally or only relatively true? Indeed they are. A preliminary point is that recent work on folk metaethics shows, perhaps surprisingly, that there is variation in relativist responses *within the ethical domain* (Goodwin and Darley 2008). People are more likely to give relativist responses concerning abortion than they are for bank robbery. More important for our purposes is that this variation in relativist responses correlates with variation in perceived consensus (Goodwin and Darley 2008). When people think there is high consensus regarding a moral claim, people are more likely to give universalist responses concerning that claim; when there is low consensus, people are more likely to give relativist responses. So far that is just a correlational claim. But Goodwin and Darley (2012) also manipulated perceived consensus about an issue by informing participants that either there was high or low consensus in the United States about the issue. They found that high consensus led to increased judgements that the action was universally wrong. Finally, in our recent work, we found that people are sensitive to consensus information in making inferences about relativism even in non-moral domains. For instance, when there is low consensus about the colour of a paint sample, people are more likely to give relativist answers (Ayars and Nichols forthcoming).

5. Philosophical implications

One familiar concern about much experimental philosophy is that it doesn't inform philosophy itself. For the research programme outlined in this chapter, one might wonder how statistical learning accounts of acquisition could inform philosophical issues. In fact, I think that providing a statistical learning account of philosophically relevant concepts and distinctions might have several significant payoffs for philosophy.

5.1. Corroborating the character of philosophical representations

One dividend of a successful statistical learning account of a philosophically relevant representation is that it can provide evidence regarding the character of the representation. This might sound a little circular, so I'll say more. For many candidate representations of philosophical significance, there is debate over what the representation is. A learning theory might help move the debate forward. I'll review two kinds of cases.

Demandingness

One might doubt a characterization of a philosophical concept because it seems excessively demanding. Take the concept of *knowledge,* discussed in Section 3. Pinillos and I gave a learning theory that explains why we have a concept of knowledge that demands infallibility. Many contemporary philosophers reject infallibilist characterizations of the lay notion of knowledge since we often ascribe knowledge to an agent when the agent's belief is manifestly fallible. One might well wonder how we would have a concept of knowledge that is so demanding that many of our uses of the concept fail. Our statistical learning account of how the evidence available to children would lead a rational learner to the acquisition of an infallibilist conception of knowledge, provides support for the idea that agents actually do possess such a representation. The mystery, 'How would we have such a demanding concept?' can be addressed by offering a specific answer based in established principles of statistical learning.

It's possible that these concepts that get acquired in childhood and are then kicked aside by grownups, such that the adult conception of knowledge is scrubbed clean of any infallibilist elements. But there are reasons to doubt this Pollyanna view. First, as noted above, many philosophers are struck by the fact – registered by grown-up philosophers! – that it's weird to say 'He knows that p, but he could be wrong'. Infallibilism gives a direct explanation for the weirdness.

A second reason to doubt the view that adults simply do away with their early conceptions comes from work on adult judgement under cognitive load. For instance, young children tend to say that plants are not alive, at least partly because plants are viewed as stationary (Richards and Siegler 1986). As adults, of course, we know that plants are alive. We have no doubts about this matter. To see whether there are still vestiges of childhood inclinations, Goldberg and Thompson-Schill (2009) had adults do a categorization task under speed-measured conditions. They were presented with words for items (including plants, animals, non-moving artefacts, moving artefacts) and told to indicate, within 1s, whether the item was alive or not. The results indicated that the adults were more likely to mistakenly deny life to plants and also slower to make correct responses, as compared to their responses for whether animals were alive.[10] Under

---

[10] Adults were also more likely to mistakenly attribute life to moving non-living things (e.g. truck, river) than to non-moving non-living things (e.g. broom, stone). This fits with a classic Piagetian finding that children tend to judge moving entities to be alive.

timed conditions, adults are more likely to exhibit the childhood inclination for 'promiscuous teleology' (e.g. agreeing that 'Earthworms tunnel underground to aerate the soil') (Kelemen and Rossett 2009). More generally, across a variety of scientific domains, including physics, astronomy and biology, children's concepts drive adult judgements and reactions even after the childhood notions have been explicitly rejected in adulthood (Shtulman and Valcarcel 2012). Childhood conceptions thus continue to have a hold over us into adulthood, and so even if we as adults come to reject an infallibilist conception of knowledge, it's far from clear that we can completely extirpate its influence on our epistemic judgements.

Complexity

Let me turn now to a second reason to doubt certain characterizations of philosophically relevant representations is that the proposed representations seem excessively complex. We see this concern arise for proposals regarding moral rules. In the familiar trolley case in which one can pull a switch to divert a train so that it will kill one on the side-track instead of five on the main track, people tend to say that this is permissible. People also judge that it would be impermissible to pull the switch if there no one on the main track. And that it would be wrong to push someone into the path of the train to stop it from killing five. How do we explain this pattern of responses? One suggestion is that people have rules against battery which conform to the 'Principle of Double Effect' (PDE), such that the rules prohibit acting intentionally to produce an outcome but do not prohibit causing that outcome as an unintended side effect (Harman 2000; Mikhail 2011). This involves rather complex representational structures, and many cognitive scientists have tried to explain the pattern of judgements with more austere representational resources like aversions and habit learning (Greene 2008; Cushman 2013). Moreover, there are significant difficulties in empirically demonstrating that moral judgement really does involve such complexly structure rules. For what we typically measure is a person's judgement about a case, and even if rules play a part, judgements also depend on several other factors, including values, emotions and heuristics (Petrinovich and Oneill 1996; Bartels 2008; Amit and Greene 2012). It is difficult to isolate the unique contribution of the rule and hence to show that the rule does indeed have the complex structure that drives judgements that conform to PDE. A successful statistical learning account can again help. For if we can show that, given the evidence and the statistical principles that are available to the child, she *should* acquire the complexly structured rule that has been posited as the acquirendum, this lends credence to the hypothesis that this is indeed what gets acquired.

5.2. Statistical learning and nativism

Let's turn now to how this work bears on debates between empiricism and nativism. As noted in Section 2, the debate between empiricism and nativism concerns the nature of the operative learning mechanisms. As also noted there, the empiricist/nativist debate must always be considered for a particular domain. In the case of moral capacities, a number of recent theorists have developed an analogy between the moral faculty and the linguistic faculty and suggested that perhaps there is an innate domain-specific learning device for the acquisition of moral principles (Dwyer 1999; Harman 2000; Mikhail 2011). The most detailed discussion is to be found in Mikhail's *The Elements of Moral Cognition*. In the concluding pages of his book, he draws out the relations between the linguistic analogy and the broader import of moral nativism:

> Linguists and cognitive scientists have argued that every normal human being is endowed with innate knowledge of grammatical principles ... Both the classical understanding of the law of nature and the modern idea of human rights ... rest at bottom on an analogous idea: that innate *moral* principles are a common human possession. (Mikhail 2011, 317)

Mikhail notes that he hasn't argued for this view, but regards it as a deep question thrown into relief by the Chomskyan programme in moral psychology. With the linguistic analogy, Mikhail suggests, 'The existence and character of a natural moral law, which ancient belief held is written in every heart, is, or can be, a problem of ordinary science' (318).

The statistical learning accounts offered here point in a less nativist direction. In Sections 3 and 4, I argued that several aspects of our philosophical outlook can be explained with statistical learning models. Statistical learning mechanisms are, of course, general purpose. For instance, one can form overhypotheses about anything from the colour of marbles in bags to the character of rules in a norm system. Thus, the statistical learning accounts are naturally affiliated with empiricism. But we should be clear that the empiricism doesn't go all the way down. The proposals in Sections 3 and 4 presuppose that the learner already has important representational resources, like representations for *agent* and *wrong*; also, insofar as the content of the rules will be framed over *harms* and *intentions*, there must also be representations for these categories. For all I've said, all of those representations could be innately specified. Indeed, there is a more general question here. Statistical learning, as I've sketched it here, starts with a hypothesis space. And it's obviously a further question how the hypothesis space gets generated (see e.g. Nichols et al. 2006, 550; Xu and Tenenbaum 2007, 251). It might be the hypothesis space itself is innately specified. Nonetheless, although the representational ingredients for moral rules might be innate, what the statistical learning approach suggests is that the moral rules that are acquired – the rules that guide judgement and action – can be explained in terms of statistical learning over the evidence. So we have less ground for thinking that there is a universal moral law written into every heart. The moral law that gets taken up is a function of the evidence that learners receive.

5.3. Statistical learning and rationality

While statistical learning accounts don't support the idea that morality is written into our hearts, it can be reassuring in other ways. One familiar strand of argument in experimental philosophy is that people's philosophically relevant judgements are based on epistemically defective factors. Perhaps most famously, Greene (2008) argues that our non-utilitarian moral judgements derive from emotional distortions of rational thought. These kinds of debunking arguments aim to expose our ordinary philosophical judgements as rationally defective.

In contrast to these kinds of debunking accounts, traditional rationalists promote a vindicatory view of philosophical judgements. For example, Descartes (whose view has obvious debts to Plato) argues for a kind of rationalism which involves the recognition of innate truths (Newman 2016, Section 1.5). We use reason to uncover the truths. While the accounts offered in Sections 3 and 4 aren't rationalist in this Cartesian sense, they are rationalist in an evidentialist sense. According to evidentialism, S's belief is rational or justified just in case it is supported by S's evidence. Of course, this is exactly what statistical learning claims to capture – the learner is using statistically appropriate principles to make inferences from the available evidence. Statistical learning is rational learning. This contrasts with the debunking accounts of philosophical judgements. But it also contrasts with Chomskyan nativism. Chomskyan arguments for innate learning devices are emphatically not rationalist in the evidentialist sense. The whole point of *poverty-of-the-stimulus* arguments is that rational inference doesn't suffice to explain how the organism transitions from the available evidence (the stimulus) to the capacity. For instance, Chomskyans maintain that in the case of grammar learning, the child jumps to conclusions about the grammar that aren't warranted by the evidence. Thus in important ways, the statistical learning accounts on offer here are more optimistic about human rationality than are the Chomskyan nativists. For the statistical learning accounts suggest that, given the evidence available to the child, she *should* infer the kinds of rules that she acquires.

6. Conclusion

The goal of this chapter has been to provide a sense of how statistical learning accounts can contribute to experimental philosophy. Part of the ambition was to show several different principles of statistical inference to illustrate the range of possibilities. Many details were left aside in favour of the breathless presentation. But the promise for experimental philosophy is considerable. For any particular philosophical concept or belief or distinction we naturally have, there is a substantive question about its history. Statistical learning accounts provide an important new tool for investigating these questions. It's a tool that I hope will be taken up by many experimental philosophers.

Suggested Readings

Perfors, A., Tenenbaum, J. B., Griffiths, T. L., and Xu, F. (2011). A tutorial introduction to Bayesian models of cognitive development. *Cognition, 120*(3), 302–321.

Shafto, P., Goodman, N., and Griffiths, T. (2014). A rational account of pedagogical reasoning. *Cognitive Psychology, 71*, 55–89.

Smith, L., Jones, S., Landau, B., Gershkoff-Stowe, L., and Samuelson, L. (2002). Object name learning provides on-the-job training for attention. *Psychological Science, 13*(1), 13–19.

Xu, F., and Tenenbaum, J. B. (2007). Word learning as Bayesian inference. *Psychological Review, 114*(2), 245–272.

References

Amit, E., and Greene, J. D. (2012). You see, the ends don't justify the means: Visual imagery and moral judgment. *Psychological Science, 23*(8), 861–868.

Ayars, A., and Nichols, S. (2017). Moral empiricism and the bias for act-based rules. *Cognition, 167*, 11–24.

Ayars, A., and Nichols, S. (forthcoming). Rational learners and metaethics.

Bartels, D. M. (2008). Principled moral sentiment and the flexibility of moral judgment and decision making. *Cognition, 108*(2), 381–417.

Berwick, R. C., Pietroski, P., Yankama, B., and Chomsky, N. (2011). Poverty of the stimulus revisited. *Cognitive Science, 35*(7), 1207–1242.

Bramley, N. R., Mayrhofer, R., Gerstenberg, T., and Lagnado, D. A. (2017). Causal learning from interventions and dynamics in continuous time. In *Proceedings of the 39th Annual Meeting of the Cognitive Science Society.* Austin, TX: Cognitive Science Society.

Cowie, F. (2008). Innateness and language. *The Stanford Encyclopedia of Philosophy* (Fall 2017 Edition), Edward N. Zalta (ed.). https://plato.stanford.edu/archives/fall2017/entries/innateness-language/.

Cushman, F. (2013). Action, outcome, and value a dual-system framework for morality. *Personality and Social Psychology Review, 17*(3), 273–292.

Dewar, K., and Xu, F. (2010). Induction, overhypothesis, and the origin of abstract knowledge evidence from 9-month-old infants. *Psychological Science, 21*(12), 1871–1877.

Dodd, D. (2011). Against fallibilism. *Australasian Journal of Philosophy, 89*(4), 665–685.

Dwyer, S. (1999). Moral competence. In K. Murasugi and R. Stainton (eds.), *Philosophy and Linguistics* (pp. 169–190). Boulder, CO: Westview Press.

Dwyer, S., Huebner, B., and Hauser, M. D. (2010). The linguistic analogy. *Topics in Cognitive Science, 2*(3), 486–510.

Fontanari, L., Gonzalez, M., Vallortigara, G., and Girotto, V. (2014). Probabilistic cognition in two indigenous Mayan groups. *Proceedings of the National Academy of Sciences, 111*(48), 17075–17080.

Gaus, G., and Nichols, S. (2017). Moral learning in the open society. *Social Philosophy and Policy, 34*(1), 79–101.

Girotto, V., and Gonzalez, M. (2008). Children's understanding of posterior probability. *Cognition, 106*(1), 325–344.

Goldberg, R. F. and Thompson-Schill, S. L. (2009). Developmental 'roots' in mature biological knowledge. *Psychological Science, 20*(4), 480–487.

Goodman, N. (1955). *Fact, Fiction, and Forecast.* Cambridge, MA: Harvard University Press.

Goodwin, G., and Darley, J. (2008). The psychology of meta-ethics: Exploring objectivism. *Cognition, 106,* 1339–1366.

Goodwin, G., and Darley, J. (2012). Why are some moral beliefs perceived to be more objective than others? *Journal of Experimental Social Psychology, 48,* 250–256.

Harman, G. (2000). Moral philosophy and linguistics. In *Explaining Value* (pp. 217–226). Oxford: Oxford University Press.

Heibeck, T. H., and Markman, E. M. (1987). Word learning in children. *Child Development,* 1021–1034.

Joyce, R. (2001). *The Myth of Morality.* Cambridge: Cambridge University Press.

Kahneman, D., and Tversky, A. (1973). On the psychology of prediction. *Psychological Review, 80,* 237–251.

Kelemen, D. and Rosset, E. (2009). The human function compunction: Teleological explanation in adults. *Cognition, 111*(1), 138–143.

Kemp, C., Perfors, A., and Tenenbaum, J. B. (2007). Learning overhypotheses with hierarchical Bayesian models. *Developmental Science, 10*(3), 307–321.

Kirkham, N. Z., Slemmer, J. A., and Johnson, S. P. (2002). Visual statistical learning in infancy: Evidence for a domain general learning mechanism. *Cognition, 83,* B35–B42.

Laurence, S., and Margolis, E. (2001). The poverty of the stimulus argument. *The British Journal for the Philosophy of Science, 52*(2), 217–276.

Mackie, J. (1977). *Ethics: Inventing Right and Wrong.* London: Penguin.

Marler, P. (2004). Science and birdsong: the good old days. In P. Marler and H. Slabbekoorn (eds.) *Nature's Music: The Science of Birdsong.* San Diego, CA: Elsevier Academic Press.

McNaughton, D., and Rawling, P. (1991). Agent-relativity and the doing-happening distinction. *Philosophical Studies, 63*(2), 167–185.

Mikhail, J. (2011). *The Elements of Moral Cognition.* Cambridge: Cambridge University Press.

Millhouse, T., Ayars, A., and Nichols, S. (forthcoming). Learnability and moral nativism: Exploring Wilde rules. In J. Suikkanen and A. Kauppinen (eds.), *Methodology and Moral Philosophy.* Routledge.

Murray, D., and Nahmias, E. (2014). Explaining away incompatibilist intuitions. *Philosophy and Phenomenological Research, 88*(2), 434–467.

Nagel, T. (1986). *The View from Nowhere.* Oxford: Oxford University Press.

Newman, L. (2016). Descartes' Epistemology, *The Stanford Encyclopedia of Philosophy* (Winter 2016 Edition), Edward N. Zalta (ed.). https://plato.stanford.edu/archives/win2016/entries/descartes-epistemology/.

Nichols, S. (2004). *Sentimental Rules: On the Natural Foundations of Moral Judgment.* New York: Oxford University Press.

Nichols, S., and Gaus, J. (forthcoming). Unspoken rules: Resolving underdetermination with closure principles.

Nichols, S., Kumar, S., Lopez, T., Ayars, A, and Chan, H. (2016). Rational learners and moral rules. *Mind & Language, 31,* 530–554.

Nichols, S., and Mallon, R. (2006). Moral dilemmas and moral rules. *Cognition, 100*(3), 530–542.

Nichols, S., and Pinillos (forthcoming). Skepticism and the acquisition of 'knowledge'. *Mind & Language.*

Parfit, D. (1984). *Reasons and Persons.* Oxford: Oxford University Press.

Pellizzoni, S., Siegal, M., and Surian, L. (2010). The contact principle and utilitarian moral judgments in young children. *Developmental Science, 13*(2), 265–270.

Perfors, A., Tenenbaum, J. B., Griffiths, T. L., and Xu, F. (2011). A tutorial introduction to Bayesian models of cognitive development. *Cognition*, *120*(3), 302–321.

Petrinovich, L., & O'Neill, P. (1996). Influence of wording and framing effects on moral intuitions. *Ethology and Sociobiology*, *17*(3), 145–171.

Pullum, G. K., and Scholz, B. C. (2002). Empirical assessment of stimulus poverty arguments. *The Linguistic Review*, *18*(1–2), 9–50.

Richards D., and Siegler R. (1986). Children's understanding of the attributes of life. *Journal of Experimental Child Psychology*, *42*, 1–22.

Rysiew, P. (2001). The context-sensitivity of knowledge attributions. *Noûs*, *35*(4), 477–514.

Saffran, J., Aslin, R., and Newport, E. (1996). Statistical learning by 8-month-old infants. *Science*, *274*(5294), 1926.

Saffran, J., Johnson, E., Aslin, R., and Newport, E. (1999). Statistical learning of tone sequences by human infants and adults. *Cognition*, *70*(1), 27–52.

Shtulman, A., and Valcarcel, J. (2012). Scientific knowledge suppresses but does not supplant earlier intuitions. *Cognition*, *124*(2), 209–215.

Shafto, P., Goodman, N., and Griffiths, T. (2014). A rational account of pedagogical reasoning. *Cognitive Psychology*, *71*, 55–89.

Smith, L., Jones, S., Landau, B., Gershkoff-Stowe, L., and Samuelson, L. (2002). Object name learning provides on-the-job training for attention. *Psychological Science*, *13*(1), 13–19.

Swain, S., Alexander, J., and Weinberg, J. (2008). The instability of philosophical intuitions. *Philosophy and Phenomenological Research*, *76*(1), 138–155.

Wright, J. C., McWhite, C., and Grandjean, P. (2014). The cognitive mechanisms of intolerance. *Oxford Studies in Experimental Philosophy*, *1*, 28–61.

Xu, F., and Denison, S. (2009). Statistical inference and sensitivity to sampling in 11-month-old infants. *Cognition*, *112*(1), 97–104.

Xu, F., and Garcia, V. (2008). Intuitive statistics by 8-month-old infants. *Proceedings of the National Academy of Sciences*, *105*(13), 5012–5015.

Xu, F., and Tenenbaum, J. B. (2007). Word learning as Bayesian inference. *Psychological Review*, *114*(2), 245–272.

2

Eyes as Windows to Minds: Psycholinguistics for Experimental Philosophy

Eugen Fischer and Paul E. Engelhardt

Much philosophical thought occurs in natural language, as thinkers read or write philosophical texts, discuss philosophical problems with each other or engage in the subvocalized speech characteristic of conscious thought (*cf.* Carruthers 2002). Philosophical thought is therefore bound to be influenced by the automatic processes that continually go on in language comprehension and production. Much philosophical reasoning proceeds from verbal descriptions of possible cases. In thought experiments, such descriptions prompt intuitions about what else is also true of the cases described, and such intuitive judgements are frequently treated as evidence for or against philosophical theories (review: Weinberg 2016; *pace* Cappelen 2012; Deutsch 2015). Many philosophical arguments involve inferences from premises that describe a possible case, to conclusions about what else must also be true of it. Such judgements and conclusions can be generated by routine comprehension inferences, which, for example, have us automatically infer from 'The secretary fell out of the window' that the protagonist is female (Atlas and Levinson 1981), was initially located in a building and was subsequently injured or killed (McKoon and Ratcliff 1980). While many inferences triggered by philosophical case descriptions may be due to domain-specific processes (like 'mindreading', which may generate intuitive knowledge attributions; see Nagel 2012; Gerken 2017), many others will be due to routine comprehension processes.

Acknowledgements

Both authors contributed to material development, study design and interpretation of results. Paul Engelhardt undertook data-collection and statistical analyses. Eugen Fischer undertook the remaining research. For comments on previous drafts, the authors thank Rachel Giora and Shaun Nichols. For comments on closely related material, we are indebted to audiences in Norwich (July 2017), Osnabrück and Reading (November 2017) and London (June 2018).

An important strand of experimental philosophy examines whether and when case intuitions have evidentiary value – and philosophers possess warrant for accepting them. Most of this work to date has proceeded by examining the sensitivity of relevant intuitions to truth-irrelevant parameters (like the order in which cases are presented) and infers lack of evidentiary value where it observes such sensitivity (reviews: Mallon 2016; Stich and Tobia 2016; Knobe and Nichols 2017). Partially in response to replication issues and theoretical challenges to the key inference from observed sensitivity to lack of evidentiary value (e.g. Horne and Livengood 2017), recent calls for an 'experimental philosophy 2.0' (Nado 2016) have suggested that research in this strand should (a) be refocused on the examination of specific cognitive processes that underpin philosophical thought (building on e.g. Nichols and Knobe 2007), (b) deploy the resulting understanding of how specific processes work, to develop 'epistemological profiles', which indicate under which conditions we may (not) trust the processes' outputs (Weinberg 2015, 2016), and (c) employ such profiles to assess a wider range of outputs: not only intuitive judgements but also inferences in arguments (Fischer and Engelhardt 2017a, 2017b).

This paper will discuss and demonstrate how experimental philosophers (and especially an 'experimental philosophy 2.0') can recruit methods from psycholinguistics to study automatic inferences in language comprehension and production, with a view to assessing philosophically relevant intuitions and arguments.[1] To demonstrate the approach and illustrate its potential philosophical usefulness, we will present a case study on the process of stereotypical enrichment (Levinson 2000) and its role in the influential 'argument from hallucination', a classical paradox about perception. This case study will contribute towards an epistemological profile of the key process: we will identify conditions under which stereotypical inferences predictably fail to lead to true conclusions, argue that these conditions obtain in formulations of the target argument, and present two studies – including one fresh study – which deploy different psycholinguistic methods to examine the hypothesis that under the conditions predicted, competent language users cannot help making inappropriate stereotypical inferences, despite knowing they are inappropriate (*sic*). The findings will support a novel resolution of the targeted philosophical paradox.

---

[1] For use of psycholinguistic methods in conceptual analysis, see Powell et al. (2014).

Section 1 will give an initial overview of the psycholinguistic methods that have been used to study automatic inferences involved in stereotypical enrichment and other comprehension/production processes. Section 2 will present our chosen philosophical application: It will identify a cognitive bias ('salience bias') besetting the generally reliable process of stereotypical enrichment, and explain how the processes' nascent epistemological profile can be used to assess philosophically relevant intuitions and arguments; a fresh analysis of the 'argument from hallucination' will suggest this classical paradox relies on stereotypical inferences which are contextually inappropriate. Section 3 will explain how we have used questionnaire-based methods and convenient 'offline' (outcome) measures to study comprehension inferences and garner first evidence of contextually inappropriate stereotypical inferences. The following sections will explore how these methods can be complemented by 'online' measures (which tap into cognitive processes as they unfold). We will explore approaches that use people's eyes as windows into their minds: Section 4 will discuss a study that employs pupillometry (measurements of pupil dilations) to provide further evidence of inappropriate automatic inferences, in speech comprehension. Section 5 will report a fresh study that measures reading times to investigate inferences in text comprehension. Section 6 will present some potential methodological lessons.

1. Automatic inferences and their psycholinguistic study

We now introduce the automatic inference process of interest and then review the psycholinguistic methods that have been used to study it.

1.1. Stereotypical enrichment

Semantic memory is our memory for facts and 'general world knowledge', as opposed to personally experienced or 'episodic' events (Tulving 2002; McRae and Jones 2013). It is commonly conceived as a semantic network which doubles as information-storage and inference engine. In first approximation, such a network consists of nodes representing concepts and links between them that can automatically pass on activation from stimuli, verbal and other, along several pathways simultaneously (Allport 1985). When a concept is *'activated'* it is more likely to be used by several cognitive processes, crucially including processes involved in utterance comprehension (from word-recognition to disambiguation) and forward inference: Links in semantic memory facilitate a plethora of probabilistic parallel inferences, in processes including language comprehension.

According to standard conceptions of semantic memory (Neely and Kahan 2001; McRae and Jones 2013),[2] the observed co-occurrence of features (things and their common properties, wholes and their common parts) and events (causes and typical effects, etc.) forges links between the respective nodes which grow stronger upon frequent activation and atrophy upon disuse. The more frequently we encounter tomatoes that are red (in the supermarket) or Germans who are nasty (in war movies), the stronger the links between the respective concepts become, the more activation gets passed on from the stimuli 'tomato' and 'German', respectively, to nodes representing 'red' and 'nasty', respectively. These concepts thus come to be *stereotypically associated* with the words: They are activated most rapidly and strongly, and come to mind first, when we encounter the words. The strength of these associations encodes information about the co-occurrence frequencies in the subject's physical and discourse environment. Such empirical knowledge is brought to bear in processes including language comprehension: While stereotypical associations do not determine the extension of words (Hampton and Passanisi 2016), they support automatic inferences from words ('tomato', 'German') to stereotypically associated features (*red* and *nasty*, respectively).

Embedded in cooperative communication (Grice 1989), such stereotypical inferences address the challenge of the 'communication bottleneck'. Articulation of speech proceeds at a slower pace than pre-articulation processes in speech production (Wheeldon and Levelt 1995) or parsing and inference processes in comprehension (Mehler et al. 1993) – 'inference is cheap, articulation is expensive' (Levinson 2000, 29). This bottleneck is mitigated by inferences which systematically draw on information encoded by stereotypes. This process of *stereotypical enrichment* is captured by the bi-partite *I-heuristic* ('What is expressed simply is stereotypically exemplified', Levinson 2000, 37): Speakers facilitate and listeners devise interpretations that are positive, stereotypical and highly specific, in line with the maxims:

---

[2] Kahneman (2011, part I) provides an elegant introduction; textbook: Harley (2014, ch.11).

(I-speaker) Skip mention of stereotypical features but make deviations from stereotypes explicit (e.g. 'male secretary').

(I-hearer) In the absence of such explicit indications to the contrary, assume that the situation talked about conforms to the relevant stereotypes, deploy (also) the most specific stereotypes relevant and fill in details in line with this knowledge about situations of the kind at issue.

We now review some of the methods psycholinguists have used to study stereotypical associations and the automatic inferences they support, alongside some of the key findings.

1.2. Psycholinguistic methods

Stereotypical associations have been examined through a variety of offline and online measures. Offline measures include plausibility ratings (How plausible/ common is it that a tomato is red?), how frequently and early features are mentioned in listing tasks (List common features of tomatoes!) and the frequency with which participants use a word to complete a sentence-frame ('Tomatoes are___') (*cloze probability*) (McRae et al. 1997).

Priming experiments then serve to examine activation: Participants are presented with a '*prime*' word, sentence (-fragment), or short text and then a '*probe*' word or letter string, and have to read out the word, decide whether the string forms a word, or judge whether the referents of prime and probe words both fall under the same category (e.g. *good* or *bad*). That the prime activates the probe concept is inferred from shorter response times, for example for 'money'-'bank' than 'honey'-'bank' (Lucas 2000). Varying the time between the presentation of prime and probe ('stimulus onset asynchrony') allows researchers to examine the time course of activation.

Priming experiments have shown that single words, presented in isolation, activate stereotypically associated features rapidly, within 250 ms (review: Engelhardt and Ferreira 2016). Event-nouns (Hare et al. 2009) and verbs (Ferretti et al. 2001) activate a particularly wide and complex range of features: Where the actions or events denoted typically involve particular kinds of agents, 'patients' acted on, instruments used, or relations between them (Tanenhaus et al. 1989), event words activate typical features pertaining to the fillers of all these different thematic roles. For example, the verb 'dig' activates the instrument *spade* (Ferretti et al. 2001), 'arrest' activates the agent *cop* and the patient *criminal* (ibid.), while telic verbs ('washing') swiftly activate both initial and resulting patient properties (*dirty, clean*) (Welke et al. 2015). Conversely, typical agent-, patient-, instrument- and location-nouns activate relevant verbs (Hare et al. 2005). Different words thus activate comprehensive stereotypes that can include perceptual features (e.g. *small, dirty*), evaluative features (e.g. *mean*) and functional features (i.e. involvement of particular instruments) (McRae et al. 1997; Ferretti et al. 2001).

Patterns of activation of features by words thus represent comprehensive 'event knowledge' about the typical features of actions like sewing and washing: who typically does it, with what instruments, what consequences are typically caused or intended, etc. Stereotypes can represent such knowledge because they are not unstructured 'bags of features' but have internal (thematic) structure. In incremental language comprehension, activation of features is sensitive to thematic roles (agent, patient, etc.): Sentence fragments ('She was arrested by the ___') activate typical agents (*cop*) in post-verbal position only when they leave the agent role blank (as above), not when they leave open the patient ('She arrested the ___'), and *vice versa* (Ferretti et al. 2001, Exp.4). These complex, internally structured stereotypes are known as *generalized situation schemas* (Rumelhart 1978; Tanenhaus et al. 1989).

Various online measures have been used to examine the inferences from words potentially supported by schemas (e.g. from 'S is sewing' to *S uses a needle*). These studies typically use a '*cancellation paradigm*' and materials where the word of interest is followed by a sequel that is inconsistent with (cancels) the hypothesized inference (e.g. ' ... the job would be easier if Carol had a needle'; see Harmon-Vukic et al. 2009). If participants make the critical inference from the prior verb, their conclusion's clash with the sequel will create comprehension difficulties, which require effort to overcome. This effort is reflected in different measures.

Pupillometry exploits the fact that cognitive effort makes our eyes widen, so that pupil dilation is an index of cognitive effort (Kahneman 1973; Laeng et al. 2012). We can therefore examine whether participants make automatic inferences from words by measuring their pupil size during and after they hear sentences with sequels that are either consistent or inconsistent with (cancel) the hypothesized inference from previous text. Dilations prompted by 'inconsistent' sentences but absent from otherwise identical 'consistent' counterparts are evidence of hypothesized inferences.

Reading-time measurements can build on the fact that when we read (rather than hear) sentences, comprehension difficulties cause us to slow down, and trigger increased backward (right-to-left) eye movements, called 'regressions'. The simplest way to detect slow-downs is to present text in small instalments of single words, sentences or lines, on a computer screen, and ask participants to read at a comfortable pace and advance the text by pressing a key on the keyboard. Studies using this '*self-paced moving window*' paradigm show that participants read the remainder of the sentence more slowly when subject and verb were followed by a patient atypical for that agent-action pairing, rather than a typical patient ('The *mechanic/journalist* checked the spelling of his latest report') (Bicknell et al. 2010, Exp.1). A similar finding was made for instruments ('Susan used the *saw/scissors* to cut the expensive paper … '), despite the absence of single-word priming of typical patients (e.g. *scissors-paper*) (Matsuki et al. 2011). These findings suggest that readers make automatic inferences supported not only by stereotypes associated with individual words ('journalist', 'mechanic', 'checking', etc.), but also by more specific situation schemas encoding knowledge (e.g. about car inspections) which are not associated with any one word but get activated by combinations of words (see also Metusalem et al. 2012).

Eye-trackers record the position of the reader's eye (up to every millisecond). They permit more fine-grained and differentiated reading-time measurements than the self-paced moving window paradigm as well as analyses of regressions. Dependent measures employed in the study of automatic inferences include[3]

- *first-pass reading time:* the sum of all fixations in a region of text, from first entering that region until leaving that region either in a forward or backward direction;
- *regression path duration:* the time from first entering a region until moving past that region forward in text (unlike first-pass reading time this also includes time spent on regressions out of the region);
- *second-pass reading time:* the sum of all fixations in a region following the initial first-pass reading; and
- *total reading times:* the sum of all fixations in a region.

A classic study (Rayner et al. 2004) examined inferences to typical patients prompted by combinations of verbs and instrument-nouns (consistent with the sequel in 1, but not 2):

1) John used a knife to chop the large carrots for dinner last night.
2) John used an axe to chop the large carrots for dinner last night.

---

[3] Unfortunately, definitions of these dependent variables differ across research labs and software packages. The present definitions are from Clifton et al. (2007) and employed in the current study (see Section 5).

Rayner and colleagues observed (marginally) increased regression-path durations on the (one-word) region of interest ('carrots') and (significantly) for the following (n+1) region ('for dinner'), in (2) than in (1). By contrast, where the word of interest was inconsistent with inferences prompted by the prior verb alone ('inflate' in 3, below), they observed also longer first-pass reading times (gaze durations) for both that word and its sequel.

3) John used a pump to inflate the large <u>carrots</u> for dinner last night.

Inferences with different support (here: schemas activated by the verb alone or by combinations of verbs and nouns) or leading to expectation violations of different magnitude (3 > 2) may thus affect different eye-tracking measures.

<u>Electroencephalogram (EEG) measurements</u> often complement approaches that rely on the responsiveness of our eyes to clashes between expectations and subsequent text: In addition to the characteristic eye responses just reviewed, there are signature electrophysiological responses in the brain, known as '*event-related brain potentials*' (ERPs) (Kutas and Federmeiner 2000, 2011). EEG measurements record electrical brain activity at the human scalp. By averaging, researchers can extract from such recordings a time series of changes in electrical brain activity before, during and after an event of interest. The amplitudes, latencies and scalp topographies of these evoked potentials were found to systematically vary not only with features of the linguistic or other stimulus but also with readers'/hearers' expectations. For example, violations of expectations based on syntactic rules (say, failure in gender agreement between pronoun and antecedent) produce positive deflections in the ERP waveform that peak 600 ms after stimulus onset (known as 'P600'). By contrast, violations of expectations based on knowledge encoded by stereotypes and schemas produce negative deflections that peak 400 ms after stimulus onset (known as 'N400').

ERP findings help us interpret the results of priming and eye-tracking studies: Semantic category violations ('Dutch trains are *sour* ... ') and conflicts with empirical knowledge ('Dutch trains are *white* ... '; when they are actually yellow) lead to the same N400 response (with similar amplitude, topography, onset and peak latency) (Hagoort et al. 2004). This suggests that lexical and empirical knowledge is deployed in the same way at the level of associative processing – and may both be encoded together as components of the same schema. Second, ERP results provide subtle further evidence that schemas are not 'bags of features' but deployed in a way sensitive to thematic roles: While N400 amplitudes indicate that the verb is expected where preceded by subject and object ('The restaurant owner forgot which *customer* the *waitress* had <u>served</u> ... '), even where their typical roles are reversed (' … which *waitress* the *customer* had <u>served</u>') (Chow et al. 2016), this reversal prompts signature responses to syntactic violations (P600) suggesting participants expected the verb in the passive voice ('which waitress the customer had been served by'), consistent with assignments of agent and patient-roles typical for the verb (Kim et al. 2016; *cf.* Kim and Osterhout 2005). Finally, ERP studies suggest that inferences prompted by combinations of verbs and preceding nouns are supported by more specific situation schemas that are activated already at the verb (Bicknell et al. 2010, Exp.2).

The research reviewed supports a *'cued schemas account'* of language comprehension and production: 'Linguistic coding is to be thought of less like definitive content and more like interpretative clue' (Levinson 2000, 29). Words and syntactic constructions (Goldberg 2003; verb aspect: Ferretti et al. 2007; Kehler et al. 2008) are used as complementary clues for indicating and accessing relevant semantic and empirical knowledge in incremental language comprehension and production (Elman 2009). Relevant knowledge is encoded by situation schemas and other stereotypes. Increasingly specific schemas are activated by words and combinations of verbs and agent-, patient-, or instrument-nouns, as well as discourse context (Metusalem et al. 2012). Activated schemas then support a multitude of rapid, parallel stereotypical inferences: At each point, receivers use the most specific inferences to flesh out utterance content, in line with the I-heuristic. The activation processes in semantic memory that support these inferences occur in language comprehension *and* production (Stephens et al. 2010; Pickering and Garrod 2013). Stereotypical enrichment will hence occur not only in interpersonal communication but also in the kind of subvocalized speech involved in philosophical thought (*cf.* Carruthers 2002).

2. Philosophical application

We can draw on psycholinguistic research to develop an epistemological profile of this important process, which tells us under what conditions we may (not) trust the stereotypical inferences we automatically make (Weinberg 2015, 2016). This profile can then be used to assess philosophical intuitions and arguments.

2.1. Epistemological profiles

With some caveats, stereotypical enrichment is generally reliable. The strength of stereotypical association gradually increases through continued observation of co-occurrence in the physical and discourse environment (seeing more red tomatoes in the supermarket, watching more war movies full of nasty Germans); it gradually decreases, as incompatible observations accumulate (seeing green tomatoes in the fields, meeting friendly Germans) (Neely and Kahan 2001; McRae and Jones 2013). Strength of stereotypical association thus encodes information about co-occurrence frequencies in the subject's physical and discourse environment. To the extent that the physical environment is stable and changes only gradually, and the discourse environment is free from systematic misrepresentation (no war propaganda), probabilistic stereotypical inferences are therefore reasonably accurate. Second, the maxim I-speaker (above) has speakers make deviations from stereotypes explicit, and where contextual cues defeat conclusions of stereotypical inferences, these conclusions typically get swiftly suppressed (Faust and Gernsbacher 1996) or simply decay for wont of reinforcement (Oden and Spira 1983). Such processing largely prevents contextually inappropriate inferences from interfering with utterance comprehension and further reasoning.

We now build up to a set of conditions under which the generally reliable process predictably leads to inappropriate inferences which go through to affect further reasoning, even so (*cf.* Fischer and Engelhardt 2016; under review). This cognitive bias arises from the way in which polysemous words are processed. Most words have more than one meaning or sense (Klein and Murphy 2001). A linguistic stimulus activates all semantic and stereotypical features associated with the expression, in *any* of its meanings or senses, regardless of contextual propriety – for example 'mint' activates the probe *candy*, even when used in a different sense (as part of the prime 'All buildings collapsed except the mint') (Simpson and Burgess 1985; Till et al. 1988). The strength of initial activation is ordered by '*salience*', understood as a function of exposure frequency (how often a language user encounters the word in that sense), modulated by prototypicality (how good examples of the relevant category the word is deemed to stand for, in that sense) (Giora 2003; Giora et al. 2015).[4] The more salient a sense is for a speaker/hearer, the more rapidly and strongly the associated situation schema is activated. The more strongly activated a schema is, the longer its activation takes to decay (Loftus 1973; Farah and McClelland 1991) and the more effort is required for its suppression (Levy and Anderson 2002; De Neys et al. 2003).

---

[4] Exposure frequency cannot be directly measured, but is inferred from occurrence frequencies in corpora, familiarity ratings, or conventionality ratings (Giora 2003). Prototypicality is usually assessed through listing, sentence-completion, or typicality-rating tasks (Battig and Montague 1969, Chang 1986).

Salience imbalances can therefore lead to an interpretation bias, where utterances employ words in less salient senses or uses – but their interpretation is unduly influenced by the schema associated with the most salient or dominant use.

Such a 'salience bias' (Fischer and Engelhardt, under review) is particularly liable to arise where less salient uses are interpreted with a *Retention/Suppression Strategy* (Giora 2003, 37; henceforward 'Retention Strategy', for short): Where utterances use a polysemous word in a less salient sense, they are often interpreted by retaining the most rapidly activated schema associated with the most salient sense and suppressing the contextually inappropriate component features of the dominant schema (Giora 2003; Giora et al. 2014). This Retention Strategy has been shown to be used in the interpretation of figurative speech (Giora et al. 2007b, 2015), crucially including (non-default) metaphor (Giora et al. 2007a). We would expect it to be involved, for example in interpreting metaphorical uses of the verb 'to see': According to a recent corpus study using the *British National Corpus* (Fischer and Engelhardt 2017b), 'see' is used far more frequently in a visual sense (68%) than in an epistemic sense ('I see your point', 12%), a doxastic sense ('as he saw fit', 10%) or a phenomenal sense ('Hallucinating, Macbeth saw a dagger', 1%).[5] The schema associated with the dominant visual sense ('*vision-schema*') includes agent-features like *S uses her eyes, S looks at X, S knows X is there* and *S knows what X is* as well as patient-features like *X is in front of S, X is near X, X is there at the same time as S*. To interpret epistemic uses in line with the Retention Strategy, most of these features get suppressed, while retaining the contextually appropriate agent-features *S knows X is there* and *S knows what X is* (yielding the interpretation, 'I know you've got a point and know what it is').[6]

Frequently co-occurring components of a situation schema activate others (McRae et al. 2005; Hare et al. 2009). Where a frequently used word has a dominant sense that is far more salient than all others (like 'see'), the components of the associated schema will frequently co-occur. Initial activation of contextually inappropriate schema components will then not only be strong (due to salience) but also complemented by lateral cross-activation from other schema components. It will then be difficult to suppress only some of the frequently co-occurring components, but not others. Where some of them are retained for utterance interpretation, the others will remain at least partially

---

[5] A production experiment using a sentence-completion task determined the same rank order for prototypicality, with even higher preponderance of the visual sense (Fischer and Engelhardt 2017b). We infer that the visual sense is more salient than the epistemic and doxastic senses and these, in turn, are more salient than the phenomenal sense.

[6] The eye-tracking study reported in Section 5 provides evidence that the Retention Strategy is applied here.

activated and support contextually inappropriate inferences that go through unsuppressed.

We have thus arrived at a first set of jointly vitiating conditions under which the generally reliable process of stereotypical enrichment is liable to lead to inappropriate conclusions that affect further judgement and reasoning:

[Salience Bias Hypothesis SBH] At least <u>when</u>

(i) one sense of a polysemous high-frequency word is much more salient than all others,

(ii) the dominant stereotype (situation schema) is deployed in interpreting utterances involving a less salient use and

(iii) some, but not all, of the stereotype's frequently co-occurring core components are contextually relevant, <u>then</u>

1) inappropriate stereotypical inferences licensed only by the dominant sense will be triggered by the less salient use and

2) these automatic inferences will influence further judgement and reasoning, even when thinkers explicitly know they are inappropriate.

We have thus obtained first components of an epistemological profile of the process of stereotypical enrichment. We now explore how they can be deployed to assess philosophically relevant intuitions and reconstructions of philosophical arguments.

2.2. Philosophical assessments

Arguably, most of the intuitive judgements about, and inferences from, case descriptions that philosophers make in thought experiments and argument rely on everyday conceptual and linguistic competencies which are also deployed in ordinary discourse (Williamson 2007, 188) – and on the routine cognitive processes underlying these competencies. Insofar as these routine processes are reliable, the philosophically relevant intuitions and inferences should be reliable as well, and the intuitions they generate should have *evidentiary value* (i.e. the fact that thinkers have them, as and when they do, should speak for the intuitions' truth). So just how far are those routine processes reliable?

To address this question, some experimental philosophers have started to develop '*GRECI explanations*' (as we have called them elsewhere, see Fischer and Engelhardt 2016). These explanations trace philosophically relevant intuitions back to cognitive processes which are generally <u>reliable</u> but subject to cognitive biases which generate cognitive illusions under specific conditions. At any rate when produced under these vitiating conditions, intuitions lack evidentiary value. For example, intuitive knowledge attributions elicited through the method of cases have been traced to a mindreading competency (Nagel 2012; Gerken 2017) that has been argued to be generally reliable (Boyd and Nagel 2014), but subject to cognitive biases including a focal bias which may assert itself, for instance, in thought experiments allegedly revealing contrast effects and supporting contrastivism about knowledge (Gerken and Beebe 2016).

Above, we built up towards a GRECI explanation that traces philosophically relevant intuitions not to a domain-specific process like mind-reading, but to a potentially complementary domain-general language comprehension/ production process, stereotypical enrichment, which we argued to be generally reliable but subject to cognitive biases including a salience bias. An empirically supported account of this bias will call into question the evidentiary value of intuitions about cases whose descriptions use familiar words in special senses for whose interpretation the dominant sense may be functional. Many philosophical thought experiments involve 'esoteric' cases involving well-behaved zombies, envatted brains, twin planets, etc. These cases are described with familiar words, given rare new uses, for whose interpretation the dominant sense will typically be functional (in the absence of explicit explanations). To the extent to which the cases deviate from dominant stereotypes (e.g. the 'zombies' behave like us), the dominant stereotypes are then liable to support contextually inappropriate inferences. Our account of salience bias thus supports a variant of the 'esotericity thesis' that intuitions about esoteric cases are less reliable than about 'normal' cases (Weinberg 2007; Williamson 2007). Prior to further psycholinguistic investigation (including examination of the precise salience structure of the relevant words), it provides an undermining defeater of the relevant intuitions (Pollock 1984). Already nascent epistemological profiles which do not (yet) go beyond arguments for general reliability of the target process and the identification of a first set of vitiating conditions thus allow us to assess the evidentiary value of at least some philosophical case intuitions.

In cognitive psychology, *intuitions* are typically conceptualized as judgements generated by automatic inferences, that is, by autonomous cognitive processes that place low demands on working memory (Evans and Stanovich 2013) and duplicate inferences with heuristic rules (Kahneman and Frederick 2005). Such processes may issue either in explicit judgements ('intuitions') or conclusions that are tacitly presupposed in further cognition (judgement and reasoning). 'Experimental philosophy 2.0' seeks to extend epistemological investigation from intuitions to arguments (Nado 2016) and we now focus on automatic inferences driving philosophical argument.

Our example is taken from the philosophy of perception, where philosophers wishing to merely describe perceivers' subjective experience systematically use familiar appearance- and perception-verbs in a rarefied 'phenomenal' sense, which lacks existential, factive and spatial implications (e.g. Ayer 1956, 90; Jackson 1977, 33–49; Fish 2010, 6; Macpherson 2013, 5; *cf.* Chisholm 1957, 44–48). We submit that it satisfies condition (i)-(iii) from Hypothesis SBH: At any rate for 'see', we have shown (i) that the verb has a clearly dominant (visual) sense and that the phenomenal sense is the least salient sense (above, Fn.5). We suggest that (ii) the dominant word schema is retained and deployed to interpret the latter (*cf.* Giora 2003; Giora et al. 2014): A situation-model that instantiates the dominant schema with specific patient-role fillers is constructed. This model contains a set of phenomenal features as a component, and these features are attributed to the target experience, in a variant of the common 'feature transfer' approach of metaphor interpretation (Ortony 1993; Bortfeld and McGlone 2001). However, (iii) what it is like to see something is strongly associated with the other features of the schema associated with the dominant use of 'see', as evidenced by embodied cognition effects associated with visual metaphors (Landau et al. 2010; Lakoff 2012). Accordingly, it is hard to retain only the phenomenal component and suppress the schema's other core components. Hypothesis SBH therefore predicts that uses of the rarefied phenomenal sense will prompt contextually inappropriate (e.g. existential and spatial) inferences supported – only – by the dominant visual sense of the verb.

The special phenomenal sense is then used to talk about unusual cases, like hallucinations. Where thinkers know little about the relevant cases (e.g. hallucinations), conclusions are yet less likely to be suppressed through integration with background knowledge (*cf.* Metusalem et al. 2012; Fischer and Engelhardt 2017a), and yet more likely to affect further cognition (judgement and reasoning). We therefore regard it as particularly likely that 'arguments from hallucination' will involve such contextually inappropriate stereotypical inferences.

In their traditional version, these arguments argue for the existence of mind-dependent objects of sense-perception ('sense-data'), which separate subjects from any physical objects in the environment (Ayer 1956, 90; Smith 2002, 194–197; Fish 2010, 12–15; Macpherson 2013, 12–13; *cf.* Crane and French 2015, Section 3.1). Here is a classic statement that explicitly marks the special phenomenal sense used:

'Let us take as an example Macbeth's visionary dagger […] There is an obvious [perceptual] sense in which Macbeth did not see the dagger; he did not see the dagger for the sufficient reason that there was no dagger there for him to see. There is another [viz., phenomenal] sense, however, in which it may quite properly be said that he did see a dagger; to say that he saw a dagger is quite a natural way of describing his experience. But still not a real dagger; not a physical object … If we are to say that he saw anything, it must have been something that was accessible to him alone … a sense-datum.' (Ayer 1956, 90)

The second half of the argument then generalizes from this special case (hallucination) to all cases of visual perception: Since subjectively indistinguishable experiences (supposedly) must be mental states of the same type, and mental states of the same type must have objects of the same kind (mind dependent or independent), actual perceptual experiences must have the same objects of awareness as possible hallucinatory experiences that are subjectively indistinguishable from them.

But note the persuasive fallacy in the first half: Macbeth is meant to have an experience just like that of seeing a physical dagger. In the phenomenal sense (where 'S sees an F' means 'S has an experience like that/as of seeing an F'), he therefore *can* be said to 'see a physical dagger' (his visual experience is, by assumption, just like that of seeing a physical dagger) – while he cannot be said, for example, to see a translucent non-physical dagger (his experience is not like that). In the quoted passage, the move from

1) 'Macbeth saw a dagger' (in the phenomenal sense)

to 'but still not a real dagger' is hence fallacious. We suggest the argument is driven by inappropriate inferences from the phenomenal use of 'see' to conclusions that typically remain implicit, but are presupposed in further reasoning, for example

2) There then was something that Macbeth saw. – But, by assumption:
3) 'There then was no physical object for Macbeth to see.' By (2) and (3):
4) 'There then was a non-physical object that Macbeth saw'.

The inference from (1) to (2) would be licensed by the dominant visual sense of 'see', but not by the contextually relevant phenomenal sense which lacks factive implications and creates an intensional context not admitting of quantification (Forbes 2013). The same conclusion can be reached also by spatial inferences, also supported only by the visual sense (Fischer and Engelhardt 2017b). Given explicit marking of the different uses in the quoted passage, more must be involved than a simple error of using the wrong sense of 'see'.

Plausible principles of charity limit the extent of irrationality and conceptual or linguistic incompetence we may attribute to competent thinkers (Adler 1994; Lewinski 2012). Empirical explanations of why competent thinkers commit fallacies are then required to validate reconstructions (Thagard and Nisbett 1983). Our proposed account of salience bias can explain why competent speakers should make contextually inappropriate stereotypical inferences from 'see' – and other verbs, used in other statements of the argument (Fischer, in preparation). But is this account correct? Do competent language users make inferences licensed only by the dominant use of a polysemous verb also from rarefied special uses, as posited by the Salience Bias Hypothesis?

2.3. Pre-study

We used a battery of complementary psycholinguistic methods to examine this hypothesis about automatic inferences which speakers/hearers are not aware of making. The hypothesis boldly claims that under certain condition (i–iii above), competent language users will go along with contextually inappropriate stereotypical inferences *even when they know the inferences at issue to be inappropriate*. In a pre-study (see Fischer and Engelhardt, under review), we identified potentially relevant inferences: Undergraduate participants rated spatial inferences from visual, epistemic, doxastic and phenomenal uses of 'see', as well as from visual and epistemic uses of 'aware' (see Table 2.1).

On a 5-point Likert scale, participants indicated their confidence that 'in situations where the first sentence [premise] is true also the second sentence [conclusion] will typically be true'. They were very confident that spatial inferences from visual uses typically lead to true conclusions (mean rating 4.7 for 'see', and 3.7 for 'aware', both significantly above neutral mid-point '3'). They were also very confident that spatial inferences from the other uses examined typically lead to conclusions that fail to be true (mean ratings significantly below 3). They were most confident that spatial inferences from epistemic uses are inappropriate (lead to conclusions that fail to be true) (mean rating 1.58 for 'see'

Table 2.1 Spatial inferences from different uses of 'see'.

	[*visual 'see'*]	[*epistemic 'see'*]
Premise	Mona sees the car on the road.	Josh sees the issues in play.
Conclusion	The car on the road is in front of Mona.	The issues in play are in front of Josh.

and 1.53 for 'aware'). To get clear on whether competent language users make contextually inappropriate stereotypical inferences under the conditions (i)-(iii) identified by our Hypothesis SBH, we therefore examined whether speakers/hearers make and deploy spatial inferences from epistemic uses of 'see', which they demonstrably know to be inappropriate.[7] We accordingly examined the verb-specific hypotheses:

H₁   Competent speakers infer spatial patient-properties (*X is in front of S*) from visual *and epistemic* uses of 'S sees X'.

H₂   Conclusions from *all* these inferences will be deployed in subsequent cognitive processing, regardless of contextual (im)propriety.

3. Plausibility ratings

3.1. Approach and predictions

Plausibility ratings offer a convenient first means for following up hypotheses about comprehension inferences from specific words that affect further cognitive processing. They thus allow us to examine at any rate the conjunction of H₁ and H₂. Participants hear or read sentences like the following, and rate their plausibility on a Likert scale:

1a. Matt sees the spot on the wall facing him. ('*s-consistent*')

2a. Chuck sees the spot on the wall behind him. ('*s-inconsistent*')

In these sentences, the expression of interest is followed by a sequel that is either consistent with the hypothesized stereotypical inference ('s-consistent') or inconsistent with it ('s-inconsistent'). Our items have post-verbal contexts that are either consistent or inconsistent with the hypothesized inference from 'S sees X' to 'X is in front of S'. If this inference is made, and not swiftly suppressed, then the clash with the sequel will render s-inconsistent items (like 2a) less plausible than s-consistent items (like 1a).

This prediction holds both on a content- and an experience-based approach to metacognitive judgements. If the plausibility judgement is based on cognitive

---

[7]   If correct assessment does not prevent inappropriate automatic inference to conclusions presupposed in further reasoning, this will provide some support for inferences from findings about undergraduates to conclusions about expert philosophers, in the light of the 'expertise objection' (reviews: Nado 2014, Machery 2015).

engagement with the content and an assessment of its probability, the clash of the s-inconsistent sequel with the conclusion of the probabilistic inference (e.g. in 2a) will make its truth less probable than that of its s-consistent counterpart (1a). According to the experience-based approach to metacognitive judgements (Koriat 2007), the subjective plausibility of a judgement results not from cognitive engagement with its content but from features of the underlying cognitive processes. Fluency or level of effort serves as a cue for a wide range of metacognitive judgements, including plausibility assessments (for a review, see Alter and Oppenheimer 2009). Perceived inconsistencies (as in 2a) reduce the degree of 'fluency' or effortlessness of the comprehension process (Carpenter and Just 1977), which in turn reduces subjective plausibility (Thompson et al. 2011). Either way, lower plausibility ratings for s-inconsistent sentences than for their s-consistent counterparts would provide evidence for automatic spatial inferences.

To show that the inferences of interest are supported by features of the verb (e.g. stereotypical features), we manipulate not only the post-verbal context but also the verb and replace 'see' in half the items by a contrast verb less strongly associated with spatial patient features. We employ 'is aware of', which is ordinarily used in an epistemic sense, to attribute knowledge that may, but need not, be acquired through the five senses (*MEDAL, WordNet*).[8] A prior production experiment with a sentence-completion task showed that this verb is paired about half the time with visual objects as patients, which agents would be aware of in virtue of looking at them – whereas 'see'-stems are provided with completions that give the verb a visual sense, over 93% of the time (Fischer and Engelhardt 2017b). We infer that 'aware' is less strongly associated with the vision-schema, and spatial patient-properties, than 'see', and include items like

1b. Matt is aware of the spot on the wall facing him.
2b. Chuck is aware of the spot on the wall behind him.

Again, content- and experience-based accounts support the same prediction from our hypotheses: The weaker association of 'aware' (than 'see') with spatial features translates into a weaker probabilistic inference (it's less probable that the patient is in front of the agent), making it more probable that the sentence is true (the agent knew all along, from previous observation or testimony, that there's a spot on the wall). Less strongly supported inferences are also easier to suppress,

---

[8] https://www.macmillandictionary.com/dictionary/british/aware, http://wordnetweb.princeton.edu/perl/webwn

leading to less disfluency. Either way, s-inconsistent 'aware'-sentences (like 2b) will be rated more plausible than their 'see'-counterparts (like 2a). If so, this will provide evidence that the inferences of interest are supported by features of the verb.

To examine whether spatial inferences are also, inappropriately, made from epistemic uses of the verb, we finally manipulate the object. In the absence of contextual cues, concrete patient-nouns ('picture', 'car') invite visual interpretations of 'see'. By contrast, epistemic readings are invited by abstract patient-nouns ('challenges', 'opportunities', henceforth 'epistemic objects', for convenience), whose referents typically cannot be literally 'seen', but known. We use items like:

3a. Joe sees the problems that lie ahead.
3b. Joe is aware of the problems that lie ahead.
4a. Jack sees the problems he left behind.
4b. Jack is aware of the problems he left behind.

In principle, perfectly plausible interpretations are readily available for s-inconsistent sentences with epistemic objects (like 4): We can complement a purely epistemic interpretation of 'see' or 'aware' with a metaphorical interpretation of the sequel (before subject=future, behind subject=past; hence 'Jack knows what problems he had in the past'). But incompletely suppressed spatial inferences from 'see' will prevent such purely metaphorical interpretation: Though conventional, the present space-time metaphors give rise to embodied cognition effects (Boroditsky and Ramscar 2002; Bottini et al. 2015) and support spatial reasoning about temporal relations (Gentner et al. 2002; Casasanto and Boroditsky 2008). We infer that these metaphorical sequels will activate spatial schemas that place objects in front of, or behind a forward/future-facing subject (Gentner et al. 2002). These schemas will be retained during comprehension of these sequels (Giora and Fein 1999; Giora 2003), and facilitate spatial reasoning from them (Casasanto and Boroditsky 2008). If spatial inferences from the prior verb are made, their conclusions will therefore engage such reasoning, and be felt to clash with s-inconsistent sequels (' … left behind'). These perceived clashes will engender comprehension difficulties and render s-inconsistent 'see'-items (like 4a) less plausible than s-consistent counterparts (like 3a).

'Aware'-items will elicit different responses: Even if the association of 'is aware of' with the vision-schema is weaker, the combination of the verb with a *visual* object-noun will activate the schema (*cf.* Bicknell et al. 2010) and prompt spatial inferences. However, their conclusions will, where contextually necessary, be easier to suppress than those from 'see' (above). Therefore s-inconsistent 'aware'-sentences with visual objects (like 2b) will be deemed less plausible than their s-consistent counterparts (like 1b), but still more plausible than otherwise identical 'see'-sentences (like 2a). By contrast, if the dominant use of 'is aware of' is epistemic, and any spatial conclusion is inappropriate where it goes with abstract epistemic objects, such conclusions should be swiftly and completely suppressed in interpreting 'aware'-sentences with *epistemic* objects. For such sentences, we therefore expect spatial inferences from 'aware' will not interfere with subsequent plausibility judgements, so that the context-manipulation will not affect plausibility ratings – which should hence again be higher than for 'see'-counterparts.

We thus derive two key predictions from our hypotheses:

[Plausibility-1] s-inconsistent 'see'-sentences, both with visual *and* with epistemic objects, will be deemed less plausible than their s-consistent counterparts.

[Plausibility-2] s-inconsistent 'see'-sentences, both with visual *and* with epistemic objects, will be rated less plausible than their 'aware'-counterparts.

To sum up, we can test the hypotheses that competent language users make inappropriate stereotypical inferences from a specific verb (here: 'see'), which influence further cognition, by using a plausibility-rating task and a $2 \times 2 \times 2$ (context [s-consistent/s-inconsistent] × verb [see/aware] × object [visual/epistemic]) design, where all variables are manipulated within-subject.

3.2. Excluding confounds

This design helps to exclude most potential confounds. The verb-manipulation helps us exclude spatial inferences from other parts of the sentence as drivers of plausibility judgements. First, patient nouns might be associated with a specific spatial orientation towards agents (e.g. 'challenges' are typically said to be 'ahead'). However, if patient-nouns were the prime source of spatial inferences, they should reduce the plausibility of s-inconsistent 'see'- and 'aware'-sentences in the same way and falsify prediction Plausibility-2.[9]

---

[9] We investigated this possibility, even so: In a norming study, participants considered the patient nouns used in Study 1 (Section 4), in 'aware'-contexts (e.g. 'Joe is aware of the problems', etc.), and rated them on whether what they stand for is typically 'ahead' (=1) or 'behind' (=−1) the agent, or one 'cannot tell' (=0). Mean ratings >0 indicate forward bias. While the overall mean for our visual nouns did not differ significantly from 0, that for our epistemic objects did. Precisely half had mean ratings significantly >0, the other half had means not significantly different from 0 (neutral). In the main study, our two key predictions were borne out by ratings for items with forward-oriented *and* 'neutral' patient nouns.

Second, in items with epistemic objects, the spatial inference might be triggered by the spatial time metaphor used by the post-verbal sequel. This metaphor activates a schema centred on a subject spatially oriented to look at things in front of her (Gentner et al. 2002). One might therefore argue that *S looks at things in front of S* is a component of this schema, and that s-inconsistent sentences like 'Jack sees the problems he left behind' will seem implausible because they clash with this component. However, in this case, both 'see'- and 'aware'-items with epistemic objects should be sensitive to the context-manipulation. If this manipulation affects the plausibility only of 'see', but not 'aware'-items, we can exclude this potential confound.

Third, lower mean plausibility ratings for s-inconsistent 'see'-sentences might be driven by existential or factive inferences from the verb, which are appropriate both with visual and epistemic objects (You can only see my point if I have one), as these too are cancelled by some of our s-inconsistent sequels with epistemic objects, like 'the problems he left behind' (which implies the problems no longer exist for the agent), though not the others (like 'the possibilities from which she has turned away'). However, existential and factive implications are shared by both 'see' and 'aware' (You can only be aware of a problem that actually exists), so that the plausibility of epistemic 'aware'-items should again be affected in the same way by the context manipulation.

Finally, plausibility judgements can be affected by word frequency, as more frequently encountered words are easier to process (Oppenheimer 2006) and higher fluency may serve as metacognitive cue anchoring plausibility judgements (Alter and Oppenheimer 2009). While this is no major issue in the present studies (since we predict lower plausibility rankings for items using the more frequent 'see', and mean frequencies for our forward and backward terms are very similar), we can assess the extent to which frequency influences participants' ratings by constructing 'frequency-congruent' filler items where the sentence with the more frequent verb is also more consistent with its associated stereotype, and 'frequency-reversed' filler items, where word-frequency and stereotype-consistency work in opposite directions. If participants make judgements predominantly in line with stereotype-consistency, and no fewer such judgements about frequency-reversed than frequency-congruent items, their plausibility judgements are unlikely to be influenced by frequency, and we can assess whether their ratings (still) bear out our predictions (Fischer and Engelhardt 2016).

Since philosophical thought takes place in both speech (oral debate) and text (reading and writing), we investigated inferences in both modalities. We employed the plausibility-rating paradigm described in two studies, in which participants heard and read the items, respectively, and the task was combined with appropriate online measures. We provide results for each after outlining the relevant further methods.

4. Pupillometry

For our purposes, the major shortcoming of offline (outcome) measures like plausibility ratings is that they allow us to examine only hypotheses about conclusions which are automatically inferred *and* maintained. To examine separately which inferences are initially made automatically at the verb of interest (as per H_1) and to what extent their conclusions are subsequently suppressed or maintained (as per H_2), we can combine plausibility ratings with pupillometry or other eye-tracking measures. Since we report the pupillometry study elsewhere (Fischer and Engelhardt, under review; see Fischer and Engelhardt 2017a for a pilot), we here focus on explaining the approach and give only an executive summary of methods and results, before reporting a fresh reading time study (Section 5).

4.1. Approach and predictions

Pupil dilations offer a window into preconscious automatic processing (reviews: Laeng et al. 2012; Sirois and Brisson 2014). Our pupils dilate when we are emotionally or cognitively aroused or expend cognitive effort. Pupil responses to task demands (as opposed to, say, changes in lighting) are highly correlated with neural activity in the *locus coeruleus*, a key node within neural circuitry that controls the muscles of the iris (Samuels and Szabadi 2008) and mediates the functional integration of the whole attentional brain system (Corbetta et al. 2008). Pupil diameter reliably increases with the 'intensity' of attention (Kahneman 1973) or cognitive load (the extent to which cognitive resources are mobilized to address a task). These pupil responses are spontaneous and impossible to suppress at will (Loewenfeld 1993); they are triggered also by subliminally presented stimuli the subject is not aware of (Bijleveld et al. 2009), and regularly commence well before any conscious task response. They thus allow us to gauge allocation of cognitive resources at pre-conscious stages of processing (Laeng et al. 2012).

In language processing, difficulties which require cognitive resources to overcome arise from several sources. While psycholinguists have only recently begun to take up pupillometry on a larger scale, pupil responses have been found sensitive to syntactic complexity and sentence length (Piquado et al. 2010) and differences in the intelligibility of speech due to interfering noise (Zekveld and Kramer 2014), where dilations peak at medium levels of interference, suggesting less resources are allocated when the task becomes too difficult. The level of difficulty is generally also dependent upon the predictability of new text in the light of old: Processing is facilitated by activation of subsequent concepts by previous words, through associative priming (based on co-occurrence of words) or semantic priming (based on activation of schemas and semantic knowledge, more generally). Accordingly, pupil responses have been found responsive to 'surprisal', that is, the predictability of the next word in a sentence, given its previous words, as estimated, for example, by recurrent neural networks, on the basis of co-occurrence frequencies (Frank and Thompson 2012). By contrast, where new text violates expectations and clashes with the conclusions of schema-based inferences, suppression is required and costs effort (Faust and Gernsbacher 1996). Accordingly, pupillometry has documented dilations in response to violations of scripts (social event/action schemas) (Raisig et al. 2012).[10]

Our study used pupillometry to garner evidence of – inappropriate – stereotypical inferences supported by event schemas. While pupil dilations are initiated rapidly, the pupil takes over one second to dilate to its maximum size, after the point of difficulty (Engelhardt et al. 2010). Since it does not respond at uniform speed to all kinds of difficulties, it may, in this period, be influenced also by difficulties preceding or following the difficulty of interest. To minimize such influence, we created the difficulty at the end of our sentence items and compare mean pupil sizes after sentence offset with mean sizes in the previous time window (rather than considering time course). In our items, the difficulty of interest arises from a clash between the last part of the sentence and inferences from the prior verb (or verb and object).

Our hypothesis H_1 thus predicts that

---

[10] Recent evidence suggests pupil responses may index conflict monitoring yet more reliably than cognitive effort (van Steenbergen and Band 2013). This would strengthen the case for using pupillometry to investigate automatic inferences through pupil responses to subsequent cancellation phrases.

[Prediction Pupil] s-inconsistent, but not s-consistent, 'see'-sentences with visual *and* with epistemic objects will prompt pupil dilations in the second after sentence offset.

Since the combination of 'is aware of' with a visual object will activate the vision-schema at the earliest possible moment (*cf.* Bicknell et al 2010), that is, at the object-noun, we also expect significant dilations in the sentence offset window for s-inconsistent 'aware' sentences with visual objects, but not with epistemic objects.

Since pupil dilations are sensitive to effort expended at pre-conscious stages of processing, they can provide evidence of inferences, even where conclusions get swiftly suppressed and fail to influence subsequent judgements. This can happen where conclusions conflict with background knowledge or beliefs that get swiftly activated in language comprehension (Metusalem et al. 2012). In a study we ran jointly with the present experiment (Fischer and Engelhardt 2017a, Exp.2), we found participants' plausibility judgements were highly sensitive to their content-related background beliefs: Two groups held opposing views on the issue at hand (whether homophobic attitudes are pathological). Pupillometry results suggested they made the same stereotypical inferences from the expression of interest ('S is homophobic'), to conclusions (*S is mentally ill*) consistent with the background beliefs of one group of participants, but not another. Even so, the groups proceeded to give opposite plausibility ratings for sentences with sequels consistent and inconsistent, respectively, with the conclusions (e.g. 'Tim is homophobic. He is mentally ill' vs 'Joe is homophobic. He is mentally healthy'). Where initial conclusions were inconsistent with their background beliefs, participants suppressed them sufficiently swiftly and comprehensively to prevent them from influencing plausibility judgements.

In our paradigm, we therefore take pupillometry and subsequent plausibility ratings to measure different things: Pupillometry picks up inferences automatically made in incremental language comprehension and allows us to examine hypothesis H_1. Plausibility ratings measure the extent to which inferences are successfully suppressed or influence subsequent cognition and let us examine hypothesis H_2.

The pupil is far more responsive to luminance variations than to changes in cognitive load (Beatty et al. 2000). Since the presentation of reading items on ordinary-sized computer screens involves luminance differences as eyes move from the beginning of the sentence (when the visual field extends beyond the screen) to the centre of the screen, only few pupillometric investigations into language processing employ reading tasks (e.g. Frank and Thompson 2012;

Raisig et al. 2012). Our study employs a listening task, which allows participants to fixate a fixation cross in the middle of a computer screen, throughout the trial.

4.2. Methods

Our 38 participants were undergraduate students. All were native speakers of English. They heard sentences including 48 critical items, viz., 6 for each of the eight conditions (examples 1a–4b, Section 3.1). Items in each category alternated post-verbal contexts (e.g. 's/he left behind' and 's/he had turned away from', for s-inconsistent items with epistemic objects) and employed the same epistemic patient nouns as the pre-study (Section 2.3).

We used a 2 × 2 × 2 (context [s-consistent/s-inconsistent] × verb [see/aware] × object [visual/epistemic]) design and manipulated all variables within subject. Participants were seated at the eye tracker, given a set of verbal task instructions, and placed their chins on a chinrest. After a calibration procedure, they completed practice and experimental trials. On each trial, a fixation cross appeared for 1500 ms prior to sentence onset. The pre-recorded sentence was played out on the computer speakers, and after sentence offset the fixation cross remained on the screen for 1000 ms. After the cross disappeared, a plausibility rating prompt appeared, and participants rated sentences' plausibility from 1 to 5, using the corresponding key on the keyboard. Mean pupil diameter was measured with an SR Research Eyelink 1000 in time windows including the second half of the sentence and the offset period.[11] We baseline corrected the pupil diameter based on the preceding time window: We divided the mean size of the pupil during offset by the mean size during the second half of the sentence, for each condition. This allowed us to assess whether the pupil size was changing between time windows. To do so, we conducted one-sample *t*-tests with a test value of 1. A value of 1 would indicate that mean pupil diameter remained the same.

4.3. Results and discussion

All our predictions were confirmed. Pupil results are shown in Figure 2.1.

The s-inconsistent items resulted in larger pupil diameters. Crucially, participants' pupil size significantly increased after hearing s-inconsistent 'see'- sentences with visual *and* with epistemic objects. As further expected, there was

---

[11] For technical specifications of this frequently used device, see https://www.sr-research.com/wp-content/uploads/2017/11/eyelink-1000-plus-specifications.pdf

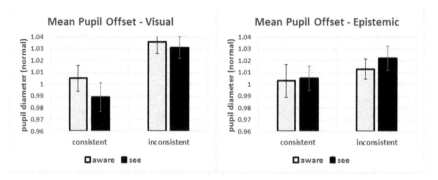

Figure 2.1 Baseline adjusted pupil diameter in 1000-ms time window following sentence offset. Error bars show the standard error of the mean.

also an increased pupil size after hearing s-inconsistent 'aware' sentences with visual objects, but no significant increases in the other conditions. However, while remaining shy of significance, pupil dilations after s-inconsistent 'aware' sentences with epistemic objects, clearly fell between the dilations observed in the other conditions with epistemic objects. We interpret this unexpected finding as evidence that the weak association of 'aware' with the vision-schema still supports initial spatial inferences.

Plausibility results are given in Figure 2.2. S-inconsistent 'see' sentences with visual *and* with epistemic objects were deemed less plausible than s-consistent counterparts (as per prediction Plausibility 1) and s-inconsistent 'aware'-counterparts (as per Plausibility 2). Items in all s-consistent conditions were judged distinctly plausible (mean ratings significantly above mid-point 3), as were s-inconsistent 'aware'-items with epistemic objects, while s-inconsistent

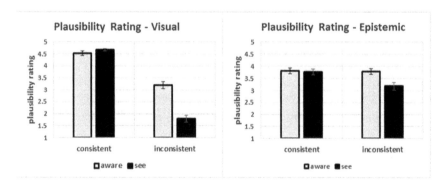

Figure 2.2 Mean plausibility ratings for each of the eight conditions in the pupillometry study. Error bars show the standard error of the mean.

'see'-items with visual objects were judged distinctly implausible (mean ratings below 3), and s-inconsistent 'see'-items with epistemic objects were deemed neutral (not significantly different from 3) (as were s-inconsistent 'aware'-items with visual objects). Predicted plausibility differences thus materialized as categorical differences.

Plausibility results cohere with pupillometry findings: Precisely in the three conditions with significant pupil dilations participants refrained from rating sentences as 'plausible'. The absence of more precise mirroring is consistent with our view that pupil dilations and plausibility ratings measure different things (Section 4.1): We interpret the observed pupil dilations as evidence of automatic inferences – including contextually inappropriate spatial inferences from epistemic uses of 'see'. We regard subsequent plausibility ratings as evidence of the extent to which suppression is successful. A purely epistemic interpretation of s-inconsistent items with epistemic objects (where 'Jack sees/is aware of/the problems he left behind' comes down to *Jack knows what problems he had in the past*, see Section 3.1), renders these items at least as plausible as their s-consistent counterparts (which then come down to, e.g. *Joe knows what problems he will have in the future* – that is notoriously hard to predict). We observe almost identical plausibility ratings for s-inconsistent and s-consistent 'aware'-sentences with epistemic objects. If the vision schema was deployed in interpreting epistemic uses of 'aware', the initial spatial inferences unexpectedly suggested by our pupillometry findings were suppressed with complete success, and a purely epistemic interpretation attained, before plausibility judgements were made.

This is different for items with 'see', which is more strongly associated with the vision-schema: Lower plausibility ratings for s-inconsistent 'see'-sentences with epistemic objects than for s-consistent counterparts and for analogous 'aware'-sentences suggest that inappropriate spatial conclusions inferred from epistemic uses of 'see' were not completely suppressed and prevented purely epistemic interpretation. Together with our pre-study, present findings provide evidence of contextually inappropriate inferences (from 'see') competent hearers make and presuppose in further cognition, despite knowing they are inappropriate.

5. Eye tracking: Fixation times

Philosophical thought takes place in reading and writing as well as in oral debate. We therefore followed up pupillometric investigation of automatic inferences in speech comprehension with a study that combined plausibility ratings with tracking of eye movements to examine H_1 and H_2 as hypotheses about automatic inferences in reading. This paradigm has the advantage of allowing us to localize the source of comprehension difficulties more precisely, as eye movements respond to difficulties more quickly than pupil size and display more intricate response patterns across different (early and late) eye movement measures. By exploiting these advantages, we can also close the remaining gap in the argument initially motivating our verb-specific hypotheses H_1 and H_2: We can now examine our suggestion that epistemic uses of 'see' are interpreted with the Retention Strategy (Section 2.1).

5.1. Approach and predictions

Contrary to a common folk conception, reading is *not* a sequential process where each word in a sentence is read, one after the other, as they appear in the text, at roughly the same pace. Instead, the eye moves in stops (fixations on words) and starts (saccades). Readers tend to fixate most, but not all words, as their eyes move forward (skipping the words easiest to predict from the context) *and* backwards ('regressions' at points of difficulty). According to the 'good enough processing' approach that informs much eye tracking research on reading (Ferreira et al. 2002; Ferreira and Patson 2007; *cf.* Frazier and Fodor 1978) and is consistent with broader trends in cognitive science (Ferreira and Lowder 2016), hearers/readers immediately construct local interpretations over small numbers of adjacent words; if the task at hand demands it, and only then, they subsequently integrate these local interpretations into more comprehensive interpretations of longer sentences and passages, which take long-distance dependencies into account (Swets et al. 2008). In reading, we thus need to recognize words and integrate them into local and more comprehensive interpretations. Difficulties at these different stages manifest themselves in different eye-tracking measures (*cf.* Section 1.2).

The difficulty of word recognition depends mainly on the word's frequency, length and predictability in the (local) context (Rayner 1998; Clifton et al. 2016). It is reflected in first-pass reading (fixation) times. A backward eye movement (regression) upon first fixation may indicate difficulty in integrating the word into a local interpretation. The regression path duration (sum of [1] all fixations on a word or in a region before moving to the right, plus [2] all fixations made during regressions in this period) reflects the effort required to overcome this difficulty. By contrast, difficulties in integrating the local interpretation of a sentence region into a more comprehensive interpretation will show up only in increased second pass or total reading times for the region, and a higher number of saccades from it to other text (Rayner et al. 2004; Clifton et al. 2007). Difficulties arising from one sentence region may lead to longer total reading times for the next (n+1) region (Rayner et al. 2004).

We wish to examine, first, our hypothesis that epistemic uses of 'see' are interpreted by retaining the dominant vision-schema and suppressing its contextually irrelevant components. Whenever an ambiguous word with a clearly dominant meaning is used in a less salient sense and disambiguated by immediate post-verbal context (still considered in constructing local interpretations), this increases first-pass reading times and regression-path durations on the disambiguating region, as well as the number of regressions from it (Sereno et al. 2006). Use of the Retention Strategy (Giora et al. 2014), however, implies more sustained suppression effort (Faust and Gernsbacher 1996) and should translate also into longer total reading times for the disambiguating region which are not entirely driven by longer first-pass reading times. In other words, this sustained effort should translate into longer total and second pass (= total minus first pass) reading times for the disambiguating region. In our critical items, 'see' is disambiguated by the visual versus epistemic object-noun immediately following it. Higher second-pass and total reading times can also be driven by regressions from the following post-patient context, prompted by integration difficulties. In our critical items (see Table 2.2), such difficulties arise from s-inconsistent, but not s-consistent contexts. Our hypothesis thus motivates:

> [Prediction EM1] First-pass, second-pass, and total reading times for the object region will be longer for 'see' sentences with epistemic than with visual objects, across all 'see'-items *and* specifically in s-consistent sentences.

The interpretation of epistemic uses of 'aware' will show a partially distinct pattern. Epistemic objects are more abstract than visual objects, and this fact alone increases early processing effort (Binder et al. 2005) and first-pass reading times (Schwanenflugel and Shoben 1983). But differences should show up in later processing stages. According to our initial assumptions, the interpretation of epistemic uses of 'is aware of' involves discarding the weakly associated vision-schema in favour of a different, purely epistemic, situation schema. Our pupillometry findings cast first doubt on this assumption and suggest that the vision-schema may be deployed to interpret epistemic uses also of 'is aware of'. However, while salient, the visual use arguably is not as clearly dominant for 'aware' as the visual sense is for 'see'.[12] The resulting weaker association of the vision schema with 'aware' than 'see' should then translate into less effort being required to suppress contextually inappropriate schema components, and more comprehensive suppression success. The latter success would manifest in plausibility ratings, the former effort in 'late' eye-movement measures. Our initial *and* our modified assumptions about 'aware' thus both motivate

> [Prediction EM2] Second-pass and total reading times on epistemic objects will be longer for 'see' sentences than for their 'aware'-counterparts.

These measures alone will not allow us to adjudicate between assumptions concerning 'aware', but can provide evidence of the vision-schema's retention in interpreting epistemic uses of 'see'.

Second, we wish to examine our key hypothesis H_1 that competent speakers infer spatial patient-properties (*X is in front of S*) from visual *and epistemic* uses of 'S sees X'. To do so, we construct sentences where visual and epistemic objects, respectively, are followed by sequels that are s-consistent ('that lie ahead of him') or s-inconsistent ('that lie behind him'). We assume the previous text (e.g. 'Joe sees the problems') constitutes a local context, so the clash between the spatial inference and the s-inconsistent sequel will arise at the stage of integrating the local into larger interpretations (Ferreira et al. 2002). Difficulties at this stage show up in 'late' measures. From H_1 we therefore infer:

> [Prediction EM3] In 'see'-sentences with visual *and* epistemic objects, total reading times will be longer for s-inconsistent post-object contexts than for s-consistent counterparts.

The relevant clashes can also show up through increased regressions. However, to keep our items as similar as possible to the pupillometry study, we placed these post-object contexts at the end of the sentence, where regressions routinely occur as part of a 'wrap-up' process, anyway (Rayner et al. 2000). We therefore make no predictions about regressions.

---

[12]   While data on occurrence frequencies for different uses of 'aware' remains to be collected, participants in a production study used visual patient nouns about half the time to complete sentence-stems with 'aware', while providing completions resulting in a visual use of 'see' 94% of the time (Fischer and Engelhardt 2017b).

Comparing total reading times for post-object contexts (see Table 2.2) in 'see'- and 'aware'-sentences with epistemic objects can help us adjudicate between assumptions about the processing of epistemic uses of 'aware': If the vision-schema is discarded for a dominant epistemic schema, processing effort should focus on the epistemic object, rather than the post-object context, and the consistency-manipulation should affect total reading times for the context region less than when the vision-schema is retained for interpreting the utterance. We assume the latter holds for 'see'-sentences. Our initial assumption that the vision-schema is not retained to interpret epistemic uses of 'is aware of' then implies

[Prediction EM4] Total reading times of s-inconsistent post-object contexts will be longer for 'see'-sentences with epistemic objects than for their 'aware'-counterparts.

Disconfirmation of this prediction would favour our modified assumption that the vision-schema is retained in interpreting epistemic uses also of 'aware'.

Our final hypothesis H_2 is, to repeat, that, regardless of contextual (im)propriety, spatial conclusions from both visual *and* epistemic uses of 'see' will be deployed in subsequent cognitive processing beyond utterance comprehension. This hypothesis is again assessed through subsequent plausibility ratings. *Mutatis mutandis*, the above reasoning (Section 3.1) continues to apply and motivate two predictions (to repeat from above):

Table 2.2 Example stimuli and regions of interest for eye movement analysis.

	Verb	**Object**		**Context**	**Status**
Epistemic					
1 Joe	sees	the problems	that lie	ahead of him	(s-consistent)
2 Joe	sees	the problems	that lie	behind him	(s-inconsistent)
3 Joe	is aware of	the problems	that lie	ahead of him	(s-consistent)
4 Joe	is aware of	the problems	that lie	behind him	(s-inconsistent)
Visual					
1 Sheryl	sees	the picture	on the wall	facing her	(s-consistent)
2 Sheryl	sees	the picture	on the wall	behind her	(s-inconsistent)
3 Sheryl	is aware of	the picture	on the wall	facing her	(s-consistent)
4 Sheryl	is aware of	the picture	on the wall	behind her	(s-inconsistent)

[Plausibility-1] S-inconsistent 'see'-sentences, both with visual *and* with epistemic objects, will be deemed less plausible than their s-consistent counterparts.

[Plausibility-2] S-inconsistent 'see'-sentences, both with visual *and* with epistemic objects, will be rated less plausible than their 'aware'-counterparts.

5.2. Methods

Participants. Thirty-six undergraduate psychology students from the University of East Anglia, thirty women, six men, ranging in age from 18 to 26 years (M=19.83, SD=1.50), participated for course credit. All were native speakers of English with normal or corrected-to-normal vision.

Materials. The experimental items included 48 critical items, 6 for each of eight conditions. We adapted the materials from the pupillometry study (Section 4), controlling for word-frequency and length of patient-nouns ('objects') and post-patient contexts (see Table 2.2): The visual and epistemic object-nouns had very similar mean frequencies (M=84.5, SD=70.6 and M=82.5, SD=69.4, respectively)[13] and mean lengths in terms of number of characters (M=8.5, SD=2.2 and M=6.8, SD=2.2, respectively). Neither mean frequencies nor mean lengths differed significantly $t(22)$=.07, p=.95 and $t(22)$=1.82, p=.082. Similarly, the expressions used in the cancellation region were very similar in terms of mean lengths (visual-consistent M=10.7; visual-inconsistent M=10.3; epistemic-consistent M=9.0; epistemic-inconsistent M=10.0) and of the mean frequency of key words (underlined) in each (e.g. post-epistemic: 'ahead of him', 'facing him', 'before him' vs 'behind him', 'has overcome', 'turned from') (visual-consistent M=179.3; visual-inconsistent M=202; epistemic-consistent M=166; epistemic-inconsistent M=158).[14] Critical items employed the same patient nouns as the visual and epistemic items in the pre-study (Section 2.3). There were 48 filler trials.

Apparatus. Eye movements were recorded with an SR Research Ltd. EyeLink 1000 eye-tracker which records the position of the reader's eye every millisecond (see Fn.11). Head movements were minimized with a chin rest. Eye movements were recorded from the right eye. The sentences were presented in 12 pt. Arial black font on a white background.

---

[13] Here and below, frequency figures refer to occurrence frequencies in the full written and spoken British English reference corpus of Leech et al. (2001).

[14] Since our predictions do not call for comparisons between reading times for verbs, the evident differences in length and frequency between 'see' and 'is aware of' are irrelevant.

Design and Procedure. The design was a 2 × 2 × 2 (verb [see/aware] × object [visual/epistemic] × context [s-consistent/s-inconsistent]). All variables were manipulated within subject.

Participants were seated at the eye tracker and instructed verbally. They placed their chins on a chinrest. After a 9-point calibration and validation procedure, participants completed two practice trials and 96 experimental trials. These included 48 critical trials. Each participant saw 'see' and 'aware' versions of critical items in equal number, in each condition, as verbs and context-phrases were rotated across lists in a Latin-square design. At the start of each trial, participants were required to fixate a drift-correction dot on the left edge of the monitor, centred vertically. The experimenter then initiated the trial. The sentence appeared after an interval of 500 ms and the initial letter of each sentence was displayed in the same position, in terms of x and y coordinates, as the drift correction dot. The entire sentence was presented on a single line on the screen. The participant read the sentence silently and then pressed the spacebar on the keyboard. A plausibility rating prompt appeared, and participants rated sentences' plausibility on a scale from 1 to 5, by pressing the corresponding key on the keyboard. As before, endpoints were explained as 'very implausible' (1) and 'very plausible' (5) and the midpoint (3) as 'neither plausible nor implausible; the decision feels arbitrary'. The entire testing session lasted approximately 30 min.

5.3. Results and discussion

For eye movement and plausibility data, we defined outliers as means ± 3.5 SDs from the mean. There were none. Analyses were conducted with subjects ($F1$) and items ($F2$) as random effects.

Eye movements

Our eye movement findings largely confirmed our predictions, which concerned reading times for object regions [EM1 and EM2] and for post-object context regions [EM3 and EM4]. Our findings were also consistent with the 'good-enough processing' account (Ferreira et al. 2002; Ferreira and Patson 2007), according to which initial shallow processing leads to local interpretations that are subsequently integrated into more comprehensive interpretations, in more in-depth processing (see Section 5.1). Accordingly, we found regressions from the final word of the sentence, in 90% of critical trials, and observed increasing responsiveness to our manipulations, in late than in early eye-movement measures. We now first report omnibus tests mandated by the 2×2×2 design, then report how our predictions fared. In a few cases, we examine predictions where the relevant omnibus tests do not provide statistical support for the requisite comparisons.

Object Region

First pass reading times on the object region showed no three-way interaction and no main effect of verb ($p > .25$). Crucially, however, the $2 \times 2 \times 2$ (verb × object × context) repeated measures ANOVA did show a main effect of object (see Figure 2.3): The epistemic objects had longer reading times than the visual objects $F1(1,35)=16.28$, $p<.001$, $\eta^2=.32$, $F2(1,11)=9.55$, $p<.05$, $\eta^2=.47$. As predicted, they did so also specifically in 'see'-sentences (see-visual vs see-epistemic $t(35)=-2.36$, $p<.05$). Since our norming work excluded frequency and length differences between epistemic and visual objects, longer reading times for epistemic objects will be due to their more abstract character, which makes them more difficult to process (Binder et al. 2005).

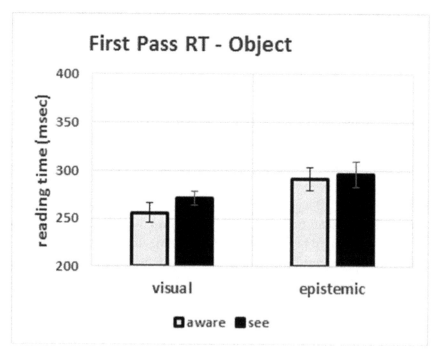

Figure 2.3 First-pass reading times on object region. Error bars show the standard error of the mean.

Second-pass reading times on the object region showed no three-way interaction ($p>.12$) but did show a main effect of verb and object, and a by-subjects interaction of object and context $F1(1,35)=5.04, p<.05, \eta^2=.13, F2(1,11)= 2.31, p=.16, \eta^2=.17$. The object-regions of 'see'-sentences had longer reading times than object-regions in 'aware'-sentences $F1(1,35)= 36.20, p<.001, \eta^2=.51, F2(1,11)= 63.77, p<.001, \eta^2=.85$. The epistemic objects had longer reading times than the visual objects $F1(1,35)= 44.22, p<.001, \eta^2=.56, F2(1,11)= 12.61, p<.01, \eta^2=.53$. As predicted by (EM1), this last point also held specifically for 'see'-sentences (irrespective of post-object context) and, yet more specifically, for 'see'-sentences with s-consistent post-object contexts: We found that epistemic objects had longer reading times than visual objects when considering 'see'-sentences irrespective of (i.e. collapsed across) contexts ($t(35)=-4.96, p<.001$), and when considering yet more narrowly only 'see'-sentences with s-consistent contexts ($t(35)=7.24, p<.001$). As predicted by (EM2), reading times for epistemic objects were longer in 'see'-sentences than in 'aware'-counterparts, irrespective of (i.e. collapsed across) context ($t(35)=-3.97, p<.001$). Mean second-pass reading times for epistemic objects were numerically almost identical when followed by s-consistent and s-inconsistent contexts, respectively, in both 'aware'-sentences and 'see'-sentences (Figure 2.4).

Total reading times on the object region showed no three-way interaction ($p>.17$) but a main effect of verb and object (see Figure 2.5). The object-regions of 'see'-sentences had longer reading times than counterparts in 'aware'-sentences $F1(1,35)=35.66, p<.001, \eta^2=.51, F2(1,11)=70.15, p<.001, \eta^2=.86,$[15] and the epistemic objects had longer reading times than the visual objects $F1(1,35)=54.25,$

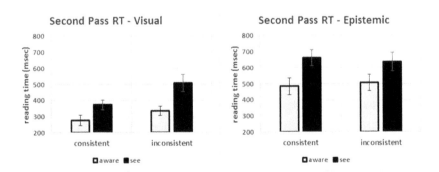

Figure 2.4 Second-pass reading times on object region. Error bars show the standard error of the mean.

---

[15] (EM2) predicts this for sentences with epistemic objects. For a possible explanation concerning sentences with visual objects, see Fn.18.

$p<.001$, $\eta^2=.61$, $F2(1,11)=11.22$, $p<.01$, $\eta^2=.51$. As predicted by (EM1), we also observed longer reading times for epistemic than visual objects when considering specifically 'see'-sentence (irrespective of context) ($t(35)=-5.67$, $p<.001$) and when focusing yet more narrowly on 'see'-sentences with s-consistent contexts ($t(35)=8.25$, $p<.001$). As predicted by (EM2), reading times for epistemic objects were longer in 'see'-sentences than in 'aware'-counterparts, irrespective of (i.e. collapsed across) context ($t(35)=-3.89$, $p<.001$). Mean total reading times for epistemic objects were numerically almost identical when followed by s-consistent and s-inconsistent contexts, respectively, in both 'aware'-sentences and 'see'-sentences (Figure 2.5).

To sum up, observed reading times for object regions were consistent with the predictions we derived from the hypothesis that the Retention Strategy is employed for interpreting epistemic uses of 'see'. As per prediction (EM1), first pass, second pass and total reading times for the object region were all longer for 'see'-sentences with epistemic than with visual objects. Crucially, this also held specifically for s-consistent sentences, where total object region reading times are not liable to be affected by difficulties to integrate post-object contexts. The fact that total reading times for epistemic objects were the same across sentences with s-consistent and s-inconsistent contexts, in sentences with either verb, further confirms that higher reading times for epistemic than visual objects were not driven by greater difficulties to integrate post-object contexts and increased regressions from such contexts. Finally, as per prediction (EM2), second pass and total reading times on epistemic objects were longer for 'see' sentences than for their 'aware'-counterparts. These findings support our hypothesis that epistemic uses of 'see' are interpreted with the Retention Strategy (Giora 2003; Giora et al. 2014).

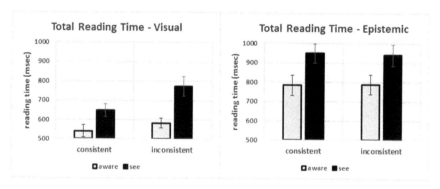

Figure 2.5 Total reading time on the object region. Error bars show the standard error of the mean.

Context region

While our predictions concerning post-object contexts only predict total reading times, we will get a better grasp of both sentence processing and eye tracking measures by considering also first-pass reading times.

First-pass reading times on the post-object context region showed no three-way interaction (p>.47) but a main effect of object and context (see Figure 2.6). Contexts following visual objects had longer reading times than contexts following epistemic objects (*sic*) $F1(1,34)=11.57$, p<.01, η^2=.25, $F2(1,11)=4.53$, p=.057, η^2=.29 and s-inconsistent contexts had longer reading times than s-consistent contexts $F1(1,34)=15.29$, p<.001, η^2=.31, $F2(1,11)=20.55$, p<.01, η^2=.65. The remaining main effect of verb and the interactions were not significant (p's>.45).

While striking *prima facie*, these findings are broadly consistent with the 'good enough processing' account (Ferreira and Patson 2007): At least in first-pass reading, readers only expend effort up to a threshold, tend to process subordinate clauses (like those containing our post-object context regions) superficially (Ferreira and Lowder 2016), and only attempt to integrate information beyond local interpretations where integration is easy enough. Abstract epistemic objects are more difficult to process than concrete visual objects (see above). Our findings suggest that participants made efforts to integrate information from post-object contexts already in first-pass reading only when reading sentences with visual, but not with epistemic objects – where integration efforts get deferred to later processing stages and show up only in second-pass or total reading times. This would account for the observed *longer* reading times for contexts following visual objects.

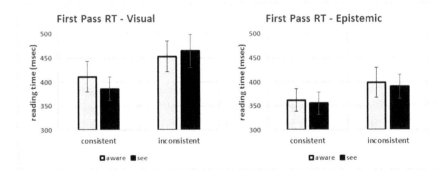

Figure 2.6 First-pass reading times on the context region. Error bars show the standard error of the mean.

Total reading times on the context region showed a significant three-way interaction between the variables $F1(1,34)=4.88$, $p<.05$, $\eta^2=.13$, $F2(1,11)=6.64$, $p<.05$, $\eta^2=.38$ as well as a main effect of object and of context (see Figure 2.7). Context regions following epistemic objects had longer total reading times (significant in the by-subjects analysis) than regions following visual objects $F1(1,34)=5.33$, $p<.05$, $\eta^2=.14$, $F2(1,11)=2.15$, $p=.17$, $\eta^2=.16$ and s-inconsistent contexts had longer reading times than their s-consistent counterparts $F1(1,34)=21.67$, $p<.001$, $\eta^2=.31$, $F2(1,11)=26.92$, $p<.001$, $\eta^2=.71$.

To decompose the three-way interaction, we considered items with visual and epistemic objects separately. For sentences with visual objects, there was a marginal (by subjects) interaction between context and verb $F1(1,35)=3.32$, $p=.077$, $\eta^2=.09$, though this failed to be confirmed by item analysis $F2(1,11)=1.97$, $p=.19$, $\eta^2=.15$. Neither of the main effects were significant ($p's >.10$). The marginal interaction arose from the fact that total reading times for s-inconsistent contexts in 'see'-sentences were significantly longer than reading times for s-consistent counterparts $t(35)=-2.20$, $p<.05$ and marginally longer than for s-inconsistent contexts in 'aware'-sentences $t(35)=-1.86$, $p=.07$. For sentences with epistemic objects, there was only a main effect of context $F1(1,34)=20.80$, $p<.001$, $\eta^2=.38$, $F2(1,11)=39.17$, $p<.001$, $\eta^2=.78$. Paired comparisons confirmed that s-inconsistent contexts had significantly longer reading times than s-consistent counterparts, both in 'see'-sentences $t(35)=-2.43$, $p<.05$ and in 'aware'-sentences $t(35)=-4.47$, $p<.001$.

These findings are consistent with our key prediction (EM3) about 'see' sentences: As predicted, total reading times for s-inconsistent contexts were longer than for s-consistent contexts, in 'see'-sentences with visual *and* with epistemic

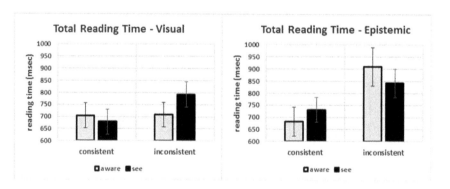

Figure 2.7 Total reading time on the context region. Error bars show the standard error of the mean.

objects. Reading times for 'aware'-sentences, however, did not conform to our expectations: In 'aware'-sentences with visual objects, total reading times for post-object contexts were not affected by the consistency manipulation, resulting in (marginally) shorter reading times for s-inconsistent contexts than in analogous 'see'-sentences. By contrast, the consistency manipulation greatly affected context reading times in 'aware'-sentences with epistemic objects. As a result, total reading times for s-inconsistent contexts were not significantly different for 'see'- and 'aware'-sentences with epistemic objects ($t(35)=1.37$, $p=.18$), *pace* (EM4). Indeed, reading times for s-inconsistent contexts in 'aware'-sentences were even numerically higher than for counterparts in 'see'-sentences. This finding favours our modified over our initial assumptions about the processing of epistemic uses of 'is aware of' (Section 5.1): it suggests that the vision schema is retained also for interpreting such uses of this verb.

To better understand the processing of post-object contexts, we considered the different cancellation phrases separately. After epistemic objects, we used three different phrases to create s-inconsistent contexts which clash with inferences that *X is in front of S*: 'behind him', 'has overcome' and 'turned from' all place patients behind agents in spatial schemas (*cf.* Section 3.1). Since each phrase was used in just two items, insufficient for a by-items (*F2*) inferential analysis, we only conducted a by-subject (*F1*) analysis. A 2 × 3 (verb × cancellation) repeated measures ANOVA revealed a main effect of cancellation $F(2,70)=4.94$, $p<.05$, $\eta^2=.12$. Contexts with cancellation phrase 'has overcome' were read marginally and significantly more quickly than contexts with, respectively, 'behind' ($t(35)=1.90$, $p=.066$) and 'turned from' ($t(35)= -3.56$, $p<.01$) (Figure 2.8). Reading times for cancellations 'behind him' and 'turned from' were not significantly different from each other ($t(35)=-1.07$, $p=.294$). The interaction between variables remained shy of marginal significance $F(2,70)=2.43$, $p=.096$, $\eta^2=.065$, depriving more detailed comparisons of statistical support. Purely exploratory comparisons between 'see'- and 'aware'-sentences suggested a marginal difference for contexts with 'turned from' $t(35)=1.83$, $p=.075$.

These differences clearly do not arise from differences in phrase length or word frequency: 'has overcome' is the longest phrase, and 'overcome' the least frequent word. Higher reading times for 'turned from' in 'aware'- than 'see'-sentences might arise from activation, by 'turned from', of the conceptual metaphor *looking-at as thinking-about* (Fischer 2018). This would lead to a perceived conflict between, for example, 'Kelly is aware of the possibilities' and the implication from 'she has turned from', namely, that she no longer

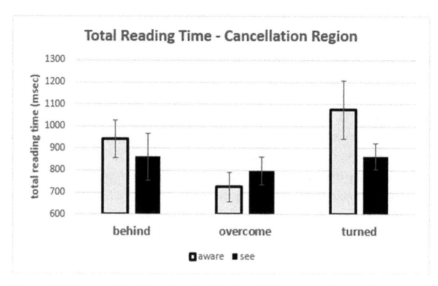

Figure 2.8 Mean total reading times for the three different cancellation phrases used after epistemic objects. Error bars show the standard error of the mean.

thinks about the possibilities. If 'is aware of' is more strongly associated with 'thinks about' than 'see' is, suppression of this previously activated stereotypical associate would lead to higher total reading times on the phrase in 'aware'- than in 'see'-sentences. If this is correct, higher reading times in these s-inconsistent 'aware'-sentences would not be (mainly) due to spatial inferences, and these sentences should be disregarded in assessing our hypotheses. Exclusion of these items yields a mean total reading time of about 800 ms for 'aware'-sentences with epistemic objects and s-inconsistent contexts. This is numerically below, though not significantly different from, the mean for the corresponding 'see'-sentences ($t(35)=-.401, p=.691$) (Figure 2.7).

The lower reading times for 'has overcome' than 'lies behind' may have an explanation in line with our Salience Bias Hypothesis (Section 2.1): According to SBH, the salience bias arises where initial activation of contextually inappropriate schema components is not only strong (due to salience) but is also complemented by lateral cross-activation from frequently co-occurring component features of the relevant schema. We now assume that the vision-schema is retained to interpret epistemic uses of both 'see' and 'is aware of'. In instantiations of the vision schema, the spatial-directional feature *X is in front of S* arguably co-occurs frequently with the spatial-vicinity feature *X is near S* (X 'is around'). Hence these features can be more readily suppressed together than selectively, as one cannot be suppressed completely as long as the other retains activation. Our cancellation phrases all activate spatial schemas serving as source-domain scenarios of conceptual space-time metaphors (Boroditsky and Ramscar 2002; Gentner et al. 2002), but the schemas differ as our phrases carry different literal (source-domain) implications. Whereas 'X is behind S' implies that X is still around in the vicinity of S, 'S has overcome X' implies X is no longer present to S or around her. The activation of these subtly different schemas therefore either reinforces or inhibits the activation of the component *X is near S* that regularly co-occurs with *X is in front of* S, and thereby hinders or helps suppression of the latter. This would translate into greater integration difficulties and longer reading times for cancellation phrases with 'behind' than 'overcome'.

This explanation can be tested against further data. The account motivates the prediction [Plausibility-3] that s-inconsistent items using the different cancellation phrases should attract different plausibility ratings: by defeating the vicinity-implication, 'has overcome' should facilitate complete suppression of the directional feature. S-inconsistent items employing it should therefore be deemed *plausible* (provided s-consistent sentences with epistemic objects are deemed plausible). By reinforcing the vicinity-implication, 'behind' should make complete suppression of the directional feature yet more difficult. S-inconsistent items with it should therefore be deemed distinctly implausible. Finally, 'S turned from X' suggests X is still around (though S redirected attention) but implies this less strongly than 'is behind' (clearly leaving open the possibility that X moved or vanished since S averted attention). Therefore, ratings of items using 'turned from' should attract ratings in between. The fact that the vision schema is more strongly associated with, and hence activated by, 'see' than 'aware', would predict that suppression effort (evidenced by total reading times) is more successful in 'aware'-sentences. As a result, 'aware'-items should be deemed more plausible than 'see'-counterparts across all cancellation phrases and are likely be placed in a higher plausibility category in the 'mid-way' condition with 'turned from'.

Plausibility

Plausibility results replicated almost perfectly those from the previous pupillometry study (see Section 4) and confirmed our predictions.

A $2 \times 2 \times 2$ (verb $\times$ object $\times$ context) repeated measures ANOVA showed a significant three-way interaction $F1(1,35)=22.81, p<.001, \eta^2=.40; F2(1,11)=20.77, p=.001, \eta^2=.65$, as well as main effects of verb $F1(1,35)=45.71, p<.001, \eta^2=.57; F2(1,11)=117.37, p<.001, \eta^2=.91$ and context $F1(1,35)=430.87, p<.001, \eta^2=.93;$

$F2(1,11) =104.72$, $p<.001$, $\eta^2=.91$. Sentences with 'see' and sentences with s-inconsistent contexts had lower plausibility ratings. To decompose the three-way interaction and examine relevant differences, we considered visual and epistemic object-conditions separately (see Figure 2.9).

There were significant 2 × 2 (context × verb) interactions in both the visual object-condition $F1(1,35)=56.89$, $p<.001$, $\eta^2=.62$; $F2(1,11)=190.17$, $p<.001$, $\eta^2=.95$ and the epistemic object-condition $F1(1,35)=14.06$, $p=.001$, $\eta^2=.29$; $F2(1,11)=4.35$, $p=.06$, $\eta^2=.28$.[16] This allowed us to follow up with paired-samples t-tests. As predicted by our first key prediction [Plausibility-1], s-inconsistent 'see'-sentences were deemed less plausible than s-consistent counterparts both when they had visual objects ($t(35) =21.87$, $p<.001$) and when they had epistemic objects ($t(35)=6.21$, $p=.001$).

Also the comparisons with 'aware'-counterparts turned out as expected: s-consistent 'see'-sentences were deemed equally plausible as 'aware'-counterparts when they took visual objects ($t(35)=-.93$, $p=.36$) and when they had epistemic objects ($t(35)=.37$, $p=.72$). The context-manipulation then significantly affected the plausibility of 'aware'-sentences when they took visual objects, but then affected it less than for 'see'-counterparts: 'aware'-sentences with visual objects were deemed less plausible when s-inconsistent than when s-consistent ($t(35)=9.66$, $p<.001$). But s-inconsistent 'see'-sentences with visual objects were still deemed less plausible than their 'aware'-counterparts ($t(35)=9.16$, $p<.001$).

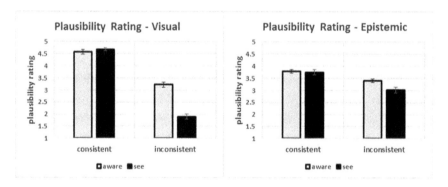

Figure 2.9 Mean plausibility ratings for each of the eight conditions in the eye-tracking study. Error bars show the standard error of the mean.

---

[16] The marginality of this by-item result is due to the fact that the by-items analysis is less powerful than the by-subjects analysis, involving fewer degrees of freedom. Lower p values are expected, and marginal by-item results do not impugn the significance of the finding (Cohen 1992).

As further expected, the plausibility of 'aware'-sentences with epistemic objects was less strongly affected by the context-manipulation. However, whereas in our previous study (Section 4), s-consistent and s-inconsistent 'aware'-sentences with epistemic objects had attracted numerically almost identical mean ratings, in the present study mean ratings were numerically lower for s-inconsistent sentences than for s-consistent counterparts, and the difference was statistically significant $t(35)=3.86$, $p<.001$. Clearly, however, the context-manipulation affected 'aware'- and 'see'-items to a different extent also when they took epistemic objects, and s-inconsistent 'see'-sentences with epistemic objects were rated less plausible than their 'aware'-counterparts ($t(35)=4.61$, $p<.001$). Findings thus confirm also our second key prediction [Plausibility-2].

As in the previous study, the predicted plausibility differences translated into categorical differences: Again, s-consistent sentences with either verb and either object were deemed distinctly plausible, that is, attracted plausibility ratings significantly above the neutral mid-point '3' (see-visual: $t(35)=27.55$, $p<.001$, aware-visual: $t(35)=16.07$, $p<.001$, see-epistemic: $t(35)=7.89$, $p<.001$, aware-epistemic: $t(35)=10.51$, $p<.001$), as were s-inconsistent 'aware'-sentences with epistemic objects ($t(35)=5.24$, $p<.001$). S-inconsistent 'see'-sentences with visual objects were deemed distinctly implausible, with a mean significantly below 3 ($t(35)=-10.40$, $p<.001$), while such sentences with epistemic objects were deemed neither plausible nor implausible, with mean ratings not significantly different from '3' ($t(35)=.292$, $p=.772$) – as were s-inconsistent 'aware'-sentences with visual objects ($t(35)=1.86$, $p=.072$).

To assess our latest prediction [Plausibility-3], we finally considered plausibility ratings for epistemic s-inconsistent items by cancellation phrase (Figure 2.10). A 2 × 3 (verb × cancellation) repeated measures ANOVA revealed a main effect of verb $F(1,35)=21.24$, $p<.001$, $\eta^2=.38$, as 'aware'-sentences had higher plausibility ratings than 'see'-counterparts, consistent with previous findings. There was also a main effect of cancellation $F(2,70)=70.67$, $p<.001$, $\eta^2=.67$, with significant differences between all three cancellation phrases (all p's <.05). The interaction between verb and cancellation was not significant ($p >.30$). Finally, all paired comparisons were significant ($p<.05$): In line with prediction [Plausibility-2], s-inconsistent 'see'-sentences of each kind were deemed less plausible than their 'aware'-counterparts. And consistent with prediction [Plausibility-3], items with 'overcome' were rated more plausible than items with 'behind', and items with 'turned from' fell between the two.

We again conducted one-sample t-tests to determine whether the means were significantly different from neutral mid-point '3'. Prediction [Plausibility-3] was

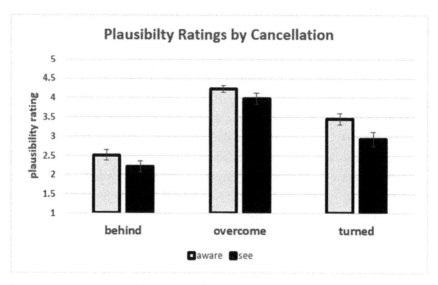

Figure 2.10 Mean plausibility ratings for see and aware for the three different cancellations. Error bars show the standard error of the mean.

fully borne out: S-inconsistent 'see'- and 'aware'-sentences with cancellation phrase 'behind' were deemed distinctly implausible (significantly below 3: $t(35)=-5.67, p<.001$ for 'see', $t(35)=-3.15, p<.01$ for 'aware'). By contrast, items with 'overcome' were deemed distinctly plausible (significantly above 3: $t(35)=6.86$, $p<.001$ for 'see', $t(35)=14.29, p<.001$ for 'aware'), and attracted mean plausibility ratings that were numerically even higher than those of s-consistent sentences with epistemic objects (*cf.* Figures 2.9 and 2.10). Items with 'turned from', finally, were placed into different categories depending upon their verb: 'aware'-items were still deemed distinctly plausible (significantly above 3: $t(35)=3.27, p<.01$). But 'see'-items were deemed neither plausible nor implausible (not significantly different from 3: $t(35) =-.39, p>.70$).

Discussion

Following on the previous pupillometry study (Section 4), present findings provide further and more detailed evidence that competent speakers make contextually inappropriate stereotypical inferences when the three conditions (i)–(iii) set out by our Salience Bias Hypothesis are met (see Section 2.1): Our prior work showed that (i) 'see' has a visual sense that is clearly dominant. The confirmation of predictions (EM1) and (EM2) suggests that (ii) epistemic uses of the verb are interpreted by retaining the situation schema associated with that dominant sense and suppressing contextually irrelevant components of this 'vision schema'. The Salience Bias Hypothesis maintains that where (iii) these irrelevant components continue to receive lateral cross-activation from frequently co-occurring schema components that are contextually relevant, suppression remains partial and contextually irrelevant schema components support contextually inappropriate inferences which influence further cognition (judgement and reasoning). Total reading times in line with prediction (EM3) provide evidence of inappropriate spatial inferences from epistemic uses of 'see'. Plausibility-judgements in line with predictions [Plausibility 1 and 2] provide evidence that inappropriate spatial conclusions influenced further cognition.

Drilling down into differences between specific cancellation phrases provided further support for our Salience Bias Hypothesis, in the shape of evidence for the relevance of condition (iii): Where the cancellation phrase (e.g. 'S has overcome X'), like the previous epistemic object, ruled out as contextually irrelevant *both* of two frequently co-occurring spatial components of the vision schema (*X is in front of S* and *X is near S*), both could be suppressed simultaneously, less suppression effort (reflected in numerically lower total context reading times) led to complete suppression, and no inappropriate inference influenced further cognition – s-inconsistent sentences were deemed as plausible as s-consistent counterparts (as per prediction Plausibility 3). By contrast, where the cancellation phrase failed to rule out one of the two regularly co-occurring spatial features as irrelevant ('turned from') or even implied its relevance ('lies behind'), this feature continued to pass on lateral activation to its regular companion. Accordingly, we observed numerically higher total context reading times and lower plausibility ratings, which provide evidence of inappropriate directional inferences and their influence on subsequent judgement.

The comparison of 'see'- and 'aware'-conditions suggests that salience imbalances (as per condition (i)) are relevant for the cognitive efficacy of the inappropriate inferences examined. Total reading times for object and context regions displayed the same pattern for 'see'- *and* 'aware'-sentences with epistemic objects.[17] This suggests that, contrary to our initial expectations, epistemic uses

---

[17] For both, s-inconsistency of context increased the reading times for context regions (Figure 2.7), but not for prior object regions (Figure 2.5) or verb regions ('see': s-consistent = 350 ms, s-inconsistent = 340 ms, no difference $t(35)=.34$, $p=.74$; 'aware': s-consistent = 730 ms, s-inconsistent = 755 ms, no difference $t(35)=-.65$, $p=.52$).

of 'is aware of' are interpreted, like epistemic uses of 'see', by retaining the vision schema and suppressing its contextually irrelevant component features.[18] The visual use (where the verb takes visual objects) may be the most salient use of 'aware', but is not clearly dominant (see Fn.12), resulting in weaker association of the vision schema with 'aware' than with 'see'. Accordingly, similar amounts of suppression effort (as evidenced by increased total reading times for s-inconsistent contexts) led to more complete suppression (evidenced by plausibility ratings), in 'aware'-sentences: we observed higher plausibility ratings for s-inconsistent epistemic items with 'aware' than 'see' and, second, found that such 'aware'-items were deemed distinctly plausible not only when contexts supported suppression through cancellations ('has overcome') that explicitly ruled out the contextual relevance of a schema component cross-activating the spatial component. Rather, s-inconsistent 'aware'-items were also deemed distinctly plausible when the cancellation ('turned from') remained silent on the relevance of 'cross-activators'. Only when the cancellation phrase ('lies behind') reinforced the activation of this schema component (X *is near* S), and thereby the lateral cross-activation of the spatial component of interest (X *is in front of* S), did the critical spatial inferences from 'aware' go through and affect plausibility judgements. This suggests that, beyond unhelpfully phrased contexts, the Retention Strategy does not make us generally prone to inappropriate inferences which manage to influence further cognition. Rather, our findings provide evidence that it makes us more generally prone to such inferences when the polysemous word at issue displays pronounced salience imbalances and has a dominant sense far more salient than all others.[19]

---

[18] In the visual condition, by contrast, we observed different processing patterns for 'see'- and 'aware'-items. With 'aware', the consistency-manipulation did not affect total reading times for either the object or the context region, but did so for the verb region (s-consistent = 687 ms, s-inconsistent = 804 ms, a significant difference $t(35)=-2.51$, $p<.05$). With 'see', all three regions had higher total reading times, in s-inconsistent items (verb: s-consistent = 338 ms vs s-inconsistent = 416 ms $t(35)= -2.38$, $p<.05$). This suggests that the extra processing effort prompted by s-inconsistency was, in 'aware'-items, devoted to switching to a less salient interpretation of the verb (e.g. from 'is visually aware' to 'has seen and now knows'), but is spread across all three regions in 'see'-items, where no re-interpretation of the verb alone does the trick.

[19] In line with the adaptive behaviour and cognition programme (Gigerenzer et al. 2011; *cf.* Ferreira and Patson 2007), future research could fruitfully examine to what extent this apparent defect results from a system design that strikes the best balance overall between processing effort and accuracy of information inferred and retained, given real-world task demands.

6. Conclusion

6.1. Main findings

Two studies combining plausibility ratings with pupillometry and eye tracking, respectively, provided evidence of a salience bias in speech and text comprehension: Where a polysemous word has a clearly dominant sense (like 'see'), utterances that use the word in a less salient sense may trigger contextually inappropriate stereotypical inferences that are licensed only by the dominant sense – and go through to influence further judgement and reasoning, even when hearers/readers know they are inappropriate. Our studies documented inappropriate spatial inferences from epistemic uses of 'see' and (to a lesser extent) 'aware'; in a pre-study, participants drawn from the same population deemed such inferences inappropriate. Our findings suggest that inappropriate stereotypical inferences occur at least where pronounced salience imbalances coincide with an interpretation strategy ('Retention Strategy') whereby less salient uses of words are interpreted by retaining the situation schema associated with the most salient sense and attempting to suppress their contextually inappropriate conclusions. Where such attempts remain unsuccessful, for example due to continued lateral cross-activation from contextually relevant schema components, competent language users go along with contextually inappropriate stereotypical inferences, despite knowing they are inappropriate.

6.2. Philosophical relevance

Our findings contribute towards an epistemological profile (Weinberg 2015, 2016) of the key process of stereotypical enrichment. This generally reliable process of automatic inference routinely goes on in language comprehension. It is bound to generate many intuitions thinkers have when considering verbal descriptions of possible cases, in philosophical thought experiments, and to drive many inferences they draw from such descriptions in philosophical arguments. The epistemological profile helps us assess the evidentiary value of these intuitions and to reconstruct such arguments.

Philosophers often take familiar words which ordinary discourse may have endowed with a dominant sense and use them in a special sense (Section 2.2). They may do so to talk about unusual cases which deviate from the stereotype associated with the dominant sense (as envisaged in philosophical thought experiments, for example, about hallucinations or well-behaved zombies).

Where this happens, thinkers are liable to make stereotypical inferences which are contextually inappropriate. When the conclusions of such inappropriate inferences strike thinkers as obvious, these intuitions lack evidentiary value. To acquire the right to treat intuitions about unusual (stereotype-divergent) cases as evidence, philosophers need to engage in psycholinguistic investigation at least into the salience structures of the relevant words. Already the first set of jointly vitiating conditions we have identified in the process helps us assess at least some philosophical case intuitions.

Where inappropriate conclusions are not explicitly endorsed but implicitly presupposed, the finding of the salience bias helps us reconstruct the relevant lines of thought and vindicate reconstructions in the light of plausible principles of charity which permit the attribution of fallacies to competent thinkers only in the presence of empirically supported explanations of why thinkers commit the relevant fallacies under relevant conditions (Thagard and Nisbett 1983). Our specific finding that competent speakers make inappropriate inferences from less salient uses of the verb 'to see' helps vindicate our proposed reconstruction of the 'see-version' of the 'argument from hallucination', which we took to rely on such an inference in its opening step (Section 2.2). Elsewhere (Fischer in prep) we explain how related salience effects can account for fallacies in versions of the argument that employ the verb 'is aware of', instead. These empirically supported reconstructions help resolve this classical paradox and the 'problem of perception' (Smith 2002; Crane and French 2015) it engenders together with a parallel paradox (the 'argument from illusion', examined in Fischer and Engelhardt, 2016).

6.3. Methodological lessons

Our studies hopefully provide a useful model of how to study automatic inferences by combining plausibility ratings with pupillometry or reading time measurements, in the cancellation paradigm. In conclusion, we stress three methodological points these studies may help to illustrate. In the cancellation paradigm, inferences are studied by manipulating the consistency of subsequent text with hypothesized inferences and measuring indices of cognitive effort. Due to 'good enough' processing strategies with initial focus on local interpretations (Ferreira and Patson 2007), increased processing effort engendered by inconsistencies may show up only in later reading time measures (second-pass and total reading times). It may also materialize with delay, namely, on the post-conflict sentence region (Rayner et al. 2004) and at the likely ultimate source of difficulty (on the region regressed to from the conflict region, as with our s-inconsistent visual 'aware'-sentences; see Fn.18).

Second, different eye tracking methods (reading time measurements and pupillometry) can provide complementary evidence but need not yield equivalent results on any specific measure. Our pupillometry study examined increases in mean pupil size between two time windows, namely, the second half of the sentence and the 1000 ms window after sentence offset. Such pupil dilations are indicative of cognitive effort involved in processing the second half of the sentence (Section 4.1). The reading time measure that comes closest to capturing this effort would be summed total reading times for object and context regions (Figure 2.11). We observed pupils dilations for s-inconsistent 'see'- and 'aware'-sentences with visual objects and s-inconsistent 'see'-sentences with epistemic objects. Despite the replication of plausibility results across studies, these dilations are not mirrored in these summed reading times, which are significantly different for visual s-inconsistent 'see'- and 'aware'-sentences, while s-inconsistent 'see'-sentences with epistemic objects do not have significantly longer summed reading times than analogous 'aware'-sentences and s-consistent counterparts.[20] Some differences arise from the facts that in speech comprehension (involved in the pupillometry study) there is no 'going back' to sources of difficulty (as in reading) and that pupil dilations and reading times are affected by overlapping but distinct

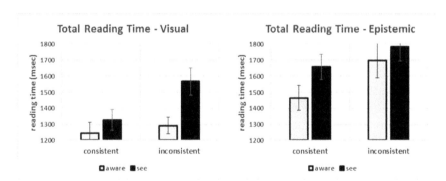

Figure 2.11 Total reading time on object and cancellation regions. Error bars show standard error of the mean.

---

[20] Full analysis of summed total reading times showed a significant three-way interaction $F(1,35)=7.66$, $p<.01$, $\eta^2=.18$. Follow up 2×2 analyses, considering visual and epistemic object conditions separately, showed a significant interaction for visual objects $F(1,35)=5.02$, $p<.05$, $\eta^2=.13$, and for epistemic objects, main effects of context $F(1,35)=8.19$, $p<.01$, $\eta^2=.19$ and verb $F(1,35)=4.48$, $p<.05$, $\eta^2=.04$. We observed significant differences between visual-inconsistent see- and aware-items $t(35)=-3.82$, $p<.01$, but not between epistemic-inconsistent see-items and aware-counterparts $t(35)=-.91$, $p>.36$ or epistemic-consistent see-items $t(35)=-1.49$, $p>.14$.

factors (Sections 4.1 and 5.1). More generally, more fine-grained measures need not 'add up' to a global measure and require independent derivation of predictions.

Finally, plausibility ratings and online measures measure different things: In the cancellation paradigm, higher total reading times are indicative of *extent* of suppression and integration effort, at different points. Plausibility ratings reflect *success* of this effort. The two measures hence need not pattern together, since similar effort may lead to more complete suppression of irrelevant schema components, where associations are weaker (as we observed for s-inconsistent epistemic items with 'aware' vs 'see'). Only the plausibility ratings tell us whether an inappropriate inference gets completely suppressed or goes on to influence further cognition. The moment we turn from psycholinguistic questions about sentence processing to experimental philosophy's questions about how automatic inferences affect our judgements and reasoning for better or worse, we need to complement 'online' (process) measures with 'offline' (outcome) measures.

Suggested Readings

Clifton, C., Staub, A., and Rayner, K. (2007). Eye movements in reading words and sentences. In R. P. G. van Gompel et al. (eds.), *Eye Movements. A Window on Mind and Brain* (pp. 341–371). Amsterdam: Elsevier.

Fischer, E., and Engelhardt, P. E. (2017). Stereotypical inferences: Philosophical relevance and psycholinguistic toolkit. *Ratio, 30*, 411–442.

Laeng, B., Sirois, S., and Gredebäck, G. (2012). Pupillometry: A window to the preconscious? *Perspectives on Psychological Science, 7*, 18–27.

Raney, G. E., Campbell, S. J., and Bovee, J. C. (2014). Using eye movements to evaluate the cognitive processes involved in text comprehension. *Journal of Visualized Experiments, 83*, e50780, doi:10.3791/50780 (with video).

Sirois, S., and Brisson, J. (2014). Pupillometry. *WIREs Cognitive Science, 5*, 679–692.

References

Adler, J. E. (1994). Fallacies and alternative interpretations. *Australasian Journal of Philosophy, 72*, 271–282.

Allport, D. A. (1985). Distributed memory, modular subsystems and dysphasia. In S. K. Newman and R. Epstein (eds.), *Current Perspectives in Dysphasia* (pp. 207–244). Edinburgh: Churchill Livingstone.

Alter, A.L. and Oppenheimer, D.M. (2009). Uniting the tribes of fluency to form a metacognitive nation. *Personality and Social Psychology Review, 13*, 219–235.

Atlas, J., and Levinson, S. C. (1981). It-clefts, informativeness and logical form: Radical pragmatics. In P. Cole (ed.), *Radical Pragmatics* (pp. 1–62). New York: Academic Press.

Ayer, A. J. (1956/1990). *The Problem of Knowledge*. London: Penguin.

Battig, W. F., and Montague, W. E. (1969). Category norms for verbal items in 56 categories: A replication and extension of the Connecticut category norms. *Journal of Experimental Psychology, 80*, 1–46.

Beatty, J., and Lucero-Wagoner, B. (2000). The pupillary system. In Cacioppo J. T., Tassinary L. G., and Berntson G . (eds.) *Handbook of Psychophysiology* (pp. 142–162). Cambridge: Cambridge University Press.

Bicknell, K., Elman, J. L., Hare, M., McRae, K., and Kutas, M. (2010). Effects of event knowledge in processing verbal arguments. *Journal of Memory and Language, 63*, 489–505.

Bijleveld, E., Custers, R., and Aarts, H. (2009). The unconscious eye opener: Pupil dilation reveals strategic recruitment of resources upon presentation of subliminal reward cues. *Psychological Science, 20*, 1313–1315.

Binder, J. R., Westbury, C. F., McKiernan, K. A., Possing, E. T., and Medler, D. A. (2005). Distinct brain systems for processing concrete and abstract concepts. *Journal of Cognitive Neuroscience, 17*, 905–917.

Boroditsky, L., and Ramscar, M. (2002). The roles of body and mind in abstract thought. *Psychological Science, 13*, 185–188.

Bortfeld, H., and McGlone, M. S. (2001). The continuum of metaphor processing. *Metaphor and Symbol, 16*, 75–86.

Bottini, R. et al. (2015). Space and time in the sighted and blind. *Cognition, 141*, 67–72.

Boyd, K., and Nagel, J. (2014). The reliability of epistemic intuitions. In E. Machery and E. O'Neill (eds.), *Current Controversies in Experimental Philosophy* (pp. 109–127). London: Routledge.

Cappelen, H. (2012). *Philosophy without Intuitions*. Oxford: Oxford University Press.

Carpenter, P.A. and Just, M.A. (1977). Integrative processes in comprehension. In D. LaBerge and S.J. Samuels (eds.), *Basic Processes in Reading: Perception and Comprehension* (pp. 217–241).Washington, DC: Erlbaum.

Carruthers, P. (2002). The cognitive functions of language. *Behavioral and Brain Sciences, 25*, 657–674.

Casasanto, D., and Boroditsky, L. (2008). Time in the mind: Using space to think about time. *Cognition, 106*, 579–593.

Chang, T. M. (1986). Semantic memory: Facts and models. *Psychological Bulletin, 99*, 199–220.

Chisholm, R. (1957). *Perceiving*. Ithaca: Cornell University Press.

Chow, W., Smith, C., Lau, E., and Phillips, C. (2016). A 'bag-of-arguments' mechanism for initial verb predictions. *Language, Cognition, and Neuroscience, 31*, 577–596.

Clifton, C., Ferreira, F., Henderson, J. M., Inhoff, A. W., Liversedge, S. P., Reichle, E. D., and Schotter, E. R. (2016). Eye movements in reading and information processing: Keith Rayner's 40 year legacy. *Journal of Memory and Language, 86*, 1–19.

Clifton, C., Staub, A., and Rayner, K. (2007). Eye movements in reading words and sentences. In R. P. G. van Gompel et al. (eds.), *Eye Movements. A Window on Mind and Brain* (pp. 341–371). Amsterdam: Elsevier.

Cohen, J. (1992). A power primer. *Psychological Bulletin, 112*, 155–159.

Corbetta, M., Patel, G., and Shulman, G. L. (2008). The reorienting system of the human brain: From environment to theory of mind. *Neuron, 58*, 306–324.

Crane, T., and French, C. (2015). The problem of perception. In N. Zalta (ed.), *The Stanford Encyclopedia of Philosophy*. Summer 2015. http://plato.stanford.edu/entries/perception-problem/

De Neys, W., Schaeken, W., and d'Ydewalle, G. (2003). Inference suppression and semantic memory retrieval: Every counterexample counts. *Memory & Cognition, 31*, 581–595.

Deutsch, M. (2015). *The Myth of the Intuitive*. Cambridge, MA: MIT Press.

Elman J.L. (2009). On the meaning of words and dinosaur bones: Lexical knowledge without a lexicon. *Cognition, 33*, 547–582.

Engelhardt, P. E., and Ferreira, F. (2016). Reaching sentence and reference meaning. In P. Knoeferle, P. Pyykkonen, and M. W. Crocker (eds.), *Visually Situated Language Comprehension*. Amsterdam: John Benjamins.

Engelhardt, P. E., Ferreira, F., and Patsenko, E. G. (2010). Pupillometry reveals processing load during spoken language comprehension. *Quarterly Journal of Experimental Psychology, 63*, 639–645.

Evans, J. S. B. T., and Stanovich, K. E. (2013). Dual-process theories of higher cognition: Advancing the debate. *Perspectives on Psychological Science, 8*, 223–241

Farah, M. J., and McClelland, J. L. (1991). A computational model of semantic memory impairment: Modality specificity and emergent category specificity. *Journal of Experimental Psychology: General, 120*, 339–357.

Faust, M., and Gernsbacher, M. A. (1996). Cerebral mechanisms for suppression of inappropriate information during sentence comprehension. *Brain and Language, 53*, 234–259.

Ferreira, F., Ferraro, V., and Bailey, K. (2002). Good-enough representations in language comprehension. *Current Directions in Psychological Science, 11*, 11–15.

Ferreira, F., and Lowder, M. W. (2016). Prediction, information structure, and good-enough language processing. *Psychology of Learning and Motivation, 65*, 217–247.

Ferreira, F., and Patson, N. (2007). The 'good enough' approach to language comprehension. *Language and Linguistics Compass, 1*, 71–83.

Ferretti, T. R., Kutas, M., and McRae, K. (2007). Verb aspect and the activation of event knowledge. *Journal of Experimental Psychology: Learning, Memory, and Cognition, 33*, 182–196.

Ferretti, T., McRae, K., and Hatherell, A. (2001). Integrating verbs, situation schemas, and thematic role concepts. *Journal of Memory and Language, 44*, 516–547.

Fischer, E. (2018). Two strategies for analogical reasoning: The cases of mind metaphors and introspection. *Connection Science, 30*, 211–243.

Fischer, E., and Engelhardt, P. E. (2016). Intuitions' linguistic sources: Stereotypes, intuitions, and illusions. *Mind & Language*, *31*, 65–101.

Fischer, E., and Engelhardt, P. E. (2017a). Stereotypical inferences: Philosophical relevance and psycholinguistic toolkit. *Ratio*, *30*, 411–442.

Fischer, E., and Engelhardt, P. E. (2017b). Diagnostic experimental philosophy. *Teorema*, *36*(3), 117–137.

Fischer, E., and Engelhardt, P. E. (under review). Lingering stereotypes: Salience bias in philosophical argument.

Fischer, E. (in prep). Empirically supported argument analysis: Revisiting the argument from hallucination.

Fish, W. (2010). *Philosophy of Perception*. London: Routledge.

Forbes, G. (2013). Intensional transitive verbs. In E.N. Zalta (ed.), *The Stanford Encyclopedia of Philosophy* (Fall 2013 Edition). http://plato.stanford.edu/archives/fall2013/entries/intensional-trans-verbs/

Frank, S., and Thompson, R. (2012). Early effects of word surprisal on pupil size during reading. *Proceedings of the Annual Meeting of the Cognitive Science Society*, *34*, 1554–1559.

Frazier, L., and Fodor, J. D. (1978). The sausage machine: A new two-stage parsing model. *Cognition*, *6*, 291–325.

Gentner, D., Imai, M., and Boroditsky, L. (2002). As time goes by: Evidence for two systems in processing space time metaphors. *Language and Cognitive Processes*, *17*, 537–565.

Gerken, M. (2017). *On Folk Epistemology*. Oxford: Oxford University Press.

Gerken, M., and Beebe, J. (2016). Knowledge in and out of contrast. *Nous*, *50*, 133–164.

Gigerenzer, G., Hertwig, R., and Pachur, Th. (2011). *Heuristics: The Foundations of Adaptive Behaviour*. Oxford: Oxford University Press.

Giora, R. (2003). *On Our Mind. Salience, Context, and Figurative Language*. Oxford: Oxford University Press.

Giora, R., and Fein, O. (1999). On understanding familiar and less-familiar figurative language. *Journal of Pragmatics*, *31*, 1601–1618.

Giora, R., Fein, O., Aschkenazi, K., and Alkabets-Zlozover, I. (2007a). Negation in context: A functional approach to suppression. *Discourse Processes*, *43*, 153–172.

Giora, R., Fein, O., Laadan, D., Wolfson, J., Zeituny, M., Kidron, R., Kaufman, R. and Shaham, R. (2007b). Expecting irony: Context vs. salience-based effects. *Metaphor and Symbol*, *22*, 119–146.

Giora, R., Givoni, S., and Fein, O. (2015). Defaultness reigns: The case of sarcasm. *Metaphor and Symbol*, *30*, 290–313.

Giora, R., Raphaely, M., Fein, O., and Livnat, E. (2014). Resonating with contextually inappropriate interpretations: The case of irony. *Cognitive Linguistics*, *25*, 443–455.

Goldberg, A. E. (2003). Constructions: A new theoretical approach to language. *Trends in Cognitive Sciences*, *7*, 219–224.

Grice, H. P. (1989). Logic and conversation. In his *Studies in the Ways of Words* (pp. 22–40). Cambridge, MA: Harvard University Press.

Hagoort, P., Hald, L., Bastiaansen, M., and Petersson, K. M. (2004). Integration of word meaning and world knowledge in language comprehension. *Science, 304*, 438–441.

Hampton, J. A., and Passanisi, A. (2016). When intensions do not map onto extensions: Individual differences in conceptualization. *Journal of Experimental Psychology: Learning Memory and Cognition, 42*, 505–523.

Hare, M., Jones, M., Thomson, C., Kelly, S., and McRae, K. (2009), Activating event knowledge. *Cognition, 111*, 151–167.

Harley, T. A. (2014). *The Psychology of Language*, 4th edition. London: Psychology Press.

Harmon-Vukić, M., Guéraud, S., Lassonde, K. A., and O'Brien, E. J. (2009). The activation and instantiation of instrumental inferences. *Discourse Processes, 46*, 467–490.

Horne, Z., and Livengood, J. (2017). Ordering effects, updating effects, and the spectre of global scepticism. *Synthese, 194*, 1189–1218.

Jackson, F. (1977). *Perception. A Representative Theory*. Cambridge: Cambridge University Press.

Kahneman, D. (1973). *Attention and Effort*. Engelwood Cliffs, NJ: Prentice Hall.

Kahneman, D. (2011). *Thinking Fast and Slow*. London: Allen Lane.

Kahneman, D., and Frederick, S. (2005). A model of heuristic judgment. In K. J. Holyoak and R. Morrison (eds.), *The Cambridge Handbook of Thinking and Reasoning* (pp. 67–293). Cambridge: Cambridge University Press.

Kehler, A., Kertz, L., Rohde, H., and Elman, J. L. (2008). Coherence and coreference revisited. *Journal of Semantics, 25*, 1–44.

Kim, A. E., Oines, L. D., and Sikos, L. (2016). Prediction during sentence comprehension is more than a sum of lexical associations: The role of event knowledge. *Language, Cognition, and Neuroscience, 31*, 597–601.

Kim, A. E., and Osterhout, L. (2005). The independence of combinatory semantic processing: Evidence from anticipatory eye-movements. *Journal of Memory and Language, 52*, 205–225.

Klein, D. E., and Murphy, G. L. (2001). The representation of polysemous words. *Journal of Memory and Language, 45*, 259–282.

Knobe, J., and Nichols, S. (2017). Experimental philosophy. In E. N. Zalte (ed.), *The Stanford Encyclopedia of Philosophy*, Winter 2017. https://plato.stanford.edu/archives/win2017/entries/experimental-philosophy

Koriat, A. (2007). Metacognition and consciousness. In P. D. Zelazo, M. Moscovitch and E. Thompson (eds.), *The Cambridge Handbook of Consciousness* (pp. 289–326). Cambridge: Cambridge University Press.

Kutas, M., and Federmeier, K. T. (2000). Electrophysiology reveals semantic memory use in language comprehension. *Trends in Cognitive Sciences, 4*, 463–460.

Kutas, M., and Federmeier, K. T. (2011). Thirty years and counting: Finding meaning in the N400 component of the event-related brain potential (ERP). *Annual Review of Psychology, 62*, 621–647.

Laeng, B., Sirois, S., and Gredebäck, G. (2012). Pupillometry: A window to the preconscious? *Perspectives on Psychological Science, 7*, 18–27.

Lakoff, G. (2012). Explaining embodied cognition results. *Topics in Cognitive Science, 4*, 773–785.

Landau, M. J., Meier, B. P., and Keefer, L. A. (2010). A metaphor-enriched social cognition. *Psychological Bulletin, 136*, 1045–1067.

Leech, G., Payson, P., and Wilson, A. (2001). *Word Frequencies in Written and Spoken English: Based on the British National Corpus*. London: Longman.

Levinson, S. C. (2000). *Presumptive Meanings. The Theory of Generalized Conversational Implicature*. Cambridge, MA: MIT Press.

Levy, B. J., and Anderson, M. C. (2002). Inhibitory processes and the control of memory retrieval. *Trends in Cognitive Sciences, 6*, 299–305.

Lewinski, M. (2012). The paradox of charity. *Informal Logic, 32*, 403–439.

Loewenfeld, I. (1993). *The pupil: Anatomy, physiology, and clinical applications*. Detroit, MI: Wayne State University Press.

Loftus, E. F. (1973). Activation of semantic memory. *The American Journal of Psychology, 86*, 331–337.

Lucas, M. (2000). Semantic priming without association: A meta-analytic review. *Psychonomic Bulletin and Review, 7*, 618–630

Machery, E. (2015). The illusion of expertise. In E. Fischer and J. Collins (eds.), *Experimental Philosophy, Rationalism, and Naturalism* (pp. 188–203). London: Routledge.

Macpherson, F. (2013). The Philosophy and Psychology of Hallucination. In F. Macpherson and D. Platchias (eds.), *Hallucination: Philosophy and Psychology* (pp. 1–38). Cambridge, MA: MIT Press.

Mallon, R. (2016). Experimental philosophy. In H. Cappelen, T. Szabo Gendler, and J. Hawthorne (eds.), *Oxford Handbook of Philosophical Methodology* (pp. 410–433). Oxford: Oxford University Press.

Matsuki, K., Chow, T., Hare, M., Elman, J. L., Scheepers, C., and McRae, K. (2011). Event-based plausibility immediately influences on-line language comprehension. *Journal of Experimental Psychology: Learning, Memory and Cognition, 37*, 913–934.

McKoon, G., and Ratcliff, R. (1980). Priming in item recognition: The organization of propositions in memory for text. *Journal of Verbal Learning and Verbal Behavior, 19*, 369–386.

McRae, K., Ferretti, T. R., and Amyote, I. (1997). Thematic roles as verb-specific concepts. *Language and Cognitive Processes, 12*, 137–176.

McRae, K., Hare, M., Elman, J. L., and Ferretti, T. R. (2005). A basis for generating expectancies for verbs from nouns. *Memory & Cognition, 33*, 1174–1184.

McRae, K., and Jones, M. (2013). Semantic memory. In D. Reisberg (ed.), *Oxford Handbook of Cognitive Psychology*, Oxford: Oxford University Press.

Mehler, J., Sebastian, N., Altmann, G., Dupoux, E., Christophe, A., and Pallier, C. (1993). Understanding compressed sentences: The role of rhythm and meaning. *Annals of the New York Academy of Sciences, 682*, 272–282.

Metusalem, R., Kutas, M., Urbach, T. P., Hare, M., McRae, K., and Elman, J. L. (2012). Generalized event knowledge activation during online sentence comprehension. *Journal of Memory and Language, 66,* 545–567.

Nado, J. (2014). Philosophical expertise. *Philosophy Compass, 9,* 631–641.

Nado, J. (2016). Experimental philosophy 2.0. *Thought, 5,* 159–168.

Nagel, J. (2012). Intuitions and experiments: A defence of the case method in epistemology. *Philosophy and Phenomenological Research, 85,* 495–527.

Neely, J. H., and Kahan, T. A. (2001). Is semantic activation automatic? A critical re-evaluation. In H. L. Roediger, J. S. Nairne, I. Neath and A. M. Surprenant (eds.), *The Nature of Remembering* (pp. 69–93). Washington, DC: APA..

Nichols, S., and Knobe, J. (2007). Moral responsibility and determinism: The cognitive science of folk intuitions. *Noûs, 41,* 663–685.

Oden, G. C., and Spira, J. L. (1983). Influence of context on the activation and selection of ambiguous word senses. *Quarterly Journal of Experimental Psychology, 35A,* 51–64.

Oppenheimer, D. M. (2006). Consequences of erudite vernacular utilized irrespective of necessity: Problems with using long words needlessly. *Applied Cognitive Psychology, 20,* 139–156.

Ortony, A. (1993). The role of similarity in similes and metaphors. In A. Ortony (ed.), *Metaphor and Thought*, 2nd edition (pp. 342–356). Cambridge: Cambridge University Press.

Pickering, M. J., and Garrod, S. (2013). An integrated theory of language production and comprehension. *Behavioral and Brain Sciences, 36,* 329–347.

Piquado, T., Isaacowitz, D., and Wingfield, A. (2010). Pupillometry as a measure of cognitive effort in younger and older adults. *Psychophysiology, 47,* 560–569.

Pollock, J. (1984). Reliability and justified belief. *Canadian Journal of Philosophy, 14,* 103–114.

Powell, D., Horne, Z., and Pinillos, A. (2014). Semantic integration as a method for investigating concepts. In J. Beebe (ed.), *Advances in Experimental Epistemology* (pp. 119–144). London: Bloomsbury.

Raisig, S., Hagendorf, H., and Van der Meer, E. (2012). The role of temporal properties on the detection of temporal violations: Insights from pupillometry. *Cognitive Processing, 13,* 83–91.

Rayner, K. (1998). Eye movements in reading and information processing: 20 years of research. *Psychological Bulletin, 124,* 372–422.

Rayner, K., Kambe, G., and Duffy, S. A. (2000). The effect of clause wrap-up on eye movements during reading. *Quarterly Journal of Experimental Psychology, 53,* 1061–1080.

Rayner, K., Warren, T., Juhasz, B. J., and Liversedge, S. P. (2004). The effect of plausibility on eye movements in reading. *Journal of Experimental Psychology: Learning, Memory, and Cognition, 30,* 1290–1301.

Rumelhart, D. E. (1978). Schemata: The building blocks of cognition. In R. Spiro, B. Bruce, and W. Brewer (eds.), *Theoretical Issues in Reading Comprehension*. Hillsdale, NJ: Erlbaum.

Samuels, E. R., and Szabadi, E. (2008). Functional neuroanatomy of the noradrenergic locus coeruleus: Its role in the regulation of arousal and autonomic function Part I: Principles of functional organisation. *Current Neuropharmacology, 6,* 1–19.

Schwanenflugel, P. J., and Shoben, E. J. (1983). Differential context effects in the comprehension of abstract and concrete verbal materials. *Journal of Experimental Psychology: Learning, Memory, and Cognition, 9,* 82–102.

Sereno, S. C., O'Donnell, P. J., and Rayner, K. (2006). Eye movements and lexical ambiguity resolution: Investigating the subordinate bias effect. *Journal of Experimental Psychology: Human Perception and Performance, 32,* 335–350.

Simpson, G. B., and Burgess, C. (1985). Activation and selection processes in the recognition of ambiguous words. *Journal of Experimental Psychology: Human Perception and Performance, 11,* 28–39.

Sirois, S., and Brisson, J. (2014). Pupillometry. *WIREs Cognitive Science, 5,* 679–692.

Smith, A. D. (2002). *The Problem of Perception.* Cambridge, MA: Harvard University Press.

Stephens, G. J., Silber, L. J., and Hasson, U. (2010). Speaker-listener neural coupling underlies successful communication. *Proceedings of the National Academy of Sciences, 107,* 14425–14430.

Steenbergen, H. van, and Band, G. P. (2013). Pupil dilation in the Simon task as a marker of conflict processing. *Frontiers of Human Neuroscience, 7,* 215.

Stich, S., and Tobia, K. (2016). Experimental philosophy and the philosophical tradition. In J. Sytsma and W. Buckwalter (eds.), *Blackwell Companion to Experimental Philosophy* (pp. 5–21). Wiley Blackwell: Malden.

Swets, B., Desmet, T., Clifton, C., and Ferreira, F. (2008). Underspecification of syntactic ambiguities: Evidence from self-paced reading. *Memory and Cognition, 36,* 201–216.

Tanenhaus, M. K., Carlson, G. N., and Trueswell, J. T. (1989). The role of thematic structures in interpretation and parsing. *Language and Cognitive Processes, 4,* SI 211–234.

Thagard, P., and Nisbett, R. E. (1983). Rationality and charity. *Philosophy of Science, 50,* 250–267.

Thompson, V. A., Prowse Turner, J. A., and Pennycook, G. (2011). Intuition, reason, and metacognition. *Cognitive Psychology, 63,* 107–140.

Till, R. E., Mross, E. F., and Kintsch, W. (1988). Time course of priming for associate and inference words in a discourse context. *Journal of Verbal Learning and Verbal Behaviour, 16,* 283–298.

Tulving, E. (2002). Episodic memory: From mind to brain. *Annual Review of Psychology, 53,* 1–25.

Weinberg, J. (2007). How to challenge intuitions empirically without risking scepticism. *Midwest Studies in Philosophy, 31,* 318–343.

Weinberg. J. (2015). Humans as instruments, on the inevitability of experimental philosophy. In E. Fischer and J. Collins (eds.), *Experimental Philosophy, Rationalism, and Naturalism* (pp. 171–187). London: Routledge.

Weinberg, J. (2016). Intuitions. In H. Cappelen, T. Szabo Gendler, and J. Hawthorne (eds.), *Oxford Handbook of Philosophical Methodology* (pp. 287–308). Oxford: Oxford University Press.

Welke, T., Raisig, S., Nowack, K., Schaadt, G., Hagendorf, H., and van der Meer, E. (2015). Semantic Priming of Progression Features in Events. *Journal of Psycholinguistic Research*, *44*, 201–214.

Wheeldon, L. R., and Levelt, W. J. M. (1995). Monitoring the time course of phonological encoding. *Journal of Memory and Language*, *34*, 311–334.

Williamson, T. (2007). *The Philosophy of Philosophy*. Oxford: Blackwell.

Zekveld A. A., and Kramer S. E. (2014). Cognitive processing load across a wide range of listening conditions: Insights from pupillometry. *Psychophysiology*, *51*, 277–284.

3

Judge No Evil, See No Evil: Do People's Moral Choices Influence to Whom They Visually Attend?

Jennifer Cole Wright, Evan Reinhold,
Annie Galizio and Michelle DiBartolo

Introduction

Eye-tracking technology: Tracking gaze

Eye-tracking technology allows researchers to record and analyse a range of information about what people visually attend to and how they process visual information. For example, eye-tracking technology can be used to document the order in which people attend to different features of a visual image, whether they gaze at (i.e. fixate on) particular elements of an image (or completely avoid them), and, if so, the frequency and duration of these gazes. It can also be used to track more basic processing information, such as pupil dilation (see Chapter 2 in this volume) and gaze 'directionality' (i.e. whether people's eyes tend to gaze in particular directions first or most dominantly).

There are a variety of ways that researchers can track people's eye movements and gaze direction. For example, there are free-standing systems that are typically placed in front of the person – and, thus, require that the person remain still in one location, typically while viewing visual stimuli on a screen – as well as systems that can be secured to a person's head, and are thus more mobile, able to move with the person and track eye movement and gaze more organically, during motion (see suggested readings for reviews).

Acknowledgements:

Special thanks to Dr Martin Jones, College of Charleston, for assistance with the ERICA system and protocol design.

Eye-tracking technology has been employed to study a range of phenomena and has been used to investigate several specific domains (e.g. see Lai et al. 2013; Tien et al. 2014; Blondona et al. 2015; Kredel et al. 2017; Ashraf et al. 2018). It provides an exciting window into the cognition of individuals who are not verbal, or cannot verbalize certain experiences, such as infants (Franchak et. al. 2011; Yeung et al. 2016; De Pascalis et al. 2017), children on the autism spectrum (Thorup et al. 2016; Nyström et al. 2017) and people otherwise unable to communicate (Galdi et al. 2016; Hong et al. 2017; Lee 2017). It has been used to study differences in sensory processing between experts versus novices in a given area, such as art production and evaluation (Zhiwei and Qiang 2004; Rosenberg and Klein 2015; Mitschke et al 2017; Bauer and Schwan 2018), athletic activities (Krzepota et al. 2016; Decroix et al. 2017; Vickers et al. 2017) and other physically related skills (van Leeuwen et al. 2017), teaching (McIntyre et al. 2017a; McIntyre et al. 2017b), and medical diagnoses (Södervik et al. 2017). It has also been used to investigate ways in which pre-existing dispositional attitudes and attitudinal states influence socially relevant sensory processing (Kawakami et al. 2014; Flechsenhar and Gamer 2017; Frazier et al. 2017).

In this chapter, we will discuss one way this technology could be used to explore questions of interest to experimental philosophers – in particular, questions related to their research into 'folk morality'.

Why is this useful for experimental philosophy? Exploring 'folk morality'.

Of much interest to philosophers and psychologists alike is the question of 'folk morality'. For instance, do people, as the age-old model holds, form moral judgements on the basis of impartially evaluating all of – and only – the relevant information (Piaget 1932; Kohlberg 1969)? Or can their judgements be influenced by seemingly irrelevant information? The last few decades of experimental research have largely toppled the former view in favour of the latter. Indeed, we now know that people's moral judgements can be biased by emotional processing (Haidt 2001; Greene and Haidt 2002; Greene 2007), but such biasing can be triggered by the presence of seemingly irrelevant sensory information, such as certain types of smells (Schnall et al. 2008a; Schnall et al. 2008b; Tobia et al. 2013), tastes (Eskine et al. 2011; Eskine et al. 2012), sounds (Prinz and Seidel 2012; Seidel and Prinz 2013a, 2013b) and visual imagery (Amit and Greene 2012; Zarkadi and Schnall 2013).

People's moral judgements are apparently also influenced by how they attend to and process this sensory information. People who were better at processing visual information formed moral judgements that were more strongly influenced by its presence than people who were better at processing verbal information (Amit and Greene 2012). And when viewing a social interaction, the judgements people formed about it were found to be influenced by what part of the interaction they had attended to the most. Specifically, when viewing videotapes of suspects' police interviews, the viewpoint the video takes (i.e. facing either the suspect or the police officer) influenced people's judgements about the degree to which a confession was given voluntarily, the likelihood that the suspect was guilty and the severity of the punishment s/he should receive (Lassiter and Irvine 1986; Lassiter et al. 2001; Lassiter et al. 2002) – even when it involved actual suspect interview footage (Lassiter et al. 2009) in real-life trial situations (Lassiter et al. 2002) and even when the viewers were judges and police interrogators (Lassiter et al. 2007). Later research revealed that this effect was largely a function of which person (the suspect or the police officer) the viewer had attended to the most – when allowed to view both equally (from the side), which of the two people the viewer looked at the longest was a complete mediator of the effect (Ware et al. 2008). It is thus clear that – at least in some situations – the way we attend to and process sensory information impacts the moral judgements we form. But what if the moral judgements we have already formed also influence the way we attended to and processed subsequent sensory information? Such a finding would further complicate the story surrounding moral judgements, and would be of relevance to both philosophers and psychologists interested in better understanding the causes and consequences of people's moral cognition.

There is plenty of research to suggest that our moral judgements influence other kinds of 'downstream' processing. One area this has been found in is the degree to which we seek out or attend to incoming information. People frequently form judgements and then selectively seek out or attend to information that supports these judgements, ignoring or discounting disconfirming information (Kunda 1990; Ditto and Lopez 1992; Baumeister and Newman 1994; Munro and Ditto 1997; MacCoun 1998; Nickerson 1998; Jonas et al. 2001). This phenomenon is particularly apparent with moral judgements (Haidt 2001; Ditto et al. 2009) – people cling tightly to their moral judgements and are strongly motivated to defend them against disconfirming evidence and others' conflicting judgements, resulting in a tendency to discount, ignore and avoid those who have them (Skitka and Mullen 2002; Skitka et al. 2005; Wright et al. 2008; Wright 2012). Another area this has been found is in the way in which people's moral judgements inform their subsequent judgements about things for which their moral judgements should be irrelevant, such as causation (e.g.

whether someone caused a moral harm to occur: Knobe and Fraser 2008; Knobe 2010) and intentionality (e.g. whether she did so intentionally: Wright and Bengson 2009; Knobe 2010).

There is also some evidence for sensory processing being biased by people's pre-existing judgements, generally speaking. For example, people alter the way they search a visual field, depending on what they want to confirm (Jonas et al. 2001), and their pre-existing judgements about feature relevance influence what they attend to or ignore when assessing the frequency of target objects in a visual array (Goldstone 1993; Arita et al. 2012). And, apparently, people's judgements do not even have to be directly relevant to the visual scene to which they are attending. For example, Luo and Isaacowitz (2007) found that people high in 'dispositional optimism' (i.e. people who view others as generally good and think that things typically work out for the best in the end) attended to visual stimuli differently than people low in dispositional optimism. Specifically, when shown an image of a skin cancer lesion, optimistic people attended more to the healthy surrounding tissue, looking very little at the cancerous area, whereas pessimistic people focused almost exclusively on the lesion itself (see also Segerstrom 2001; Isaacowitz 2005).

Research on depression suggests that depressed individuals preferentially attend to negative information (Beck 1967) – e.g. fixating their gaze longer on dysphoric than neutral stimuli (Mathews and Antes 1992; Eizenman et al. 2003; Caseras et al. 2007; Kellough et al. 2008). And people high in anxiety have been found to visually attend more to threatening words and images, while people low in anxiety direct their visual attention elsewhere (MacLeod et al. 1986; Bradley et al. 1998; Mogg and Bradley 1999). For example, highly anxious people who were primed to think about terrorism visually attended more to Middle Eastern faces than white faces, while people low in anxiety who had received the same prime did the opposite (Horry and Wright 2009).

Together, these findings suggest that people's existing beliefs, attitudes and judgements can powerfully influence the way they attend to and process visual information – which gives us good reason to suspect that existing moral judgements will do the same. Exploring this question is where eye-tracking technology becomes useful.

Study 1

Previous research across several disciplines suggests that, just as people's processing of sensory information can influence their moral judgements, so too should their moral judgements influence the way they attend to and process sensory information. Therefore, we expected that when we asked people to form judgements about particular moral scenarios, they would be inclined to seek out or attend to (i.e. fixate their gaze upon) visual information that was most consistent with or supportive of those judgements and/or avoid visual information that was inconsistent or unsupportive.

Since both of these inclinations – i.e. the inclination to *attend to* supportive visual information and to *avoid* unsupportive visual information – generally accomplish the same objective (i.e. insulating one's judgements from challenge, maintaining integrity of one's attitudes and beliefs, etc.), we did not have any strong *prima facie* theoretical reason to expect the use of one strategy over the other, but we did expect that either one or the other (or potentially both) would reliably occur.

Method

Participants. The participants in this study were 115 students from a southern US institution (97 females; 88% Caucasian, 4% African-American, 3% Asian-American, 2% Hispanic, 3% other) taking introductory psychology classes and receiving course credit for their participation. Twenty-three participants[1] were unable to participate due to calibration issues – mostly due to the fact that they were wearing eye-glasses, though people with overly dry eyes or dark irises were also harder to calibrate.

Apparatus. We employed the ERICA eye-tracking device (Version 05.01.20, 50 mm TV Lens, 1:1.13; Eye Response Technologies 2004) with gaze fixation points set at .05 seconds (minimum) time within a gaze diameter of 40 pixels – in other words, in order to count as a discrete gaze fixation point, the participants' gaze would need to remain within a 40 pixel gaze diameter for a minimum of .05 seconds. The eye-tracker was on a desk directly in front of a computer screen (1920 by 1200 pixel resolution), which was placed approximately 2.5 feet (0.762 meters) in front of the participant. The experimenter was located at one end of the desk with a laptop computer that controlled the images on the monitor and calibrated the ERICA eye-tracker to each participant's dominant (typically right) eye.

---

[1] While 20% is a fairly high rate of disqualification, it was unavoidable given the calibration limitations of this particular eye-tracking device. Other (newer, more expensive) technologies are not as limited. We reduced our disqualification rate in Study 2 to 12% by restricting the study to participants who did not wear glasses.

Materials and Design. A text box presenting a moral scenario (two per participant) was presented to participants for 25 seconds followed by a 10 second exposure to an image depicting the characters presented in the dilemma. A three-second buffer slide containing a dark screen with centred red dot was presented between each slide to reorient the participants' eyes to the centre of the screen. In both studies we considered the first recorded fixation point on the images to be a centred focus point carried over from the buffer slide. Therefore, we counted the second recorded fixation point as the participants' first true glance at the images.

Participants were presented first with the Trolley (Thomson 1976), then the Baby/Villager scenario:

> *Trolley:* A trolley is running out of control down a track. In its path are five people working on the trolley track and they are wearing head gear designed to block out noise, so they won't hear the trolley coming in time to move out of the way. Fortunately, you can flip a switch, which will lead the trolley down a side track to safety. Unfortunately, there is a very large man working on that side track. He is listening to music on his iPod and won't hear the trolley coming in time to move out of the way. So if you pull the switch, you will move the trolley to the side track, killing him, but saving the five workers.

> *Baby/Villager:* Enemy soldiers have taken over your village. They have orders to kill all remaining civilians. You and some of your townspeople have sought refuge in the cellar of a large house. Outside, you hear the voices of soldiers who have come to search the house for valuables. Your baby begins to cry loudly. You cover his mouth to block the sound. If you remove your hand from his mouth, his crying will summon the attention of the soldiers who will kill you, your child and the others hiding out in the cellar. To save yourself and the others, you must smother your child to death.

Immediately after reading each scenario, participants read the following question: 'Would you [*Trolley:* flip the switch to save the five people][*Baby/Villager:* smother your baby to save the other villagers]?' on 7-point Likert scales, where 1 was strongly negative and 7 was strongly affirmative. They were given an answer sheet upon which to write their numerical response to the question directly after reading each scenario, before the images appeared.

After responding to the question, a split screen was presented with 5″ × 5″ (96 dpi resolution) colour grey-scaled[2] images connected to the scenario. For

---

[2] We grey-scaled (i.e. removed colour) the images from both studies to eliminate any effect that different colours might have – and also so that we did not have to control for colour deficiencies in our participants.

the Trolley scenario, they were shown an image of an overweight man on one side of the screen and an image of five workmen on the other side. For the Baby/Villager scenario, they were shown an image of a baby on one side of the screen and an image of a group of people hiding in a cave (meant to depict the villagers) on the other side. For both, the side on which the images appeared was counterbalanced between participants.

Procedure. The eye-tracking task was part of a larger study (they filled out a survey for an unrelated study about folk meta-ethics after completing the eye tracker portion) and was the first task participants engaged in after filling out an informed consent. Participants were in the laboratory for 30 minutes with roughly 5–10 minutes spent on the eye-tracker. Participants were shown into the eye-tracking room and were asked to make a triangle with their hands and focus with both eyes open on a dot hanging on the wall across the room. Once they had the dot in focus, they were asked to close one eye at a time and observe when the dot disappears. We considered the dominant eye (which was typically the right eye) the one with which they could still see the dot when it was open. After this, they were told to sit close to the desk – 2.5 feet (0.762 meters) from the monitor – in a comfortable position so that they can hold their head still during the tracking. The eye-tracker was focused on the participants' dominant eye followed by a calibration of the device, which involved having them visually follow a series of red squares as they moved across the screen, allowing the eye tracker to lock in on their pupil's location, thereby allowing it to calculate gaze fixation.

Once all this preparatory work was completed, the participants were told that they would be presented with a set of stories for them to consider and that they would be allowed to report their judgements about them after reading each scenario. They were instructed how to write their answers on the sheet in front of them without lowering their eyes from screen, so as to not disrupt the calibration with the eye-tracker for the second scenario (though a minimal amount of disruption nonetheless occurred). At this point, the program was started and the experimenter stepped out of the room until it was finished. After this, participants were instructed to leave their written responses on the desk in front of them and they accompanied the experimenter into another room to participate in the other study.

Results

For the *Trolley* scenario, the average willingness to flip the switch was $M = 5.0$, $SE = .16$. The average number of discrete gaze fixations on the workers was $M =$

10.5, SE =.39 and on the individual M = 6.2, SE =.28. Participants' first glance (measured by their second recorded gaze fixation point) was at the individual 59.3% of the time and at the workers 40.7% of the time.

For the *Baby/Villager* scenario, the average willingness to smother the baby was M = 3.0, SE =.19. The average number of discrete gaze fixations on the villagers was M = 9.4, SE =.52 and on the baby M = 7.0, SE =.39. Participants' first glance (measured by their second recorded gaze fixation point) was at the baby 50.5% of the time and at the villagers 49.5% of the time.

The first thing we examined was the relationship between participants' judgements and their pattern of first glances (measured by their second recorded gaze fixation point), using binary logistic regression.[3] We hypothesized that for both scenarios, people would gaze first *at the beneficiary of that judgement* – an action consistent with both the inclination to seek out or attend to visual information that is consistent with or supportive of the judgement and the inclination to avoid visual information that is inconsistent with or unsupportive of the judgement.

As expected, people's judgements in the *Trolley* scenario about what they would be willing to do significantly predicted the image to which they gazed at first in the Trolley scenario, $X^2(1, N = 92) = 5.3$, p =.022. For every single unit increase in participants' willingness to pull the switch, the odds of participants' first glance being at the workers increased by 143% (or a factor of 1.43). This was also true for the *Baby/Villager* scenario, $X^2(1, N = 92) = 3.6$, p =.050. For every single unit increase in participants' willingness to smother the baby, the odds of participants' first glance being at the villagers increased by 80% (or a factor of .80).

Next, we examined the relationship between participants' judgements and their gaze fixation frequencies (i.e. the number of times they fixated their gaze upon each of the image options). We hypothesized that people's judgements would predict their tendency to gaze *more* at the beneficiary of their judgement or *less* at the non-beneficiary of their judgement, but not both. Because each of these represents a distinct strategy (approach vs avoidance) we predicted that people would be inclined to employ only one of them – that is, either participants would be inclined to *gaze at the beneficiary image*, which means their gaze would fixate more frequently upon the beneficiary image than anywhere else (including

---

[3] Binary logistic regression is a statistical technique used to predict the relationship between a continuous predictor variable (in this case, participants' judgements) and a binary predicted variable (in this case, first glance, which could be either at the beneficiary or the non-beneficiary). See Hatcher (2013).

'off-image', in between or outside the range of either image) or they would be inclined to *not gaze at the non-beneficiary*, which means their gaze would fixate less frequently upon the non-beneficiary image than anywhere else. And this is what we found.

Specifically, linear regressions revealed that for the *Trolley* scenario participants' willingness to pull the switch predicted the number of discrete gaze fixations on the workers, $B = .60$, $t(90) = 2.5$, $p = .016$ – i.e. the more willing they were to pull the switch, the more frequently they gazed at the workers – but not with the number of discrete gaze fixations on the individual, $B = -.24$, $t(90) = 1.4$, $p = .171$. So, in this case, participants tended to gaze at the beneficiary of their judgement, not away from the non-beneficiary.

Interestingly, for the *Baby/Villager* scenario this flipped – participants' willingness to smother the baby predicted the number of discrete gaze fixations on the baby, $B = -.43$, $t(89) = 2.1$, $p = .042$ – i.e. the more willing they were to smother the baby, the less frequently they fixated on the baby. But it did not predict the number of discrete gaze fixations on the villagers, $B = -.15$, $t(90) = .5$, $p = .599$. Here, participants tended to *gaze away from* the non-beneficiary, rather than at the beneficiary.

Discussion

We hypothesized that for both scenarios, people would gaze first *at the beneficiary* of their judgement and this is what we found across both scenarios. We also hypothesized that people would either gaze *more* at the beneficiary of their judgement or *less* at the non-beneficiary of their judgement, but not both – which we also found. Specifically, we found that the more willing people were to kill the individual to save the group, the more they gazed at the workers (beneficiary) in the Trolley scenario – suggesting an impulse to direct their gaze *towards*, rather than *away* (i.e. the approach strategy) – and the more willing people were to kill the baby to save the group, the more they gazed away from the baby (non-beneficiary) in the Baby/Villager scenario, suggesting an impulse to direct their gaze *away*, rather than *towards* (i.e. the avoidance strategy). This finding is consistent with Amit and Greene's (2012) finding that participants who judged it wrong to kill the individual reported visualizing the individual more than the group – but the participants who judged it acceptable to kill the individual did not report visualizing the group more than the individual.

This leaves an important question unanswered, however – and that is *why* people employed one strategy for the *Trolley* scenario and another for the *Baby/*

Villager scenario. One possibility is that which strategy gets activated (approach or avoidance) depends upon certain aspects of the scenarios themselves. For example, a clear difference that stands out between the two scenarios chosen for Study 1 (*Trolley* vs *Baby/Villagers*) is the level of difficulty – in terms of the choices participants were asked to make – they represent. While many people report that it would be relatively easy to pull the switch (for discussions about why this might be, see Petrinovich et al. 1993; Petrinovich and O'Neill 1976; Greene et al. 2004; Greene et al. 2008; Greene et al. 2009), smothering a baby (even to save a village) is a much more difficult choice. And perhaps this has an influence on the strategy employed. For scenarios where the choice is relatively easy, people would be inclined to gaze at the beneficiaries, their gaze pattern functioning as a sort of visual *confirmation* of their judgement – a sort of 'feel good' impulse to gaze upon the positive outcome of their choice. On the other hand, for scenarios where the choice is difficult – and thus, discomfort and guilt would conceivably be higher – people would be inclined to gaze away from the non-beneficiary, their gaze pattern functioning as a visual *avoidance*, an impulse (likely subconscious) to avoid a source of potential guilt and regret.

Study 2

To test this possibility – namely, that people will employ different strategies for visually processing information, depending upon whether the moral choice they were asked to make was easier or more difficult – we designed another study, the results of which are reported below.

Method

Participants. The participants in this study were 120 students from a southern US institution (82 females; 86% Caucasian, 5% African-American, 4% Asian-American, 2% Hispanic, 3% other) taking Introductory Psychology classes and receiving course credit for their participation. Eye-tracking data for 14 participants was not successfully generated due to difficulties with calibration (similar to Study 1).

Apparatus. Once again, we employed the ERICA eye-tracking device with everything set up in the same way as Study 1.

Materials and Design. In order to test whether people gaze at the beneficiary of their chosen action or away from the person who does not benefit, we developed six new scenarios that required participants to make a choice that benefitted one person at the expense of another. All six scenarios had a 'strong' version, in which one of the potential beneficiaries was much more desirable/deserving of that benefit than the other (making the choice of who to benefit relatively easy and justified), and an 'ambiguous' version, in which both potential beneficiaries seemed equally desirable/deserving (making the choice between them harder because of its completely arbitrary nature). We hypothesized that for the strong scenarios, people would be inclined to gaze at the beneficiary of their judgement, while for the ambiguous scenarios people would instead be inclined to gaze away from the non-beneficiary.

All six scenarios also included written and visual information about a third person, who would neither benefit nor be negatively impacted by the participants' judgements (more on the reason for this below).

As an example, here is one of the scenarios:

Strong version: Bob has always been on time. He works well with everyone and puts in the extra effort on projects that help the company. Fred is late on a weekly basis and usually clocks out early. He doesn't really interact with anyone else in the company and never puts in extra effort to help get the job done. Joe was just recently hired and is learning the ropes.

Ambiguous version: Bob is always on time to work. He works well with all of the other employees and isn't afraid to put in the extra effort to help the company. Fred is very punctual and hasn't been late to work once. He gets along with the rest of the staff and works well with them too. Sometimes he even comes in on Saturday to help finish a project. Joe was just recently hired and is learning the ropes.

For both, participants then saw the question, 'You have the ability to promote either Bob or Fred to a position with a higher salary (being new, Joe is not eligible for a promotion) and you can <u>only</u> promote one. Which of the two would you promote?' In order to avoid the difficulties encountered with having participants write down their responses while trying *not* to look down from the screen, we asked them to respond verbally instead. Thus, they stated their decision after reading the scenario, but before the images came up on the screen. The other scenarios involved the choice of whom to (2) save from drowning, (3) help out with an errand, (4) give a job, (5) rule (in court) in favour of, (6) give an award (see Appendix for full scenarios).

A text box presenting each scenario was presented to participants for 25 seconds followed by a 10 second exposure to 3″ × 4″ grey-scaled (96 dpi resolution) images depicting the characters presented in each scenario. A three-second buffer slide containing a dark screen with centred red dot was presented between each slide to reorient the participants' eyes to the centre of the screen. Once again, we considered the first recorded gaze fixation point on the images to be a centred focus point carried over from the buffer slide. Therefore, we counted the second recorded gaze fixation point as the participants' first true glance at the images.

The key difference in Study 2 was that after each scenario, participants where shown three (rather than two) images, two of which represented the individuals they were choosing between and the third – always placed in the middle – was an irrelevant 'other' who would not benefit either way (and could therefore neither be helped nor negatively impacted by the participants' choice). This set up allowed us to better test between the two possibilities – whether people are *gazing at* the beneficiary or *away from* the non-beneficiary – because if participants' first impulse was to gaze at their beneficiary, then we should see a highest frequency of first glances at the beneficiary (as before), but no difference in frequency between the first glances at either the non-beneficiary or the 'other'. If, on the other hand, their first impulse was to gaze away from those negatively impacted by their judgements, then we would expect the lowest frequency of first glances at the person their chosen actions harm/fail to benefit, but no difference in frequency between their first glances at either the beneficiary or the 'other'.

Scenario presentation was varied along two dimensions: first, the order in which the scenarios were presented (1–6 or 6–1) and, second, which of the six scenarios were strong, which ambiguous (participants always got three of each). We did not vary the presentation of the images, or the names assigned to the images, within each scenario.

Procedure. The eye-tracking task was once again a part of a larger study (this time, they filled out a survey for an unrelated study about people's attitudes about free will after completing the eye tracker portion) and was the first task participants engaged in after filling out an informed consent. This time participants were in the laboratory for 50–60 minutes with roughly 5–10 minutes spent on the eye-tracker. Participants were shown into the eye-tracking room and told to sit close to the desk in a comfortable position where they could hold their head still during the tracking. The eye-tracker was focused on the participants' dominant (typically right) eyes followed by a preliminary calibration of the device. The participants were told that they would be presented with a set of stories to consider and that they would be allowed to verbally report their judgements about them after reading each scenario. They were instructed to verbally state their judgements, which would be recorded by a person sitting directly outside the room. At this point, the program was started and the experimenter stepped out of the room until it was finished. After this, participants accompanied the experimenter into another room to participate in the other study.

Results

In the strong (i.e. easy) versions of the scenarios, participants chose to benefit (i.e. act or judge in favour of) the clear 'deserving' individual in the vignette (e.g. Bob, Laura, Brendan, Joanna, Wayne and Ginger) far more often than the 'non-deserving' individual – on average 87.2% versus 10.8% of the time. In the ambiguous (i.e. hard) versions of the scenarios, however, participants were split between the two (48.3% vs 50.1%). Importantly, across both the strong and ambiguous cases, participants almost never chose the 'irrelevant' individual (1.9% and 1.7%, respectively).

The location of participants' first glance (measured by their second recorded fixation point) was more often the beneficiary of their choices (36.9% of the time for the strong and 32.5% for the ambiguous versions) than either the non-beneficiary (20.0% strong and 18.7% ambiguous) or the irrelevant other (17.9% strong and 30.1% ambiguous). A repeated-measures ANOVA with first glance *location* and scenario *version* (strong/ambiguous) as within-participant variables revealed a main effect for location of first glance, $F(2,182) = 14.5$, p <.001, $\eta^2 =.14$. More specifically, paired-sample t-tests revealed that – collapsing across version – participants' first glance was at the beneficiary more often (*M* = 44% of participants, *SE* =.02) than either the non-beneficiary (*M* = 25%, *SE* =.02), $t(91) = 5.0$, p <.001, or the irrelevant other (*M* = 31%, *SE* =.02), $t(91) = 3.3$, $p =.002$ and that there was no difference between participants' first glance at the non-beneficiary versus the other, $t(91) = -1.6$, $p =.096$.

Consistently with our hypothesis, however, the ANOVA revealed a more complicated story. Specifically, there was an interaction between location and version, $F(2,182) = 3.6$, $p =.029$, $\eta^2 =.04$. A series of paired-sample t-tests revealed that, for the strong version of the scenarios, the location of first glance mapped onto the general pattern mentioned above: there was a significant difference between participants' first glances towards the beneficiary (*M* = 43% of participants, *SE* =.03) versus the non-beneficiary (*M* = 21%, *SE* =.02), $t(91) = 5.0$, p <.001, as well as a significant difference between participants' first glances towards the beneficiary versus the irrelevant other (*M* = 23%, *SE* =.02), $t(91) = 4.6$, p <.001, but no difference between participants' first glances towards the non-beneficiary versus the irrelevant other, $t(91)$ = -.6, p =.528. Thus, it would appear that when the choice of who to benefit was easy and justified, people's first glance was clearly *towards the beneficiary* of their choice.

For the ambiguous version of the scenarios, however, the opposite pattern emerged. Paired-sample t-tests once again revealed a significant difference between participants' first glances towards the beneficiary (M = 36% of participants, SE =.03) versus the non-beneficiary (M = 22%, SE =.03), $t(91)$ = 2.8, p =.006. But, we also found a significant difference between participants' first glances towards the non-beneficiary versus the irrelevant other, (M = 33%, SE =.03), $t(91)$ = -2.3, p =.025 and *not* between participants' first glances towards the beneficiary versus the irrelevant other, $t(91)$ =.5, p =.636. Thus, it appears that when the choice of whom to benefit was <u>not</u> easy (but was rather difficult and arbitrary), people's first glance was *away from the non-beneficiary*, rather than towards the beneficiary of their choice (Figure 3.1).

Now turning to participants' frequency of gaze fixations, the average number of discrete gaze fixations on the beneficiary of participants' choices was M = 2.1, SE =.21 for the strong scenarios and M = 2.2, SE =.21 for the ambiguous scenarios, while the average number of discrete gaze fixations on the non-

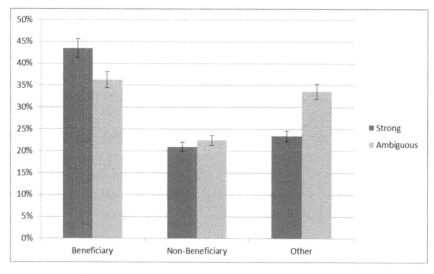

Figure 3.1 Study 2, Percentage of first-glances to the beneficiary, non-beneficiary, and 'other' individuals in the strong (easy) and ambiguous (hard) conditions. Error bars represent standard error.

Table 3.1 Study 2, Expectations of frequency of discrete fixations by condition.

	Beneficiary	Other	Non-Beneficiary
Strong (Easy) Cond	Look More	Look Less (**SONB**)	
Ambiguous (Hard) Cond	Look More (**AOB**)		Look Less

beneficiary was $M = 1.8$, $SE =.19$ for the strong scenarios and $M = 1.6$, $SE =.17$ for the ambiguous scenarios.

Given the above, our expectation should be that for the strong versions of the scenarios, people would gaze more frequently (in terms of number of discrete gaze fixations) at the beneficiary than either the non-beneficiary or the irrelevant other – yet, in the ambiguous versions of the scenarios, they would gaze less frequently away from the non-beneficiary than either the beneficiary or the irrelevant other (see Table 3.1).

In other words, we should expect that participants would gaze significantly less frequently at the other/non-beneficiary when given the strong version of the scenarios (SONB) than at the other/beneficiary when given the ambiguous version of the scenarios (AOB). And this is what we found: the number of SONB gaze fixations was significantly less ($M = 60.5\%$, $SE =.02$) than the number of AOB gaze fixations ($M = 70\%$, $SE =.01$), $t(101) = 4.3$, $p <.001$.

In addition, for the strong versions of the scenarios, people more frequently gazed at the beneficiary ($M = 37.3\%$, $SE =.02$) than either the non-beneficiary ($M = 31.6\%$, $SE =.01$) or the other ($M = 31.1\%$, $SE =.02$), t's $(107) = 2.3$ and 2.1, p's $=.027$ and.040, respectively. And for the ambiguous versions of the scenarios, they looked less frequently at the non-beneficiary ($M = 27.1\%$, $SE =.01$) than either the beneficiary ($M = 38.8\%$, $SE =.02$) or the other ($M = 33.1\%$, $SE =.01$) t's $(107) = 5.6$ and 3.1, p's $<.001$ and. 002, respectively.

Discussion

Study 2 confirmed that people's first glance was at the beneficiary of their judgement. It also confirmed that which strategy this first glance represented (approach or avoidance) depended upon whether the choice they were making was fairly easy and justified (as in the 'strong' scenarios) or hard (as in the 'ambiguous' scenarios). Specifically, in the former situation, people's first glance was *towards* the beneficiary, but in the latter case, it was *away from* the non-beneficiary. In further support, people more frequently gazed at (in terms of discrete gaze fixations) the beneficiary when given the easy choice and more frequently away from the non-beneficiary when given the hard choice.

General discussion

Taken together, these two studies provide evidence that people's moral judgements influence the way they attend to and process sensory information. Study 1 found that participants gazed first and most frequently at the beneficiary of their chosen actions. Yet, whether their gaze pattern was towards the individual benefitted or away from the individual not benefitted – that is, whether it functioned as a visual *confirmation* or a visual *avoidance* of their intended action – varied depending on the scenario presented. We hypothesized that this could be because the choice participants were being asked to make was easier in one of the scenarios (Trolley) than the other (Baby/Villager).

Study 2 was designed to test this possibility and found that when one of the two individuals in the scenario was set up to clearly deserve the benefit more than the other, making the participants' choice easier, then their gaze patterns did indeed act *confirmatory* in nature, visually gazing at the beneficiary they had chosen first and most frequently. When, on the other hand, the two individuals were on par with one other, making participants' choice about who to help/ benefit harder and more arbitrary, their gaze patterns functioned more like an *avoidance* mechanism, gazing first and most frequently at someone other than the individual their choice harmed/failed to benefit.

In sum, there is extensive evidence that people's morally relevant beliefs and judgements are influenced by the presence of seemingly irrelevant sensory information – particularly, when they attend to it – but this is the first evidence we are aware of that people's moral judgements likewise influence how they attend to and process sensory information. In conjunction with the previous literature discussed in the introduction, this suggests a strong bi-directional relationship between people's moral cognition and their attention to sensory information – in this case, visual. In other words, not only are we likely to form important moral judgements on the basis of the sensory information we have attended (or avoided attending) to, but we are likely to reinforce those judgements through our later patterns of sensory processing.

Why did participants choose one strategy (visually 'approaching' positive stimuli) when their judgements were easy – or, at least, *less* difficult – and another (visually 'avoiding' negative stimuli) when their judgements were *more* difficult? It might be that in the former situation, there are less negative emotions generated – indeed, some have argued that scenarios like *Trolley* do not elicit much emotion at all, but rather trigger more rational, utilitarian calculus (e.g. Greene et al. 2001) – and so the default (i.e. 'adaptive')[4] strategy would be for people to gain as much judgement-relevant information as they can, in this case by attending first and most frequently to the visual information relevant to the outcome of their judgements. In the latter situation, on the other hand, the presence of stronger negative emotions or conflict associated with making an 'impossible' decision could trigger an avoidance mechanism – now the main goal is to avoid gazing at anything that would exacerbate that negative state.

It is interesting to note that this seems consistent, at least on the surface, with the visual patterns observed in optimistic, non-depressed and non-anxious participants (see Introduction) – that is, their tendency to attend to positive information and avoid negative information. But this means that it is also inconsistent with the patterns observed in pessimistic, depressed and anxious people, who seemed more inclined to attend to negative information (though it is not clear that they avoided positive information). If this is correct, then it would suggest that the visual patterns we found may be consistent with more general 'low-level' (e.g. perception rather than cognition) strategies we have developed to help maintain optimal health and well-being. Of course, whether they also contribute to good social-moral decision making is a question that deserves further exploration.

Strengths and limitations of gaze-tracking technology

The research reported here explored an exciting methodological route for experimental philosophers to pursue. The use of technologies such as the eye-gaze tracking device allow for a deeper exploration into how sensory information influences and is influenced by the sorts of attitudes, beliefs and judgements philosophers care about, not only in ways that our models of rationality would

---

4   Elsewhere, people have argued for evolutionary account of 'adaptive attentional attunement', which is the view that we evolved to attend more to information that is socially/ relevant to our well-being/ survival, such as physical attractiveness in women and social status in men (e.g. Maner et al. 2007; DeWall and Maner 2008).

predict, but in surprising, counterintuitive ways – ways that help us to better map out a more accurate model of human cognition.

The strengths of this methodological approach are that it allows us to measure non-conscious processes – things people are unaware of and, thus, unable to report. This is one of the things that makes this approach so useful in research with infants, people on the autism spectrum, and people otherwise unable to communicate.

Like the Implicit Attitudes Test (IAT, Greenwald et al. 1998) and other forms of 'implicit attitude' measurement, eye-tracking technology can be used to access various qualities of people's implicit visual processing. Specifically, it can be used to measure their gaze fixation frequency and duration, as well as their scan path – i.e. how often they gaze at specific visual stimuli, for how long and in what order. This allows us to examine the role of such information in influencing, and being influenced by, people's consciously reported responses.

The downsides to using gaze-tracking technology are twofold. First, it is expensive, requiring special equipment and software to run, as well as training on how to calibrate it properly to each individual to minimize error. Such calibration can be problematic, especially for people who wear glasses or have very dark irises (making the pupil hard for the eye tracker to detect). This has a worrying side-effect of making certain demographics of people more difficult to test than others. Secondly, this technology provides us with access to a fairly limited range of information – namely, facts about how people attend to and process visual information. While there are certainly areas of cognition where this information is helpful – for example, uncovering potential sources of error or bias in people's processing of and attitudes about visual information – there is only so much we can learn by examining where and when and for how long people look (but see Chapter 2 in this volume).

Relevance to experimental philosophy

It is also worth considering why using this technology to uncover the way people's moral judgements are influenced by and influence processing of sensory information – while potentially worthwhile for social scientists, who need an accurate picture of how people form and use their judgements in order to be able to explain and predict their behaviour – would be a worthwhile endeavour for experimental philosophers.

One obvious response is because experimental philosophers are already doing this sort of research. That is, as the citations listed earlier in this chapter indicate, they already care about how people form and use their moral judgements, whether that's of deeper philosophical significance or not. And giving them a wider range of technologies and methodologies to employ only strengthens their ability to study and draw conclusions about their topic of interest, such as 'folk morality'.

But we think there are two additional reasons worth mentioning. First, as many philosophers have observed, to the extent that philosophical theories diverge from reality – to the extent that they predict judgements or behaviour that is not what we actually observe – then the onus is on the philosophical theory to justify the divergence (Knobe and Nichols 2017). And second, knowing how human beings think, feel and behave provides a valuable meta-cognitive lens for understanding how they *philosophize* (i.e. which philosophical models are likely to seem the most accurate, to be the most attractive). In other words, utilizing an expansive empirical methodology to explore the intersections between philosophical inquiry and human cognition seems like a good idea.

Suggested Readings

Chennamma and Yuan (2013). A survey on eye-gaze tracking techniques, *Indian Journal of Computer Science & Engineering*. https://arxiv.org/ftp/arxiv/papers/1312/1312.6410.pdf.

Kar and Corcoran (2017). A review and analysis of eye-gaze estimation systems, algorithms and performance evaluation methods in consumer platforms, *IEEE Access*. http://ieeexplore.ieee.org/stamp/stamp.jsp?arnumber=8003267.

References

Amit, E., and Greene, J. D. (2012). You see, the ends don't justify the means: Visual imagery and moral judgment. *Psychological Science (0956–7976)*, *23*(8), 861–868.

Arita, J. T., Carlisle, N. B., and Woodman, G. F. (2012). Templates for rejection: Configuring attention to ignore task-irrelevant features. *Journal of Experimental Psychology: Human Perception and Performance*, *38*(3), 580–584.

Ashraf, H., Sodergren, M. H., Merali, N., Mylonas, G., Singh, H., and Darzi, A. (2018). Eye-tracking technology in medical education: A systematic review. *Medical Teacher*, *40*(1), 62–69.

Bauer, D., and Schwan, S. (2018). Expertise influences meaning-making with renaissance portraits: Evidence from Gaze and thinking-aloud. *Psychology of Aesthetics, Creativity, and the Arts, 12*(2), 193–204.

Baumeister, R. F., and Newman, L. S. (1994). Self-regulation of cognitive inference and decision processes. *Personality and Social Psychology Bulletin, 20*, 3–19.

Beck, A. T. (1967). *Depression: Clinical, experimental, and theoretical aspects*. New York: Harper & Row.

Blondon, K., Wipfli, R., and Lovis, C. (2015). Use of eye-tracking technology in clinical reasoning: A systematic review. *Studies in Health Technology and Informatics, 210*, 90–94.

Bradley, B. P., Mogg, K., Falla, S. J., and Hamilton, L. R. (1998). Attentional bias for threatening facial expressions in anxiety: Manipulation of stimulus duration. *Cognition and Emotion, 12*, 737–753.

Caseras, X., Garner, M., Bradley, B. P., and Mogg, K. (2007). Biases in visual orienting to negative and positive scenes in dysphoria: An eye-movement study. *Journal of Abnormal Psychology, 116*, 491–497.

Decroix, M., Wazir, M. N., Zeuwts, L., Deconinck, F. F., Lenoir, M., and Vansteenkiste, P. (2017). Expert–Non-expert differences in visual behaviour during alpine slalom skiing. *Human Movement Science, 55*, 229–239.

De Pascalis, L., Kkeli, N., Chakrabarti, B., Dalton, L., Vaillancourt, K., Rayson, H., and Murray, L. (2017). Maternal gaze to the infant face: Effects of infant age and facial configuration during mother-infant engagement in the first nine weeks. *Infant Behavior & Development, 46*, 91–99.

DeWall, C., and Maner, J. K. (2008). High status men (but not women) capture the eye of the beholder. *Evolutionary Psychology, 6*(2), 328–341.

Ditto, P. H., and Lopez, D. F. (1992). Motivated skepticism: Use of differential decision criteria for preferred and non-preferred conclusions. *Journal of Personality and Social Psychology, 63*, 568–584.

Ditto, P. H., Pizarro, D. A., and Tannenbaum, D. (2009). Motivated moral reasoning. *Psychology of Learning and Motivation, 50*, 307–338.

Eizenman, M., Yu, L. H., Grupp, L., Eizenman, E., Ellenbogen, M., Gemar, M. et al. (2003). A naturalistic visual scanning approach to assess selective attention in major depressive disorder. *Psychiatry Research, 118*, 117–128.

Eskine, K. J., Kacinik, N. A., and Prinz, J. J. (2011). A bad taste in the mouth: Gustatory disgust influences moral judgment. *Psychological Science, 22*(3), 295–299.

Eskine, K. J., Kacinik, N. A., and Webster, G. D. (2012). The bitter truth about morality: Virtue, not vice, makes a bland beverage taste nice. *Plos ONE, 7*(7), 1–4.

Flechsenhar, A. F., and Gamer, M. (2017). Top-down influence on gaze patterns in the presence of social features. *Plos ONE, 12*(8), 1–20.

Franchak, J. M., Kretch, K. S., Soska, K. C., and Adolph, K. E. (2011). Head-mounted eye tracking: A new method to describe infant looking. *Child Development, 82*(6), 1738–1750.

Frazier, T. W., Strauss, M., Klingemier, E. W., Zetzer, E. E., Hardan, A. Y., Eng, C., and Youngstrom, E. A. (2017). A meta-analysis of gaze differences to social and nonsocial information between individuals with and without autism. *Journal of the American Academy of Child & Adolescent Psychiatry*, *56*(7), 546–555.

Galdi, C., Wechsler, H., Cantoni, V., Porta, M., and Nappi, M. (2016). Towards demographic categorization using gaze analysis. *Pattern Recognition Letters*, *82*, 226–231.

Goldstone, R. (1993). Feature distribution and biased estimation of visual displays. *Journal of Experimental Psychology*, *19*, 564–579.

Greene, J. D. (2007). Why are VMPFC patients more utilitarian? A dual-process theory of moral judgment explains. *Trends in Cognitive Science*, *11*, 322–323.

Greene, J. D., Cushman, F. A., Stewart, L. E., Lowenberg, K., Nystrom, L. E., and Cohen, J. D. (2009). Pushing moral buttons: The interaction between personal force and intention in moral judgment. *Cognition*, *111*, 364–371.

Greene, J. D., and Haidt, J. (2002). How (and where) does moral judgment work? *Trends in Cognitive Sciences*, *6*(12), 517–523.

Greene, J. D., Morelli, S. A., Lowenberg, K., Nystrom, L. E., and Cohen, J. D. (2008). Cognitive load selectively interferes with utilitarian moral judgment. *Cognition*, *107*, 1144–1154.

Greene, J. D., Nystrom, L. E., Engell, A. D., Darley, J. M., and Cohen, J. D. (2004). The neural bases of cognitive conflict and control in moral judgment. *Neuron*, *44*, 389–400.

Greene, J. D., Sommerville, R. B., Nystrom, L. E., Darley, J. M., and Cohen, J. D. (2001). An fMRI investigation of emotional engagement in moral judgment. *Science*, *293*, 2105–2108.

Greenwald, Anthony G., McGhee, Debbie E., and Schwartz, Jordan L. K. (1998). Measuring individual differences in implicit cognition: The implicit association test. *Journal of Personality and Social Psychology*, *74*(6): 1464–1480.

Haidt, J. (2001). The emotional dog and its rational tail: A social intuitionist approach to moral judgment. *Psychological Review*, *108*, 814–834.

Hatcher, L. (2013). *Advanced Statistics in Research: Reading, Understanding, and Writing Up Data Analysis Results*. Saginaw MI: Shadow Finch Media LLC.

Hong, Michael P., Guilfoyle, Janna L., Mooney, Lindsey N., Wink, Logan K., Pedapati, Ernest V., Shaffer, Rebecca C., Sweeney, John A., and Erickson, Craig A. (2017). Eye gaze and pupillary response in Angelman syndrome. *Research in Developmental Disabilities*, *68*, 88–94.

Horry, R., and Wright, D. B. (2009). Anxiety and terrorism: Automatic stereotypes affect visual attention and recognition memory for White and Middle-Eastern faces. *Applied Cognitive Psychology*, *23*, 345–357.

Isaacowitz, D. M. (2005). The gaze of the optimist. *Personality and Social Psychology Bulletin*, *31*, 407–405.

Jonas, E., Schultz-Hardt, S., Frey, D., and Thelen, N. (2001). Confirmation bias in sequential information search after preliminary decision: An expansion of dissonance theoretical research on selective exposure to information. *Journal of Personality and Social Psychology, 80,* 557–571.

Kawakami, K., Williams, A., Sidhu, D., Choma, B.L., Rodriguez-Bailón, R., Cañadas, E., and Hugenberg, K. (2014). An eye for the I: Preferential attention to the eyes of ingroup members. *Journal of Personality and Social Psychology, 107*(1), 1–20.

Kellough, J. L., Beevers, C. G., Ellis, A. J., and Wells, T. T. (2008). Time course of selective attention in clinically depressed young adults: An eye tracking study. *Behaviour Research and Therapy, 46,* 1238–1243.

Knobe, J. (2010). Person as scientist, person as moralist. *Behavioral and Brain Sciences, 33*(4), 315–329.

Knobe, J., and Fraser, B. (2008). Causal judgment and moral judgment: Two experiments. In W. Sinnott-Armstrong and W. Sinnott-Armstrong (eds.), *Moral Psychology, Vol 2: The Cognitive Science of Morality: Intuition and Diversity* (pp. 441–447). Cambridge, MA: MIT Press.

Knobe, J. and Nichols, S. (2017). Experimental philosophy. In Edward N. Zalta (ed.), *The Stanford Encyclopedia of Philosophy* (Winter 2017 Edition). https://plato. stanford.edu/archives/win2017/entries/experimental-philosophy/.

Kohlberg, L. (1969). Stage and sequence: The cognitive-developmental approach to socialization. In D. Goslin (ed.), *Handbook of Socialization Theory and Research* (pp. 347–480). Chicago: Rand McNally.

Kredel, R., Vater, C., Klostermann, A., and Hossner, E. J. (2017). Eye-tracking technology and the dynamics of natural gaze behavior in sports: A systematic review of 40 years of research. *Frontiers in Psychology, 8,* 1–15.

Krzepota, J., Stępiński, M., and Zwierko, T. (2016). Gaze control in one versus one defensive situations in soccer players with various levels of expertise. *Perceptual and Motor Skills, 123*(3), 769–783.

Kunda, Z. (1990). The case for motivated reasoning. *Psychological Bulletin, 108,* 480–498.

Lai, M. L., Tsai, M. J., Yang, F. Y., Hsu, C. Y., Liu, T. C., Lee, S. W. Y., Lee, M. H., Chiou, G. L., Liang, J. C., and Tsai, C. C. (2013). A review of using eye-tracking technology in exploring learning from 2000 to 2012. *Educational Research Review, 10,* 90–115.

Lassiter, G. D., Beers, M. J., Geers, A. L., Handley, I. M., Munhall, P. J., and Weiland, P. E. (2002). Further evidence of a robust point-of-view bias in videotaped confessions. *Current Psychology, 21,* 265–288.

Lassiter, G. D., Diamond, S. S., Schmidt, H. C., and Elek, J. K. (2007). Evaluating videotaped confessions: Expertise provides no defense against the camera perspective effect. *Psychological Science, 18,* 224–226.

Lassiter, G. D., Geers, A. L., Handley, I. M., Weiland, P. E., and Munhall, P. J. (2002). Videotaped interrogations and confessions: A simple change in camera perspective alters verdicts in simulated trials. *Journal of Applied Psychology, 87,* 867–874.

Lassiter, G. D., and Irvine, A. A. (1986). Videotaped confessions: The impact of camera point of view on judgments of coercion. *Journal of Applied Social Psychology, 16*, 268–276.

Lassiter, G. D., Munhall, P. J., Geers, A. L., Weiland, P. E., and Handley, I. M. (2001). Accountability and the camera perspective bias in videotaped confessions. *Analyses of Social Issues and Public Policy*, 53–70.

Lassiter, G., Ware, L. J., Ratcliff, J. J., and Irvin, C. R. (2009). Evidence of the camera perspective bias in authentic videotaped interrogations: Implications for emerging reform in the criminal justice system. *Legal and Criminological Psychology, 14*(1), 157–170.

Lee, J. (2017). Time Course of Lexicalization during Sentence Production in Parkinson's Disease: Eye-Tracking While Speaking. *Journal Of Speech, Language and Hearing Research, 60*(4), 924–936.

Luo, J., and Isaacowitz, D. M. (2007). How optimists face skin cancer information: Risk assessment, attention, memory, and behavior. *Psychology and Health, 22*, 963–984.

MacCoun, R. J. (1998). Biases in the interpretation and the use of search results. *Annual Review of Psychology, 49*, 259–287.

MacLeod, C., Mathews, A., and Tata, P. (1986). Attentional bias in emotional disorders. *Journal of Abnormal Psychology, 95*, 15–20.

Maner, J. K., Gailliot, M. T., and DeWall, C. (2007). Adaptive attentional attunement: evidence for mating-related perceptual bias. *Evolution and Human Behavior, 28*(1), 28–36.

Matthews, G. R., and Antes, J. R. (1992). Visual attention and depression: Cognitive biases in the eye fixations of the dysphoric and the nondepressed. *Cognitive Therapy and Research, 16*, 359–371.

McIntyre, N. A., Jarodzka, H., and Klassen, R. M. (2017a). Capturing teacher priorities: Using real-world eye-tracking to investigate expert teacher priorities across two cultures. *Learning and Instruction*.

McIntyre, N. A., Mainhard, M. T., and Klassen, R. M. (2017b). Are you looking to teach? Cultural, temporal and dynamic insights into expert teacher gaze. *Learning and Instruction*, 41–53. ISSN 0959-4752.

Mitschke, V., Goller, J., and Leder, H. (2017). Exploring everyday encounters with street art using a multimethod design. *Psychology of Aesthetics, Creativity & The Arts, 11*(3), 276–283.

Mogg, K., and Bradley, B. P. (1999). Orienting of attention to threatening facial expressions presented under conditions of restricted awareness. *Cognition and Emotion, 13*, 713–740.

Munro, G. D., and Ditto, P. H. (1997). Biased assimilation, attitude polarization, and affect in reactions to stereotyped-relevant scientific information. *Personality and Social Psychology Bulletin, 23*, 636–653.

Nickerson, R. S. (1998). Confirmation bias: A ubiquitous phenomenon in many guises. *Review of General Psychology, 2*, 175–220.

Nyström, P., Bölte, S., and Falck-Ytter, T. (2017). Responding to other people's direct gaze: Alterations in gaze behavior in infants at risk for autism occur on very short timescales. *Journal of Autism & Developmental Disorders, 47*(11), 3498–3509.

Piaget, J. (1932). *The Moral Judgment of the Child.* London: Routledge and Kegan Paul.

Petrinovich, L., and O'Neill, P. (1996). Influence of wording and framing effects on moral intuitions. *Ethology & Sociobiology, 17*, 145–171.

Petrinovich, L., O'Neill, P., and Jorgensen, M. (1993). An empirical study of moral intuitions: Toward an evolutionary ethics. *Journal of Personality and Social Psychology, 64*, 467–478.

Prinz, J., and Seidel, A. (2012). Alligator or squirrel: Musically induced fear reveals threat in ambiguous figures. *Perception, 41*(12), 1535–1539. Doi:10.1068/p7290

Rosenberg, R., and Klein, C. (2015). The moving eye of the beholder: Eye tracking and the perception of paintings. In J. P. Huston, M. Nadal, F. Mora, L. F. Agnati, C. J. Cela-Conde, J. P. Huston, and C. J. Cela-Conde (eds.), *Art, Aesthetics and the Brain* (pp. 79–108). New York: Oxford University Press.

Schnall, S., Benton, J., and Harvey, S. (2008a). With a clean conscience: Cleanliness reduces the severity of moral judgments. *Psychological Science, 19*, 1219–1222.

Schnall, S., Haidt, J., Clore, G. L., and Jordan, A. H. (2008b). Disgust as embodied moral judgment. *Personality and Social Psychology Bulletin, 34*(8), 1096–1109.

Segerstrom, S. C. (2001). Optimism and attentional bias for negative and positive information. *Personality and Social Psychology Bulletin, 2*, 1334–1343.

Seidel, A., and Prinz, J. (2013a). Mad and glad: Musically induced emotions have divergent impact on morals. *Motivation & Emotion, 37*(3), 629–637.

Seidel, A., and Prinz, J. (2013b). Sound morality: Irritating and icky noises amplify judgments in divergent moral domains. *Cognition, 127*(1), 1–5.

Skitka, L. J., Bauman, C. W., and Sargis, E. G. (2005). Moral conviction: Another contributor to attitude strength or something more? *Journal of Personality and Social Psychology, 88*, 895–917.

Skitka, L. J., and Mullen, E. (2002). The dark side of moral conviction. *Analyses of Social Issues and Public Policy, 7*, 35–41.

Södervik, I., Vilppu, H., Österholm, E., and Mikkilä-Erdmann, M. (2017). Medical students' biomedical and clinical knowledge: Combining longitudinal design, eye tracking and comparison with residents' performance. *Learning & Instruction, 52*, 139–147.

Thomson, J. J. (1976). Killing, letting die and the trolley problem. *The Monist, 59*(2), 204–217.

Thorup, E., Nyström, P., Gredebäck, G., Bölte, S., and Falck-Ytter, T. (2016). Altered gaze following during live interaction in infants at risk for autism: An eye tracking study. *Molecular Autism, 7*(12), 71–10.

Tien, T., Pucher, P. H., Sodergren, M. H., Sriskandarajah, K., Guang-Zhong, Y., and Darzi, A. (2014). Eye tracking for skills assessment and training: A systematic review. *Journal of Surgical Research, 191*(1), 169–178.

Tobia, K., Chapman, G., and Stich, S. (2013). Cleanliness is next to morality, even for philosophers. *Journal of Consciousness Studies*, *20*(11–12), 195–204.

van Leeuwen, P. M., de Groot, S., Happee, R., and de Winter, J. F. (2017). Differences between racing and non-racing drivers: A simulator study using eye-tracking. *Plos ONE*, *12*(11), 1–19.

Vickers, J. N., Causer, J., Stuart, M., Little, E., Dukelow, S., Lavangie, M., and Emery, C. (2017). Effect of the look-up line on the gaze and head orientation of elite ice hockey players. *European Journal Of Sport Science*, *17*(1), 109–117.

Ware, L. J., Lassiter, G., Patterson, S. M., and Ransom, M. R. (2008). Camera perspective bias in videotaped confessions: Evidence that visual attention is a mediator. *Journal of Experimental Psychology: Applied*, *14*(2), 192–200.

Wright, J. C. (2012). Children's and adolescents' tolerance for divergent beliefs: Exploring the cognitive and affective dimensions of moral conviction in our youth. *British Journal of Developmental Psychology*, *30*(4), 493–510.

Wright, J. C., and Bengson, J. (2009). Asymmetries in judgments of responsibility and intentional action. *Mind & Language*, *24*(1), 24–50.

Wright, J. C., Cullum, J., and Schwab, N. (2008). The cognitive and affective dimensions of moral conviction: Implications for tolerance and interpersonal behaviors. *Personality and Social Psychology Bulletin*, *34*(11), 1461–1476.

Yeung, H. H., Denison, S., and Johnson, S. P. (2016). Infants' looking to surprising events: When eye-tracking reveals more than looking time. *Plos ONE*, *11*(12), 1–14.

Zarkadi, T., and Schnall, S. (2013). 'Black and White' thinking: Visual contrast polarizes moral judgment. *Journal of Experimental Social Psychology*, *49*(3), 355–359.

Zhiwei, Z., and Qiang, J. (2004). Eye and gaze tracking for interactive graphic display. *Machine Vision & Applications*, *15*(3), 139–148.

Appendix to Chapter 3

Example of images:

Study 1, Baby/Villagers

Study 2, Case 1

| Fred | Joe | Bob |

Study 2, Scenarios

Case 1 – described above

Case 2

Strong: Jack and Laura are two seven-year-olds making snow angels on a hill near a river (and you are babysitting Jill, Jack is one of her neighbours). Jack gets jealous because Laura's snow angle is better than his and so he shoves her really hard and she tumbles down the hill into the river. Shoving her causes him to slip on the ice and so he tumbles into the river directly behind her. Pat is a friend of theirs who was also making a snow angel nearby and saw what happened. You were standing down by the river and witnessed the event, as well as both children falling in.

You jump in to save them (Pat was too far away to respond), but the current is very strong and is pulling them in two different directions. Therefore, you only have time to save <u>one</u> of them. Who do you save?

Ambiguous: Jack and Laura are two seven-year-old twins that you are babysitting. They are making snow angels on a hill near a river. Jack tries to help Laura stand up out of her snow angel but pulls too hard and they both tumble down the hill into the river. Pat is a friend of theirs who was also making a snow angel nearby and saw what happened. You were standing down by the river and witnessed the event, as well as both children falling in.

You jump in to save them (Pat was too far away to respond), but the current is very strong and is pulling them in two different directions. Therefore, you only have time to save <u>one</u> of them. Who do you save?

Case 3

Strong: You are the only college student in your dorm with a car. Brendan needs you to take a letter to the post office before they close; Richard needs you to take a package to the FedEx store. Brendan has always been there for you when you needed something. Richard is very flaky with everything, especially when you need a favour. Matt is your other friend you often go drinking with.

These two stores are in different parts of town and you only have time to make it to one. Which friend (Brendan or Richard – at this point Matt is not asking you for a favour) do you help out?

Ambiguous: You are the only college student in your dorm with a car. Brendan needs you to take a letter to the post office before they close; Richard needs you to take a package to the FedEx store. Both of them have been there for you when you needed a favour and are very dependable. Matt is your other friend you often go drinking with.

These two stores are in different parts of town and you only have time to make it to one. Which friend (Brendan or Richard – at this point Matt is not asking you for a favour) do you help out?

Case 4

Strong: Joanna has volunteered at numerous community centres focused on helping the elderly. She has just received her BS in Psychology from the College of Charleston. She has numerous awards, was an executive member of the Psychology Club and received an overall GPA of 3.8. Sarah got the interview because her father knew the director. She dropped out of college after her junior year. Her cumulative GPA until that point was a 2.0 and on a background check, an arrest came up for shoplifting. Zoey is another former CofC student who was thinking about applying for the job, but decided not to.

You only have one job opening at your nursing home and so you have to choose who to give the job to (either Joanna or Sarah – Zoey decided not to apply for the job), so who do you hire?

Ambiguous: Joanna has a lot of relevant experience working with the elderly. She is a recent CofC graduate with a BS in Psychology. Her cumulative GPA is 3.8. Sarah has volunteered at a lot of nursing homes and has gained relevant experience from doing so. She graduated from college last year with a BS in Psychology. She has received many awards and honours for volunteering and helping the elderly community. Her cumulative GPA is a 3.5. Zoey is another former CofC student who was thinking about applying for the job, but decided not to.

You only have one job opening at your nursing home and so you have to choose who to give the job to (either Joanna or Sarah – Zoey decided not to apply for the job), so who do you hire?

Case 5

Strong: Wayne has accused Patrick of beating him up at a local bar and stealing his wallet. Patrick has had numerous arrests in the past. He once almost beat another man to death. His fingerprints were all over Wayne's wallet when they found it. Wayne's injuries are consistent with his story and the bruises on Patrick's hand provide evidence as well. Plus, there was a security camera in an ATM in the bar that caught Patrick confronting Wayne. Nonetheless, Patrick maintains his innocence. Ralph was the bartender at the bar, but he had stepped away from the bar at the time and so didn't see anything.

You are the judge in this case and you have to decide who to rule in favour of (Patrick or Wayne). Ralph can't help you decide since he didn't see anything. If you rule in favour of Patrick, the charges will be dropped and he will be free to go. If you rule in favour of Wayne, Patrick will be charged with battery and theft. Who do you rule in favour of?

.**Ambiguous**: Wayne has accused Patrick of beating him up at a local bar and stealing his wallet. Both men were pretty battered. Patrick says that Wayne was the one who started the fight by throwing a beer bottle at him and that Wayne's wallet must have fallen out of his pocket during the fight. Wayne, on the other hand, says that Patrick was trying to go for his wallet during the fight. The wallet was found on the bar floor with no cash and a missing debit card. Patrick's prints were found on the outside of the wallet, but he might have touched it during the fight. Ralph was the bartender at the bar, but he had stepped away from the bar at the time and so didn't see anything.

You are the judge in this case and you have to decide who to rule in favour of (Patrick or Wayne). Ralph can't help you decide, since he didn't see anything. If you rule in favour of Patrick, the charges will be dropped and he will be free to go. If you rule in favour of Wayne, Patrick will be charged with battery and theft. Who do you rule in favour of?

Case 6

Strong: Two students are in the running for 'best overall female student of the year'. Ginger is a senior at CofC with a cumulative GPA of 3.9 and has been involved in many campus clubs and organizations, including one she started to have a positive impact on child literacy rates in the local community. She also has an on-campus job helping incoming freshmen adjust to college and is considered by many to be a fabulous mentor. Amber is also a senior at CofC. Her application is also very strong – cumulative GPA of 3.75, lots of community and campus involvement, in addition to a few strong leadership positions in student affairs. However, right before the decision is to be made, you discover that Amber is suspected of cheating on her final exams and dealing drugs on campus. Last year's award winner, Rachel, will be the one to deliver the award to the new winner this year.

You are on the committee determining who (Ginger or Amber) should receive the award that Rachel will deliver to the winner. Which of these students would you give award to?

Ambiguous: Two students are in the running for 'best overall female student of the year'. Ginger is a senior at CofC with a cumulative GPA of 3.9 and has been involved in many campus clubs and organizations, including one she started to have a positive impact on child literacy rates in the local community. She also has an on-campus job helping incoming freshmen adjust to college and is considered by many to be a fabulous mentor. Amber is also a senior at CofC. Her application is also very strong – cumulative GPA of 3.75, lots of community and campus involvement, in addition to a few strong leadership positions in student affairs. Also, right before the decision is to be made, you discover that Amber volunteers at a local senior centre with patients with Alzheimer's disease. Last year's award winner, Rachel, will be the one to deliver the award to the new winner this year.

You are on the committee determining who (Ginger or Amber) should receive the award that Rachel will deliver to the winner. Which of these students would you give award to?

4

Using fMRI in Experimental Philosophy: Exploring the Prospects

Rodrigo Díaz

1. Introduction

Experimental philosophy is an interdisciplinary approach that consists in investigating traditional philosophical questions using experimental methods from the social sciences (Knobe and Nichols 2017). After a first period in which experimental philosophers basically relied on the use of questionnaires, the field has grown to adopt a wide variety of methods from other disciplines such as linguistics or neuroscience (see other chapters in this volume for several examples). The present chapter will analyse the possibilities of neuroimaging methods, fMRI in particular, within the practice of experimental philosophy.

1.1. fMRI

Functional Magnetic Resonance Imaging (fMRI) is probably the most used neuroimaging method in cognitive neuroscience. Most research in cognitive neuroscience is committed to the goal of localizing mental processes in the brain,[1] and fMRI serves this purpose well. Researchers use fMRI to measure activity in the brain of the participants in their experiments. Used while participants perform cognitive tasks, it provides information about the neural correlates of those tasks, which is used to make links between brain structure and mental function. The localization of the brain activity found in an experiment is usually

Acknowledgements

I would like to thank Jonas Blatter, Eugen Fischer, Guillermo del Pinal, Lena Kästner, Colin Klein, Kevin Reuter and an anonymous reviewer for their comments on early drafts of this chapter.

[1] But see Section 4.1.

represented through colouring superimposed on brain pictures. The resulting 'neuroimages' are not like photographs; however, there are a considerable number of inferential steps from the acquisition of raw fMRI data to the attainment of brain activity maps (Roskies 2008).

The details of fMRI technology are complex and I can only explain a few key points here (Logothetis 2008; Klein 2010; Parens and Johnston 2014). First, fMRI is an *indirect* measure of brain activity. The fMRI signal measures changes in the oxygenation of blood in the research subject's brain. These changes are detected taking advantage of the different magnetic properties of deoxygenated haemoglobin in comparison to oxygenated haemoglobin in cerebral blood flow. As brain activity consumes oxygen, it is inferred that more deoxygenated haemoglobin in a particular brain area equals activity in that area. Second, the signal has limitations in terms of *temporal* and *spatial* resolution: changes in blood oxygenation are slower than brain activity, and the signal is recorded in volume units called *voxels* that include millions of neurons. Finally, and most importantly, fMRI results are *qualitative*. The final neuroimage is the result of complex mathematical and statistical processing of the signal, and it does not map the intensity of the signal, but the statistical significance of differences in the signal between experimental conditions. However, despite these limitations, fMRI is generally assumed to be suitable for the purpose of associating mental processes with their neural substrates, mapping mental function to brain structure.

1.2. Experimental philosophy

To evaluate the prospects of using fMRI within experimental philosophy, it is necessary to explain the ways in which this movement employs experimental methods to address philosophical questions. For this purpose, it is useful to differentiate two different projects within the practice of experimental philosophy based on their relation with the analytical philosophy tradition (Rose and Danks 2013; Fischer and Collins 2015; Knobe and Nichols 2017): The *intuitions project*, and the *psychology project*.

The intuitions project builds on the use – widespread in analytic philosophy – of the method of cases: using hypothetical scenarios to trigger intuitions about a particular topic (e.g. knowledge, moral responsibility, free will). But instead of relying on 'armchair reflection', experimental philosophers use surveys and questionnaires to investigate people's intuitions, as well as the factors that underlie those intuitions: psychological mechanisms, cultural background, etc. The ultimate goal behind this investigation differentiates between two programmes within the intuitions project: the positive programme and the negative program. Work in the positive program collects data to make progress on the understanding of the concept at hand, while work on the negative program tries to raise questions about the validity of the method of cases by showing that intuitions depend on unreliable mental processes or are affected by putatively irrelevant factors (Machery 2017).

The psychology project, in contrast to the intuitions project, is not primarily concerned with the study of intuitions and does not necessarily engage with the method of cases. Instead of collecting data on people's thoughts and feelings to inform some further philosophical issue, the issue is people's thoughts and feelings themselves. Work on this project blurs the boundaries between disciplines, as most questions involved are both questions of philosophy and questions of psychology. For example, the question of how people come to attribute mental states to others is of interest to both psychologists (mentalizing processes) and philosophers (the problem of other minds).

Within the intuitions project, using fMRI can help to uncover the mental processes underlying people's intuitions about hypothetical cases. An example of this approach is Greene et al.'s 2001 study on the impact of emotion on responses to moral dilemmas. In this study, fMRI is used to identify the mental processes underlying different moral intuitions to assess their epistemological warrant.

Within the psychology project, fMRI data can *also* be used in a different way, to inform questions about cognitive architecture. An example of this approach is Lindquist et al.'s (2012) meta-analysis of studies investigating the brain basis of emotion. In this study, fMRI data is used to elucidate the nature of emotion and examine whether emotions are 'basic' or 'constructed'.

In the following, I will use Greene et al. (2001) and Lindquist et al. (2012) as case studies to present and discuss the above two ways in which fMRI could be used to inform philosophical work. These two influential papers[2] are selected because while both use fMRI to study the same mental process, emotion, they use different approaches. By looking at fMRI-recorded patterns of brain activity, Greene et al. aim to infer the *engagement* of emotion, while Lindquist et al. use fMRI to infer the *nature* of emotions. The former strategy is an instance of reverse inference (from brain activity to *what* the mind is doing), the latter constitutes an example of ontology testing (from brain activity to *how* the mind is organized).

---

[2] According to Google Scholar, Greene et al. (2001) has been cited 3656 times, while Lindquist et al. (2012) has been cited 1067 times (date of the search: 14.03.2018).

Table 4.1 Project, strategy and method of the studies discussed. Note that there is no one-to-one correspondence between method, strategy and project. For example, subtraction is not only used for reverse inference, and reverse inference is not only used within experimental philosophy's intuitions project.

Study	Project	Strategy	Method
Greene et al. (2001)	Intuitions project	Reverse inference	Subtraction
Lindquist et al. (2012)	Psychology project	Ontology testing	Meta-analysis

The problems associated with reverse inference, in particular the lack of one-to-one mappings between mental function and brain structure, is one of the aspects that motivate the use of ontology testing approaches. The discussion of Greene et al. (2001) and Lindquist et al. (2012) will provide a concrete illustration of how information about the neural substrates of a particular mental process (e.g. emotion) cannot support reverse inferences but can tell us something about the nature of that mental process. Furthermore, each study exemplifies one of the most used methods within each strategy: subtraction within reverse inference, and meta-analysis within ontology testing (see Table 4.1). The discussion of these two studies will serve to argue against reverse inference and in favour of ontology testing approaches within experimental philosophy.

2. Greene et al. (2001): Intuitions project and reverse inference

2.1. Research context

Greene et al.'s 2001 study uses a *subtractive design* common in cognitive neuroscience. This design usually involves two tasks, which differ in one specific task component. For example: task T_1 (hear a 'beep' sound), consisting in one component C_1 (auditory), and task T_2 (tap a finger when hear a 'beep' sound) involving components C_1 (auditory) and C_2 (motor). The design tries to *isolate* the differing component (C_2), which is the target of investigation. Neuroimaging is used to record participants' brain activity while they perform each task, and the recorded pattern of brain activity in the contrast task (T_1) is *subtracted* from the pattern of brain activity in the task that involves the component of interest (T_2). This subtraction consists on performing a statistical test for each voxel to see if it shows a significant difference in activity between the two tasks. The result is the identification of differential activity in certain areas of the brain, which is taken to be the neural correlate of the investigated task component (C_2).

This design can be used for two different goals: to identify the neural basis of a specific mental process, or to identify the mental process(es) involved in a task.

First, a subtractive design can aim at identifying the neural basis of a specific mental process. If the isolated task component is assumed to correspond to the involvement of a specific mental process, the differential pattern of activity is interpreted as the neural basis of that mental process. For example, on this approach, the differential brain activity between listening to emotionally un-evocative music and listening to emotionally evocative music is interpreted as the neural basis of emotion (Mitterschiffthaler et al. 2007). This strategy is known as forward inference, and it serves to establish 'structure-function links', links between localized brain activity and mental processes.

Second, subtractive designs can be used to identify the mental process(es) involved in a task. In some cases, the mental process(es) associated with the isolated task component are unknown, and subtraction is used to identify them. For example, contrasting an evaluative judgement task with a factual judgement task can serve to investigate the mental processes underlying evaluation (Moll et al. 2001). In these cases, the differential pattern of brain activity between one task and the other is interpreted in the light of previously established structure-function links. The logic is the following: (1) In task T, activation is found in region R; (2) In previous studies, activity in region R has been associated with mental process M; therefore (3) T recruits M. This is known as reverse inference: inferring the involvement of a particular mental process by looking at the localization of brain activity. Greene et al.'s study follows this strategy: It uses subtraction to study the mental processes underlying moral intuitions.

2.2. Greene et al.'s study

Participants in Greene et al.'s study were presented with a battery of dilemmas while their brain activity was recorded using an fMRI scanner. There were three different types of dilemmas: personal moral dilemmas, impersonal moral dilemmas and non-moral dilemmas (which served as a control condition). Non-moral dilemmas involve decisions such as which of two coupons to use at a store. The paradigmatic example of an impersonal moral dilemma is the switch dilemma: A runaway trolley is about to run over five persons, and you have to decide whether to hit a switch that turns the trolley onto a different set of tracks in which it will only kill one person instead of five. Its personal counterpart is the footbridge dilemma: A runaway trolley is about to run over five persons, and you have to decide whether to push a fat person onto the tracks in front of the trolley in order to stop it, saving the other five people but killing the one that you push.

Both personal and impersonal dilemmas involve the same trade-off of lives: sacrifice one person to save five; but they differ in the kind of action required: pulling a switch versus pushing the person to sacrifice. Performing these actions is interpreted as a 'utilitarian' response (choosing the better outcome), while inaction is interpreted as a 'deontological' response (not violating someone's rights). It is a consistent finding that, despite both types of dilemmas involving the same trade-off of lives, most people give utilitarian responses in impersonal dilemmas but deontological responses in personal dilemmas. Greene and colleagues hypothesize that this asymmetric pattern of responses is due to the higher emotional salience of personal dilemmas in comparison to impersonal ones. The deontological intuitions triggered by personal dilemmas would be grounded in people's emotional reactions, while utilitarian responses would be triggered in more 'cognitive' settings. They used fMRI to test this hypothesis via reverse inference.

When comparing the fMRI data relative to each category of dilemmas, Greene et al. found that personal moral dilemmas, in comparison to impersonal and non-moral dilemmas, are more likely to trigger activity in brain areas associated with emotion (Brodmann's Areas 9–10, 31 and 39). Conversely, personal dilemmas are less likely to trigger activity in areas associated with working memory (Brodmann's Areas 46 and 7/40).

2.3. Discussion

The results seem to support Greene and colleagues' hypothesis: personal dilemmas differ from impersonal ones in that they generate emotional reactions in participants, what would in turn explain the prevalence of 'deontological' responses to this type of dilemma. This evidence is supposed to undermine the epistemological warrant of deontological intuitions (Greene 2014). Moral dilemmas are unfamiliar situations, and in unfamiliar situations (e.g. driving a car for the first time) we shouldn't rely on 'automatic' but rather 'controlled' processes. As the results suggest that deontological responses to moral dilemmas are associated with 'automatic' emotional reactions, this would undermine deontological intuitions. However, it is important to take a closer look at the evidence that supports Greene et al.'s reverse inference from the fMRI data to the engagement of emotion. The inference has the following structure:

(1) When responding to personal moral dilemmas, activity in areas 9–10, 31 and 39 is found.
(2) Activity in areas 9–10, 31 and 39 has been associated with emotion.
(3) Therefore, responding to personal moral dilemmas involves emotion.

In order for (3) to follow from (1) and (2), there must be a one-to-one correspondence between brain structure and cognitive function (Poldrack 2006): emotion should consistently recruit activity in areas 9–10, 31 and 39; and activity in areas 9–10, 31 and 39 should not be consistently recruited by other mental processes. Deducing the engagement of a mental process from activity in a particular brain area is justified *only if* no other mental processes have been associated with activity in that area. However, as Klein (2011) has already noted, none of the areas reported by Greene and colleagues are *selective* for emotion. Instead, they are *pluripotent*: they are involved in the realization of multiple different mental processes. Thus, we cannot know whether activity in those areas means that emotion or one of the other processes that have been associated with those areas are engaged. Activity in areas 9–10, 31 and 39 is *not sufficient* to infer emotion engagement.

However, selectivity might be an excessive requirement for reverse inference. Although activity in areas 9–10, 31 and 39 is not *selective* for emotion, it might be that activity in those areas is *preferentially* associated with emotion. That is, when there is activity in those areas, the probability that emotion is being engaged is higher than the probability that other mental processes are being engaged. By reformulating reverse inference in probabilistic terms, it is possible to use fMRI data to provide some support for cognitive hypotheses (Poldrack 2006). To calculate the probability of emotion engagement given activation in areas 9–10, 31 and 39, we need to know how many times those regions were active when emotion was engaged, and how many times those regions were active when emotion was not engaged. Although the latter information is difficult to obtain, fMRI databases can be used to estimate the probability of a mental process being engaged given activation in a particular brain area.

Neurosynth (Yarkoni et al. 2011) is an automated database of fMRI studies that can help determine the *selectivity* and *consistency* of structure-function links. Neurosynth uses text-mining techniques to provide information about the probability of finding a term (e.g. 'emotion', 'memory', 'attention' …) in the abstract of a paper when activation in a specific brain area is reported in that paper (reverse inference/selectivity) and, conversely, the probability of finding activation across different brain areas given the presence of a specific term (forward inference/consistency). A search in Neurosynth[3] revealed that, within the areas that Greene and colleagues reported, only area 9–10 is consistently associated with emotion (z = 6.93), and it is thus the only candidate to provide evidence for the engagement of emotion. There was a high probability of finding the term 'emotional' (.71) or 'affective' (.68) given activity in this area.[4] However, this probability was as high as the probability of finding other terms such as 'mentalizing' (.85), 'theory of mind' (.83), 'social cognition' (.83), 'self-referential' (.79), 'inferences' (.83), 'evaluation' (.76), 'intentions' (.82), 'belief' (.84), 'autobiographical' (.80), 'default mode' (.71), 'judgements' (.73), 'moral' (.82) or 'economic' (.81) (see note 5 for a complete list). The data suggests that the selectivity of area 9–10 for emotion is low. Thus, the presence of activity in this area is not able to provide strong support to Greene et al.'s hypothesis.

Nevertheless, one could argue that the possibility of mentalizing, theory of mind, social cognition, etc. being involved should not be taken into account. The only mental processes that should be considered are those relevant in the case at hand. Although this possibility is not systematically investigated by Greene and colleagues, some have argued that their study could help deciding whether deontological responses to personal moral dilemmas are generated by (H1) emotional reactions to the vignette or (H2) the application of an abstract moral rule, for example, the doctrine of double effect (Del Pinal and Nathan 2013; Machery 2013). In this case, the only mental processes to take into account would be (1) emotion and (2) rule application. Although fMRI evidence cannot provide strong support to a cognitive hypothesis in isolation, it could help selecting between two competing hypotheses. Our search in Neurosynth for areas 9–10 and 31 revealed that the use of the term 'rule' was not consistently associated with activity in any of these areas (see note 5). Thus, in this comparative framework, Greene et al.'s results would support H1 over H2, as activity in area 9–10 is more likely to be found when reacting emotionally than when applying an abstract rule.

However, the hypothesis that participants are applying an abstract rule when responding to personal moral dilemmas is compatible with those subjects experiencing emotion (Huebner et al. 2009; Mole and Klein 2010). Emotion and

---

[3] The Talairach coordinates for peak activations reported by Greene et al. were transformed to MNI space (Lacadie et al. 2008) to enable the search in Neurosynth (date of the search: 14.03.2018). Full results can be found in the following links: BA 9–10 (http://neurosynth.org/locations/2_56_20_6/) BA 31 (http://neurosynth.org/locations/-6_-58_38_6/). The coordinates for BA 39 are not provided in Greene et al.'s paper.

[4] Some terms related to emotion such as 'empathy' (.78) or 'unpleasant' (.79) could arguably support Greene et al.'s hypothesis too, while others like 'positive' (.67) don't.

rule application can occur alongside each other. Thus, even if it is the case that activity in area 9–10 is due to emotional engagement, we should not prefer H1 over H2. In order for the neuroimaging data to help us select between these two competing hypotheses, we need strong structure-function links for both of them. This is necessary to contrast the predictions of H1 and H2 one against the other. H1 posits that personal moral dilemmas, in contrast to impersonal ones, engage emotional processes. Thus, it predicts differential activity in areas associated with emotion for personal dilemmas. Conversely, H2 posits that responding to both personal and impersonal moral dilemmas involve the application of a rule: a deontological rule in the former, and a consequentialist rule in the latter. Thus, it predicts activity in areas associated with rule application for both types of dilemmas.

We might say that the neuroimaging data was in line with H1 predictions,[5] but was it *against* H2 predictions? For this, we need to know the link between rule application and localized brain activity. Although emotions have been the target of much neuroimaging work, this is not the case for rule application.[6] In Greene et al.'s study, rule application processes would putatively be supported by areas 46 and 7/40, which showed diminished activity in personal dilemmas in comparison to impersonal ones. This pattern of activity would favour H1 over H2. However, the consistency and selectivity of these areas for rule application is important at this point. If the link is weak, a proponent of H2 could argue that other areas which didn't show differential activity, and not areas 46 and 7/40, are the ones supporting rule application. A search in Neurosynth[7] revealed that activity in areas 46 and 7/40 was not consistently associated with the term 'rule'. Thus, the evidence does not undermine H2, and H1 should not be preferred over it.

The use of Neurosynth and Greene et al.'s study has shown the problems that the different formulations of reverse inference face when they are to be applied. It is important to note that the problems exposed in this section are not exclusive to Greene et al.'s study, but of reverse inference in general. Lack of

---

[5] One could still question why differential activity was only found in 9–10 but not in other areas that are also consistently associated with emotion.

[6] Using Neurosynth, 790 studies were found to be associated with the term 'emotion' (http://neurosynth.org/analyses/terms/emotion/). There were no results for 'rule application', and only 141 studies associated with the term 'rule' (http://neurosynth.org/analyses/terms/rule/)

[7] The Talairach coordinates for peak activations reported by Greene et al. were transformed to MNI space (Lacadie et al. 2008) to enable the search in Neurosynth (date of the search: 14.03.2018). Full results can be found in the following links: BA 46 (http://neurosynth.org/locations/46_36_24_6/), left BA 7/40 (http://neurosynth.org/locations/-48_-68_26_6/) right BA 7/40 (http://neurosynth.org/locations/50_-60_18_6/)

selectivity is not an anomaly of some brain areas, such as the ones that Greene and colleagues associate with emotion. Decades of fMRI research have provided a structure-function mapping that is far from selective, with most brain areas involved in many different mental processes (Poldrack 2010; Anderson et al. 2013). When differential activity is found in a specific brain area, it is usually not possible to determine which of the many mental processes that are (to a similar extent) likely to be engaged is actually engaged. Thus, although the probabilistic approach makes reverse inference viable in principle, in practice it is of limited use in most cases. The same seems to be true for extant comparative approaches to reverse inference, which also need strong structure-function links in order to avoid the 'compatibility problem'.

3. Lindquist et al. (2012): Psychology project and ontology testing

3.1. Research context

Lindquist et al.'s 2012 study is an example of 'science by synthesis'. Instead of conducting a new fMRI experiment to investigate emotion, Lindquist and colleagues conducted a meta-analysis of the neuroimaging literature on emotion. A meta-analysis is a statistical technique that allows to obtain a formal synthesis of the results across different studies. In neuroimaging research, the difficulty of establishing selective structure-function links is one of the main motivations for conducting these formal syntheses of the existing evidence (Yarkoni et al. 2010). Meta-analyses of neuroimaging studies can serve to evaluate both the consistency and the selectivity of associations between localized brain activity and mental function. First, by aggregating data across experiments investigating the neural correlates of the *same* mental process, meta-analyses can provide information about the brain areas that are *consistently* associated with that mental process. Second, when used to aggregate data across studies investigating *different* mental processes, meta-analyses can also allow to evaluate the *selectivity* of brain areas for those mental processes.

The lack of selectivity of structure-function associations has also prompted a debate about whether we should rethink our cognitive ontology, our theory about the organization of the mind (Price and Friston 2005; Poldrack et al. 2009; Lenartowicz et al. 2010). Some authors have suggested that mental processes might not map well onto brain structures because the cognitive ontology that has guided the design of neuroimaging experiments is not adequate. That is, the task components isolated in forward inferences might not correspond to basic operations of the mind. Neuroimaging research assumes that psychology is *not* independent from neuroscience, so it is to be expected that basic operations of the mind are also basic operations of the brain. Thus, when tasks that are supposed to isolate different mental processes produce overlapping patterns of brain activity, the nature of those processes as basic operations or building blocks of our mind is called into question.

Following this 'same pattern of brain activity equals same mental processes' logic, neuroimaging results can be used to test the adequacy of our cognitive ontology, and possibly guide a reformulation of the categories we use to understand the organization of the mind. Meta-analyses are specially well suited for this ontology testing approach, as they can tell us whether tasks that are supposed to involve similar (distinct) mental processes, recruit activity in the same (distinct) brain structures. This approach is used by Lindquist and colleagues to inform the debate around the nature of emotions. In particular, they investigate the question of whether emotions are 'basic' or 'constructed'.

3.2. Lindquist et al.'s study

Lindquist and colleagues conducted a meta-analysis of the results from almost a hundred neuroimaging studies. They selected studies investigating the neural correlates (forward inference) of the perception and experience of five discrete emotions: anger, sadness, fear, disgust and happiness. These are the basis of the 'basic emotion' theory (Ekman 1999), according to which these emotions are irreducible building blocks of the mind, which are the result of evolutionary pressures, and have their roots in hard-wired mechanisms in the brain and the body. On this view, it is to be expected for each basic emotion to be associated with activity in a specific region or network of regions in the brain.[8] That is, that there are brain regions *selectively* associated with each basic emotion (but see Scarantino and Griffiths 2011). In contrast to basic emotion theory, a constructionist view about emotion posits that all emotions emerge from the combination of the same domain-general mental processes: core affect and conceptualization (Lindquist and Barrett 2008). Thus, constructivism predicts

---

[8] It is necessary to posit emotion-specific central nervous system (CNS) activity in my account of basic emotions. […] There must be unique physiological patterns for each emotion, and these CNS patterns should be specific to these emotions not found in other mental activity. (Ekman 1999, 50)

that the patterns of brain activity associated with each emotion will overlap. In other words, that there are no selective associations between brain regions and each basic emotion.

The results of Lindquist et al.'s meta-analysis show that it is not possible to distinguish each basic emotion by its associated pattern of brain activity. Each emotion was associated with activity across a number of different brain regions, and none of those regions were selective for a particular emotion. For example, the amygdala, which has been taken to be the 'fear area', or at least the most important hub in a fear circuit, is shown to be consistently recruited by the experience and perception of all the emotions in the analysis. Similar results were found for other areas such as the anterior insula and the lateral orbitofrontal cortex, among others.

3.3. Discussion

Lindquist and colleagues' results show that there are no 'basic emotion circuits' in the brain. Instead, different basic emotions share (to some extent) the same neural substrates. Contrary to what basic emotion theory posits, this suggests that emotions are not basic operations of the mind. Furthermore, the brain regions consistently activated by emotions have also been associated with non-emotional processes. This seems to support the constructivist view, in which emotions are built from the combination of operations that are not specific to emotion, but rather domain-general. These conclusions, however, do not remain unchallenged. It is possible to question both the negative claim against basic emotions (Scarantino 2012), and the positive claim in favour of constructivism (Sander 2012; Scherer 2012). Regarding the second objection, it has been argued that the results of Lindquist et al.'s meta-analysis do not provide clear support for constructivism. The evidence is merely consistent with this theory, and thus also compatible with other theories of emotion such as appraisal theory. However, the evidence is *inconsistent* with basic emotion theories. Thus, the negative claim about the status of emotion categories as basic operations of the mind still holds.

Even proponents of basic emotions theory have accepted the evidence regarding the absence of basic emotion circuits in the brain (Scarantino 2012; Adolphs 2016). Similar to Lindquist and colleagues, they have used the data to argue in favour of a reformulation of our cognitive ontology, although not a constructivist one.

Scarantino (2012) agrees that there are no specific biological markers for each emotion category. However, he argues that this does not prove basic emotion theory wrong. He claims that basic emotions exist, but they do not match our traditional folk-psychological categories of emotion. The reason that there are no selective associations between brain activity and emotion categories is that neuroimaging studies of emotion have been guided by the wrong cognitive ontology. Like Lindquist and colleagues, Scarantino argues that the evidence should make us rethink the organization of the mind. However, instead of considering that emotions are not part of an adequate cognitive ontology, he argues that *our current emotion categories* are not part of an adequate cognitive ontology.

Adolphs (2016) has argued that the main problem with the cognitive neuroscience of emotion to date has been the lack of conceptual clarity. On his view, in order to find selective associations between emotions and brain circuits, neuroimaging studies need to employ more rigorous designs. In particular, it is important to distinguish between emotion states and the conscious experience, attribution, conceptual knowledge and expression of emotions. Only emotion states would constitute building blocks of the mind, while the others are abilities derivative from them. Neuroimaging studies usually conflate these different dimensions, and do not properly isolate or control for them. This is why neuroimaging studies have not found specific neural correlates for emotions, and not because basic emotion theory is wrong. On Adolphs' view, in order for the neuroscientific study of emotion to progress, researchers should take into account these conceptual distinctions when designing neuroimaging experiments.

In both examples above, the fMRI evidence is taken to be relevant to assess hypotheses about our cognitive ontology, and the categories we use to capture the organization of the mind are reformulated in the light of this evidence. Further meta-analysis of neuroimaging data regarding constructs other than emotion, such as working memory (Lenartowicz et al. 2010), have been used for the same ontology-testing purpose.

4. General discussion

The two case studies presented in this chapter exemplify different ways in which fMRI evidence can be used to inform philosophical work, and the problems associated with them.

The discussion of Greene et al.'s (2001) results brought out the problems associated with using fMRI to identify the mental processes involved in an experimental task. This strategy, known as reverse inference, faces a problem because associations between brain structure and mental processes are always one-to-many. There are two main proposals to make reverse inference a viable strategy: (1) to reformulate it in probabilistic terms and (2) to use it to decide between competing hypotheses. Greene et al.'s example showed that (1) usually provides little support for cognitive hypotheses and (2) is prone to fail because cognitive hypotheses are often compatible with multiple different neuroimaging results (see Russell A. Poldrack 2006; and G. Del Pinal and Nathan 2017 for similar conclusions).

Lindquist et al. (2012) provided an example of how the many-to-many character of structure-function links, which limited the use of reverse inference, can be used to inform questions about cognitive architecture. Their approach consists in looking at the degree of overlap between patterns of brain activity across tasks, to assess whether those tasks involve the same or distinct mental processes. Lindquist et al.'s example is especially powerful because their results are used against a theory about the organization of the mind that *makes predictions* about brain activity. However, it has been argued that this strategy can inform debates about the organization of the mind even when the theories involved make no predictions about brain activity (Mather and Kanwisher 2013). Although this approach implies assumptions which are debatable, such as the correspondence between brain and mental function, and the relevance of neuroscientific data for psychological science, it has greater potential than reverse inference.

4.1. Network-based approaches

A fundamental difference between Greene et al.'s and Lindquist et al.'s approaches is that the latter does not rely on localizing mental processes in particular brain structures. Lindquist et al.'s conclusions against basic emotions theory depend just on the degree of overlap between emotions' neural correlates, in whichever part of the brain this overlap happens. The 'localizationist paradigm' in cognitive neuroscience has been heavily criticized. In particular, it has been argued that the subtractive method (see Section 2.1) relies on certain theoretical assumptions about the modular and serial organization of the mind and the brain which are likely to be false (Orden and Paap 1997; Uttal 2001). In consequence, many researchers in cognitive neuroscience have switched from location-based to network-based approaches. On the network-oriented view, mental function does not depend on the activity of isolated brain modules, but on complex patterns of interaction between anatomically separated neural populations. This switch of perspective requires refining our methods, to be better able to detect these kinds of interactions.

A method of increasing popularity is multivariate decoding (Hebart and Baker 2017). Multivariate decoding differs from the subtractive method in two substantial ways. First, while subtractive designs use *univariate* methods of analysis, which consist in running a separate analysis on each voxel (aggregation of neurons, see Section 1.1), *multivariate* methods consist on the joint analysis of multiple voxels. This allows us to analyse distributed patterns of activity across separate neural populations. Second, while subtractive designs use methods of *encoding*, which aim to predict the neural data from the experimental task, *decoding* aims to predict the experimental task from the neural data. Multivariate decoding strategies use tools from machine learning to create classifiers which, after being 'trained' (or 'fed') with fMRI data for different tasks, can predict (decode) which task is being performed, just by looking at the associated pattern of brain activity. Multivariate decoding has been proposed as a methodological improvement on both strategies discussed here: reverse inference and ontology testing.

Some authors have claimed that the use of multivariate decoding could provide a viable alternative to traditional location-based reverse inference (Poldrack 2011; Del Pinal and Nathan 2017). For example, if we want to know whether a task T involves mental process M or mental process M', we can design a task that uncontroversially engages M, a second task that uncontroversially engages M', and train a classifier with fMRI data for both tasks. Then, we collect data for task T, and use the classifier to determine whether T involves M or M' based on the decoding accuracy. That is, based on whether the pattern of brain activity in T resembles the one in the task that involves M or the one in the task that involves M'. This pattern-based reverse inference is an interesting possibility. However, at least to date, it does not seem suitable for use within experimental philosophy's intuitions project. Multivariate decoding is especially informative with simple tasks, in which we are certain about the mental processes involved. For example, it is possible to decode whether a subject is viewing a shoe or a bottle by looking at the pattern of brain activity associated with each task (Norman et al. 2006). However, the possibilities of multivariate decoding as a base for reverse inferences are limited when using complex tasks, which are likely to involve several different processes (not only M or M', but also M_1, M_2, M_n). In these cases, it is difficult to determine whether the decoding accuracy is due to the tasks being similar in terms of the mental process of interest (M or M'), or similar in terms of the other mental processes involved (M_1, M_2, M_n).

Multivariate decoding has also been proposed as an improvement on Lindquist et al.'s methodology and has led to doubts about their conclusions. Although Lindquist and colleagues' claim against basic emotions theory does not rely on the localization of mental states, their meta-analysis uses data coming from location-based subtractive designs. If the neural basis for each basic emotion is to be found in distributed patterns of brain activity, then their study is fundamentally incapable of finding distinct neural correlates for each basic emotion (Hamann 2012). Using multivariate decoding strategies, other studies have shown that it is possible to discriminate between basic emotions based on their associated patterns of brain activity (see Kragel and LaBar 2016 for a review). However, it is important to note that the success of multivariate decoding has to do with predictive power, and predictive power does not necessarily imply neurobiological reality (Poldrack 2011; Hebart and Baker 2017; Ritchie et al. 2017). That a distributed pattern of brain activity can be used to predict, for example, the engagement of an emotion (*decode* the emotion), doesn't mean that that pattern of activity has the function of generating the emotion (*encode* the emotion). That pattern is not necessarily present in any of the individual emotion instances, so it cannot be considered as an emotion circuit in the brain (Clark-Polner et al. 2016). In fact, the pattern of distributed neural activity associated with each emotion differs across studies (Kragel and LaBar 2016). This is not an argument against the success of multivariate decoding in distinguishing between different emotions. But it is an argument against its significance. It suggests that successfully decoding an emotion is not the same as finding a neural circuit for that emotion. Thus, successful decoding of emotions will not provide evidence in favour of basic emotion theory.

5. Conclusion

In this chapter, I discussed the possibilities of fMRI for the purposes of experimental philosophy. In the introduction, I briefly described fMRI and the two main projects within experimental philosophy: the intuitions project and the psychology project. This allowed us to identify two specific ways in which fMRI can be used in experimental philosophy: reverse inference within the intuitions project and ontology testing within the psychology project. I used two examples to discuss the methods associated with each of these approaches.

The pluripotency of brain areas, that is, the one-to-many character of associations between brain structure and mental function, was shown to limit the possibilities of reverse inference. Although methodological advances might overcome this problem, the ones available to date do not seem suitable for the purposes of experimental philosophy's intuitions project. Similar to Greene et al.'s hypothesis regarding deontological intuitions, other researchers in experimental philosophy have advanced explanations in terms of emotional biases for compatibilist intuitions (Nichols and Knobe 2007) or intuitions about the intentionality of bringing about negative side-effects (Nadelhoffer 2006). The latter hypothesis has already been tested with fMRI (Ngo et al. 2015). However, as in the case of Greene et al.'s study, their neuroimaging results are open to alternative explanations (Díaz, Viciana and Gomila 2017). One of the main conclusions of this chapter is that experimental philosophers should not use fMRI to test these kinds of hypotheses.

Ontology testing, on the other hand, was shown to have the potential to address philosophically relevant issues about the organization of the mind. Although this approach itself raises a series of philosophical questions about the relationship between neuroscience and psychology, this should be an *additional* reason for philosophers to engage in the debate. Building a proper cognitive ontology is a project that calls for the interdisciplinary approach characteristic of experimental philosophy's Psychology Project. Furthermore, questions about the nature of mental states such as pain are at the centre of much philosophical debate. Here, I suggest that fMRI evidence can be used to inform these debates. For example, it could be possible to use fMRI to investigate whether pain should be understood as a bodily sensation (like a tickle) or a complex emotional state (like disgust) by looking at the degree of overlap between the patterns of brain activity recruited by each of these processes.

To sum up, I argue that (1) fMRI is currently not suitable for the intuitions project's goal of discovering the mental processes underlying intuitions, but (2) experimental philosophers should explore the potential of fMRI to inform questions about cognitive architecture within the psychology project.

Suggested Readings

Hanson, S. J., and Bunzl, M. (2010). *Foundational Issues in Human Brain Mapping*. Cambridge MA: MIT Press.

Hebart, M. N., and Baker, C. I. (2017). Deconstructing multivariate decoding for the study of brain function. *NeuroImage* (April), 1–15. Doi: 10.1016/j. neuroimage.2017.08.005.

Klein, C. (2010). Philosophical Issues in Neuroimaging. *Philosophy Compass*, 5(2), 186–198. Doi: 10.1111/j.1747-9991.2009.00275.x.

Logothetis, N. K. (2008). What we can and what we cannot do with fMRI. *Nature*, 453(7197), 869–878. Doi: 10.1038/nature06976.

Poldrack, R. A., and Yarkoni, T. (2016). From brain maps to cognitive ontologies: Informatics and the search for mental structure. *Annual Review of Psychology*, 67(1). Doi: 10.1146/annurev-psych-122414-033729.

References

Adolphs, R. (2016). How should neuroscience study emotions? By distinguishing emotion states, concepts, and experiences. *Social Cognitive and Affective Neuroscience*, 12(1). doi: 10.1093/scan/nsw153.

Anderson, M. L., Kinnison, J., and Pessoa, L. (2013). Describing functional diversity of brain regions and brain networks. *NeuroImage*, 73, 50–58. Doi: 10.1016/j.neuroimage.2013.01.071.

Clark–Polner, E., Johnson, T. D., and Barrett, L. F. (2016). Multivoxel pattern analysis does not provide evidence to support the existence of basic emotions. *Cerebral Cortex*, 27(3). doi: 10.1093/cercor/bhw028.

Díaz, R., Viciana, H., and Gomila, A. (2017). Cold side-effect effect: Affect does not mediate the influence of moral considerations in intentionality judgments. *Frontiers in Psychology*, 08(February), 295. Doi: 10.3389/fpsyg.2017.00295.

Ekman, P. (1999). Basic emotions. In T. Dalgleish, and M. Power (eds.), *Handbook of Cognition and Emotion*. Sussex, U.K.: John Wiley & Sons.

Fischer, E., and Collins, J. (2015). Rationalism and naturalism in the age of experimental philosophy. In E. Fischer and J. Collins (eds.), *Experimental Philosophy, Rationalism, and Naturalism. Rethinking Philosophical Method* (pp. 3–33). Routledge. Available at: https://books.google.es/books?hl=en&lr=&id=l1uhCAAAQBAJ&oi=fnd&pg=PT10&dq=rationalism+and+naturalism+in+the+age+of+experimental+philosophy&ots=jHJ1zwkG4s&sig=Q3AVbXBHOcIPaEwrC1RDxShIfns. Accessed 21 July 2017.

Greene, J. D. (2014). Beyond point-and-shoot morality : Why cognitive (Neuro)science matters for ethics. *Ethics*, 124(4), 695–726. Doi: 10.1086/675875.

Greene, J. D. et al. (2001). An fMRI investigation of emotional engagement in moral judgment. *Science*, 293(5537), 2105–2108. Doi: 10.1126/science.1062872.

Hamann, S. (2012). What can neuroimaging meta-analyses really tell us about the nature of emotion? *Behavioral and Brain Sciences*, 35(03), 150–152. Doi: 10.1017/S0140525X11001701.

Hebart, M. N., and Baker, C. I. (2017). Deconstructing multivariate decoding for the study of brain function. *NeuroImage* (April), 1–15. Doi: 10.1016/j.neuroimage.2017.08.005.

Huebner, B., Dwyer, S., and Hauser, M. (2009). The role of emotion in moral psychology. *Trends in Cognitive Sciences*, *13*(1), 1–6. Doi: 10.1016/j.tics.2008.09.006.

Klein, C. (2010). Philosophical issues in neuroimaging. *Philosophy Compass*, *5*(2), 186–198. Doi: 10.1111/j.1747–9991.2009.00275.x.

Klein, C. (2011). The dual track theory of moral decision–making: A critique of the neuroimaging evidence. *Neuroethics*, *4*(2), 143–162. Doi: 10.1007/s12152–010–9077–1.

Knobe, J., and Nichols, S. (2017). Experimental philosophy. *Stanford Encyclopedia of Philosophy*. Available at: https://plato.stanford.edu/archives/win2017/entries/experimental–philosophy/. Accessed 6 January 2018.

Kragel, P. A., and LaBar, K. S. (2016). Decoding the nature of emotion in the brain. *Trends in Cognitive Sciences*, *20*(6), 444–455. Doi: 10.1016/j.tics.2016.03.011.

Lacadie, C. M. et al. (2008). More accurate Talairach coordinates for neuroimaging using non–linear registration. *NeuroImage*, *42*(2), 717–725. Doi: 10.1016/j.neuroimage.2008.04.240.

Lenartowicz, A. et al. (2010). Towards an ontology of cognitive control. *Topics in Cognitive Science*, *2*(4), 678–692. Doi: 10.1111/j.1756–8765.2010.01100.x.

Lindquist, K. A., and Barrett, L. F. (2008). Constructing emotion: The experience of fear as a conceptual act. *Psychological Science*, *19*(9), 898–903. Doi: 10.1111/j.1467–9280.2008.02174.x.

Lindquist, K. et al. (2012). The brain basis of emotion: a meta–analytic review. *Behavioral and Brain Sciences*, *35*(3), 121–143. Doi: 10.1017/S0140525X11000446.

Logothetis, N. K. (2008). What we can and what we cannot do with fMRI. *Nature*, *453*(7197), 869–878. Doi: 10.1038/nature06976.

Machery, E. (2013). In defense of reverse inference. *British Journal for the Philosophy of Science*, *65*(2), 251–267. Doi: 10.1093/bjps/axs044.

Machery, E. (2017). *Philosophy within Its Proper Bounds*. Oxford: Oxford University Press.

Mather, M., and Kanwisher, N. (2013). How can fMRI inform cognitive theories. *Psychological Science*, *8*(1), 108–113. Doi: 10.1177/1745691612469037.

Mitterschiffthaler, M. T. et al. (2007). A functional MRI study of happy and sad affective states induced by classical music. *Human Brain Mapping*, *28*(11), 1150–1162. Doi: 10.1002/hbm.20337.

Mole, C., and Klein, C. (2010). *Confirmation, Refutation, and the Evidence of fMRI*. Available at: http://philpapers.org/rec/MOLCRA. Accessed 8 July 2015.

Moll, J., Eslinger, P. J., and De Oliveira–Souza, R. (2001). Frontopolar and anterior temporal cortex activation in a moral judgment task: Preliminary functional MRI results in normal subjects. *Arquivos de Neuro–Psiquiatria*, *59*(3 B), 657–664. Doi: 10.1590/S0004–282X2001000500001.

Nadelhoffer, T. (2006). Bad acts, blameworthy agents, and intentional actions: Some problems for juror impartiality. *Philosophical Explorations*, *9*(2), 203–219. Doi: 10.1080/13869790600641905.

Ngo, L. et al. (2015). Two distinct moral mechanisms for ascribing and denying intentionality. *Nature Scientific Reports*, 5, 1–11. Doi: 10.1038/srep17390.

Nichols, S., and Knobe, J. (2007). Moral responsibility and determinism: The cognitive science of folk intuitions. *Nous*, 41(4), 663–685. Doi: 10.1111/j.1468–0068.2007.00666.x.

Norman, K. A. et al. (2006). Beyond mind-reading: Multi-voxel pattern analysis of fMRI data. *Trends in Cognitive Sciences*, 10(9), 424–430. Doi: 10.1016/j.tics.2006.07.005.

Orden, G. C. Van, and Paap, K. R. (1997). Functional neuroimages fail to discover pieces of mind in the parts of the brain. *Philosophy of Science*, 64(S1), S85. Doi: 10.1086/392589.

Parens, E., and Johnston, J. (2014). Neuroimaging: Beginning to appreciate its complexities. *The Hastings Center Report*, Spec No, pp. S2–S7. Doi: 10.1002/hast.293.

Del Pinal, G., and Nathan, M. J. (2013). There and up again: On the uses and misuses of neuroimaging in psychology. *Cognitive Neuropsychology*, 30(4), 233–252. Doi: 10.1080/02643294.2013.846254.

Del Pinal, G., and Nathan, M. J. (2017). Two kinds of reverse inference in cognitive neuroscience. *The Human Sciences after the Decade of the Brain*, 121–139. Doi: 10.1016/B978–0–12–804205–2.00008–2.

Poldrack, R. A.(2006). Can cognitive processes be inferred from neuroimaging data? *Trends in Cognitive Sciences*, 10(2), 59–63. Doi: 10.1016/j.tics.2005.12.004.

Poldrack, R. A. (2010). Mapping mental function to brain structure: How can cognitive neuroimaging succeed? *Perspectives on Psychological Science: A Journal of the Association for Psychological Science*, 5(6), 753–761. Doi: 10.1177/1745691610388777.

Poldrack, R. A. (2011). Inferring mental states from neuroimaging data: From reverse inference to large-scale decoding. *Neuron*, 72(5), 692–697. Doi: 10.1016/j.neuron.2011.11.001.

Poldrack, R. A., Halchenko, Y. O., and Hanson, S. J. (2009). Decoding the large-scale structure of brain function by classifying mental states across individuals. *Psychological Science*, 20(11), 1364–1372. Doi: 10.1111/j.1467–9280.2009.02460.x.

Price, C. J., and Friston, K.J. (2005). Functional ontologies for cognition: The systematic definition of structure and function. *Cognitive Neuropsychology*, 22(3–4), 262–275. Doi: 10.1080/02643290442000095.

Ritchie, J. B., Kaplan, D., and Klein, C. (2017). Decoding the brain: Neural representation and the limits of multivariate pattern analysis in cognitive neuroscience. bioRxiv, p. 127233. Doi: 10.1101/127233.

Rose, D., and Danks, D. (2013). In defense of a broad conception of experimental philosophy. *Metaphilosophy*, 44(4), 512–532. doi: 10.1111/meta.12045.

Roskies, A. L. (2008). Neuroimaging and inferential distance. *Neuroethics*, 1(1), 19–30. Doi: 10.1007/s12152–007–9003–3.

Sander, D. (2012). The role of the amygdala in the appraising brain. *Behavioral and Brain Sciences*, *35*(03), 161. Doi: 10.1017/S0140525X11001592.

Scarantino, A. (2012). How to define emotions scientifically. *Emotion Review*, *4*(4), 358–368. Doi: 10.1177/1754073912445810.

Scarantino, A. (2012). Functional specialization does not require a one-to-one mapping between brain regions and emotions. *Behavioral and Brain Sciences*, *35*(03), 161–162. Doi: 10.1017/S0140525X11001749.

Scarantino, A., and Griffiths, P. (2011). Don't give up on basic emotions. *Emotion Review*, *3*(4), 444–454. Doi: 10.1177/1754073911410745.

Scherer, K. R. (2012). Neuroscience findings are consistent with appraisal theories of emotion; but does the brain 'respect' constructionism? *Behavioral and Brain Sciences*, *35*(03), 163–164. Doi: 10.1017/S0140525X11001750.

Uttal, W. R. (2001). *Life and Mind: Philosophical Issues in Biology and Psychology. The New Phrenology: The Limits of Localizing Cognitive Processes in the Brain.* Cambridge, MA: The MIT Press, 221–228.

Yarkoni, T. et al. (2010). Cognitive neuroscience 2.0: Building a cumulative science of human brain function. *Trends in Cognitive Sciences*, *14*(11), 489–496. Doi: 10.1016/j.tics.2010.08.004.

Yarkoni, T., Poldrack, R.A., and Nichols, T. (2011). Large-scale automated synthesis of human functional neuroimaging data. *Nature Methods*, *8*(8), 665–670. Doi: 10.1038/nmeth.1635.Large–scale.

5

Using VR Technologies to Investigate the Flexibility of Human Self-Conception

Adrian J. T. Alsmith and Matthew R. Longo

1. Introduction

This chapter will focus on the prospects of using virtual reality to study our pattern of use of the self-concept. A frequently discussed claim about the self-concept (or the first-person concept, as it is often called) is that when a subject employs it in thought, that thought is guaranteed to refer to the subject (Shoemaker 1968). But a less frequently discussed claim is that our thoughts about ourselves are nevertheless *flexible*, insofar as they involve a great deal of indeterminacy. That is, even if a subject is guaranteed to refer to herself when using the self-concept, she does not thereby ultimately determine what kind of thing she is.

To illustrate, suppose that Louis XIV once thought 'I am the State' and then around lunchtime thought 'I am hungry'.[1] If the celebrated referential guarantee of the concept expressed by 'I' obtains, then, in each use of the concept, Louis XIV would refer to himself. Yet he would refer to very different kinds of entity in each case. In the first case, he would (somewhat absurdly) be referring to himself *qua* the French government; in the second, he would (more plausibly) be referring to himself *qua* organism in need of sustenance. The claim that thoughts employing the self-concept are flexible implies that despite referring to very different kinds of thing, it might seem to Louis XIV that he refers to a single thing, namely himself. If true, this would be a highly peculiar feature. For in

Acknowledgements

This work was supported by a grant from the Volkswagen Foundation (Grant no. 89434: *Finding Perspective: Determining the embodiment of perspectival experience*) to AA and MRL. We would also like to thank an anonymous reviewer and the editors for insightful comments and suggestions.

[1] We take a little poetic licence here, as Louis XIV is actually credited with the statement 'L'etat c'est moi' (the state is me) rather than 'Je suis l'etat' (I am the state).

other cases in which multiple thoughts seem to refer to different kinds of entity, one might rightly judge that these thoughts do not all refer to a single thing. So, what is odd about our pattern of use of the self-concept, if it does exhibit such flexibility, is that it can seem to encompass reference to different kinds of entity whilst also seeming to refer to the very same thing in each case, namely oneself.

In the next section, we will clarify what exactly is meant by the claim that we will refer to as *flexibility*. The rest of the chapter will focus on assessing flexibility as an empirical claim, and, especially, the prospects of using virtual reality technology to investigate it. In Section 3, we will review virtual reality research that seems promising in this regard. In Section 4, we will raise certain key methodological issues with this research insofar as it might serve to demonstrate flexibility.

2. Flexibility and self-conception

2.1. Concepts and conceptions

To clarify the target of our discussion, we will operate with a distinction between concepts and conceptions. This distinction is commonly associated with Rawls' (1971) discussion of justice: Disagreement about justice is rife. But if that disagreement is genuine, it requires a common subject matter – the concept of justice – which the parties to the disagreement might apply very differently. These differences in application correspond to differences in conception, each of which may be wrong or partial with respect to the true reference of the concept. We distinguish, then, the concept of the self – whatever its true referent may be – from an individual's self-conception.

The distinction requires some way of cashing out the assumption that concepts have correctness conditions that are fixed independently of an individual's conceptions. For instance, an externalist theory of concepts might treat them as abstract objects, the correctness conditions for which are individuated independently of any given subject who might possess them (Putnam 1975). In a similar respect, an essentialist account of the psychology of concepts might hold that individuals believe that, for any member of a natural kind, there is an essence, the possession of which is necessary and sufficient for an entity to be a member of that kind (Keil 1989, Ch. 8). This essence would serve as an independent criterion according to which the accuracy of the individual's application of the concept (her conception) of that kind may be judged. For the purpose of our discussion, we will merely assume that there is some independent means of individuating concepts such that we can rightly distinguish them from conceptions. For those that do not embrace the distinction between concepts and conceptions, we ask only that they bear in mind that our discussion concerns how individuals apply a concept (and the underlying cognitive processes involved) not that to which the concept should be applied (and thus its correct reference).

Discerning an individual's self-conception is an empirical matter. One way of investigating it is to trace out the pattern of an individual's self-ascriptions, that is, ascriptions of properties and processes to herself in statements employing the self-concept, such as 'I am x', 'I am y-ing' and 'z is mine'. This approach assumes that the statistics of linguistic usage are a guide to the structure of an individual's conception. Though the research we will review later can only be understood as revelatory of individuals' self-conceptions on this assumption, we do not mean to suggest that other approaches are invalid.

It is also worth distinguishing flexibility, as a feature of conceptions, from similar features of language, such as polysemy and metonymy. Consider, for instance, the sentence 'The White House got a new paint job' and the sentence 'The White House issued a press release'. In these two sentences, very different properties are attributed to the grammatical subject of each sentence because the noun phrase 'The White House' is polysemic: it can refer to a group of buildings or a group of people. However, if it seemed to someone that she were referring to the very same thing by 'The White House' when ascribing properties to a group of buildings and to a group of people, then her conception of 'The White House' would be flexible in this regard. The hypothesis we will consider is that individuals possess a self-conception that is flexible, such that an individual's self-ascriptions might encompass properties and processes attributable to various kinds of entity, whilst she nevertheless refers to a single entity – herself.

2.2. Self-conceptions and the concept of the self

Much philosophical discussion of self-consciousness involves responding to Cartesian claims about the concept of the self; in particular, the claims that the self is not a part of the objective world and that the self is indivisible.[2] These claims are inherently difficult to reconcile with the idea that selves can be identified with human bodies, as these are usually conceived to be objectively

---

[2] For exposition of these claims in Descartes' (1642/1984) 2nd and 6th meditations, as well as other writings, see Wilson (1978, Chs 2 and 6).

existing, divisible entities. This reflects a broader tension that constitutes one of the most ancient and enduring problems in the study of the mind, namely the reconciliation of our apparent mental and material natures.

One reaction to this tension is to attempt to find its roots in the contrast between two compelling *conceptions* of ourselves, as mental and material entities (cf. Papineau 2002; Bloom 2003).[3] What is compelling about the Cartesian conception of the self is the certainty of the *cogito* – the certainty with which one can infer 'I exist' from 'I am thinking', where the latter is grounded in an occurrent conscious mental event. Self-ascriptions of bodily properties are also compelling, especially when made on the basis of perceptual information concerning our bodies. Thus, we are disposed (on the basis of the right kinds of information) to make self-ascriptions such as 'I am hot and sweaty' and 'I am in front of a tree' (Evans 1982, Ch 7).

Evans concluded that because these bodily self-ascriptions exhibit the referential guarantee distinctive of the self-concept, they can serve as a 'powerful antidote to a Cartesian conception of the self' (Evans 1982, 220).[4] Leaving aside the issue of whether these cases do have such a guaranteed reference, it is clear that if an individual is disposed to make such bodily self-ascriptions, then she conceives of her body as herself. Evans' conclusion is then that the Cartesian theorist's conception of the self as potentially disembodied does not map onto our self-conception.

But it is not clear that this move is valid. Certainly, there is a manifest incoherence in the idea that we could be incorporeal minds and yet possess such properties as being hot and sweaty. And to claim that the Cartesian conception maps onto a *part* of our self-conception, would be to thereby render that conception as a whole potentially incoherent. But why should we assume that the question 'How do individuals conceive of themselves?' must return a metaphysically coherent answer?[5] Indeed, the distinction between concepts and conceptions is effectively a distinction between the metaphysical principles that hold true of a domain and an individual's capacity to apply concepts concerning

---

[3] Accordingly, flexibility might not be unique to our use of the self-concept, but a more general feature of our conception of entities to which we attribute psychological properties, such as persons (cf. Perry 1978), animals (Clark 2003) and even groups (List and Pettit 2011).

[4] Evans claim, it should be noted, is that these perceptually based self-ascriptions are immune to error through misidentification relative to the first-person pronoun. Here Evans is following the common practice of taking immunity to error of this kind as a guide to an account of self-consciousness. For a discussion of Evans strategy on these terms, see Brewer (1995, 291–297).

[5] This is characteristic of the descriptive approach to metaphysics, the aim of which in this case would be to lay bare the core components that any individual's self-conception must possess in order that she might have the capacities for thought the she does (cf. Strawson 1959/2003, especially Ch 3).

that domain. Thus, even if a metaphysical theory of the self may be in the business of determining the nature of the referent of the self-concept, individuals might not need to meet this standard to engage in self-ascription (Peacocke 2014, 140–141). As Campbell puts it, it might just be that 'our ordinary use and understanding of the first person leaves it open what kinds of things we are' (2004, 476). In short, it is not clear that our self-conception is so determinate that it cannot abide incoherence.

The notion of a *flexible* self-conception expresses the idea that indeterminacy in our self-conception can potentially yield incoherence. We suggest that flexibility can be more precisely stated in terms of two dimensions of higher-order difference/similarity between pairs of self-ascriptions. Dimension 1 concerns the degree of *difference* in the range of entities to which self-ascriptions are made. Thus, compare the following two pairs of thoughts:

Pair 1: "I am a physical body" and "I am a non-physical soul"
Pair 2: "I am a human" and "I am a cyborg"

One of the ways in which pairs 1 and 2 differ is that there is a greater difference between physical and non-physical entities than there is between humans and cyborgs. In the framework we propose, this amounts to greater flexibility in pair 1 than pair 2.

Dimension 2 concerns the degree of *similarity* in context between self-ascriptive thoughts. For instance, consider:

Pair 3: "I will die one day" and "I am immortal"

According to the framework we propose, when the thoughts in pair 3 occur in a single context, such as a Sunday sermon, they exhibit greater flexibility than when they each occur in different contexts, such as a biology class and a Sunday sermon.

We can illustrate a case of extreme flexibility by recalling a great conflict within Descartes' work. Whilst he famously conceived of the subject of thought as potentially disembodied, he also conceived of the subject of bodily sensations as united with the body as a whole (Wilson 1978, 181 ff.), apparently admitting this tension (in his correspondence with Princess Elizabeth of Bohemia) as follows:

[It] does not seem to me that the human mind is capable of conceiving very distinctly, and at the same time, the distinction between the soul and the body and their union, since to do so it is necessary to conceive them as one single thing and at the same time to conceive them as two, which is contradictory. (Descartes 1643/2007, 70)

With a slight liberty of interpretation, we can translate Descartes' failure of imagination here into a claim to be stated in the framework we are suggesting: there cannot be a self-conception so flexible that the subject can, in the very same context, conceive of themselves as an indivisible and a divisible entity.[6]

It would be unwise to test the general claim of flexibility by focusing on an extreme case such as this. Flexibility in more moderate forms will consist of self-ascriptions in slightly different contexts, involving properties attributable to more similar, but nevertheless distinct entities. Of particular interest to us are self-ascriptions involving properties attributable to the human body and entities similar to the human body. Since the 19th century at least, theorists have suggested that humans' relationship to technology calls into question whether we ought to identify the bodily self with the human body, or the human body and some technological complement (see e.g. Lotze 1888, 587–90). A philosopher in the grip of metaphysical theory might insist that we identify the human bodily self with the human body. But this leaves open the question of whether such an assumption – or indeed general assumptions about determinacy – are in any way built into our self-conception, rather than merely built into contemporarily popular accounts of the reference of the self-concept (Martin 1997, 133–134).

If our self-conception is indeed flexible, then it ought to be possible to induce individuals to self-ascribe properties attributable to a range of distinct entities in somewhat similar contexts. Philosophers have, as we have seen, expressed a variety of intuitions on this subject, but this is clearly an issue that ought to be subject to systematic empirical research. The remainder of this chapter will focus on the question of whether and how empirical research using virtual reality technology might establish this, by inducing subjects to make self-ascriptions to not only their actual bodies but also distinct entities, such as virtual bodies.

3. Flexibility and VR

In this chapter, we operate with a fairly inclusive definition of virtual reality (VR) technology as ranging over a variety of sensor display and tracking technologies. Perhaps the most well-known technologies are *visual displays*, such as light-weight, head-mounted displays (HMDs), or ultra-high resolution,

---

[6]   It is not clear that Descartes is right on this point: I might conceive of myself as divisible in the sense that, for example, a guillotine could conceivably 'divide' me into a head and torso. But I might also think that were that to happen, I would stop existing, and that I am therefore indivisible in this respect.

large-screen immersive displays (sometimes known as CAVE systems). Either of these can be combined with wireless motion-tracking to enable exploration of a virtual environment (a computer-generated simulation) or a video feed from another location in a real environment. Visual displays can also be combined with headphones to provide a realistic *acoustic* dimension to the environment, by implementing functions (known as head-related transfer functions) that characterize how the human head filters sound (Bergström et al. 2017, Berger et al. 2018). *Haptic* feedback can also be integrated with tracking and other displays to simulate physical encounters with objects in the environment by means of force-feedback devices, pressure devices, vibrotactile devices or even low frequency audio (Spanlang et al. 2014).

When correctly combined, these technologies have the capacity to enable subjects to feel 'present' in virtual worlds through synthetic sensory stimulation. At the heart of the phenomenon of feeling present in a virtual world is the experience of virtual embodiment, resulting from the use of a virtual body (known as an 'avatar') to regulate sensory and motor engagement with the environment (Slater 2009). In this section, we review recent work on virtual embodiment to show how it provides *prima facie* empirical support for flexibility.

3.1. Virtual embodiment

Contemporary research in VR aims not merely to simulate previous experiences of physical reality, but rather to provide fundamentally different forms of experience using the unique possibilities of the medium (Slater and Sanchez-Vives 2016). These unique possibilities depend upon inducing in users the feeling of being 'present', not in their actual environment, but in a virtual environment. As it is currently understood, presence is a twofold illusion, consisting of both a 'place illusion' – in which it seems to a subject that she is placed within a real scene – and a 'plausibility illusion' – in which it seems to a subject that she is participating in real events (Slater 2009). Recent work has demonstrated that users' 'embodiment' of a virtual character involved in unfolding events is not only essential to place illusions, it is also a powerful contributor to plausibility (Skarbez et al. 2017).

'Embodiment' is a notion that means many things to many people (Alsmith and de Vignemont 2012). In the VR literature, it is commonly used as an umbrella term that encompasses various ways in which information concerning an entity's properties is processed in a manner that is similar to information concerning an individual's actual body (cf. Kilteni et al. 2012). Accordingly, it is a graded notion, with degrees of embodiment according to degrees of similarity in the relevant information processing (de Vignemont 2011, 88).

Paradigms for inducing and measuring embodiment were originally developed using physical props, especially rubber hands, and multisensory stimulation protocols. In a typical setup, such as the 'rubber hand illusion', a participant would observe a brush making physical contact with a rubber hand in a stroking motion, and their actual hand would also be stroked, whilst obscured from view (Botvinick and Cohen 1998). When the stroking motion participants saw on the rubber hand was kept in spatial congruence and temporal synchrony with the stroking motion they felt on the rubber hand, participants would exhibit a variety of behavioural and physiological responses that indicate their embodiment of the rubber hand. These include introspective reports about subjective experiences of body ownership (Longo et al. 2008), perceptual judgements of the location of their own hand as closer to the rubber hand after stimulation (Tsakiris and Haggard 2005); reduced temperature of the participant's own hand after stimulation (Moseley et al. 2008); increased electrodermal activity (Armel and Ramachandran 2003) and a distinctive cortical anxiety response (Ehrsson et al. 2007) when the observed hand is subjected to violence after stimulation.

In recent years, these paradigms have been adapted by researchers using VR technology to embody virtual objects. For instance, using motion-tracking of the head and a stereoscopic image projection system (similar to that used in 3D cinemas) researchers created a virtual analogue of the rubber hand illusion, the virtual arm illusion (Slater et al. 2008). On the display, participants would see a virtual arm that would appear (from the participant's perspective) to be extending out from the position of the shoulder of their right arm. Their right arm itself was hidden from view, extending in a slightly different direction from the virtual arm. A virtual ball, motion-tracked to an actuator wand held by the experimenter, was used to tap the virtual ball on the hand of the virtual arm, by tapping the wand on the participant's actual hand. Participants' responses indicated that they had embodied the virtual arm: After stimulation, the location of the hand of their actual (hidden) arm was judged to be closer to the hand of the virtual arm. Also, when the virtual arm was programmed to rotate slowly, electromyography revealed increased corresponding muscle activity in the participant's actual arm during the virtual arm's rotation.

Perhaps the majority of research on the embodiment of virtual objects studies users' embodiment of virtual avatars. Virtual avatars are (typically humanoid) virtual objects whose shape, position and movement are highly congruent with the shape, position and movement of the participant's actual body. In an early study, Lenggenhager et al. (2007) used a head-mounted display to present participants with a body projected two meters in front of their perspective in virtual space, which they observed being stroked on its back whilst they were stroked on their own back. Again, participants' behavioural responses indicated that they had embodied the virtual body: after stimulation, participants were moved backwards and asked to walk forwards to their original position, and they judged it to be closer to the position of the avatar.

Complex VR systems can involve a range of spatially and temporally integrated sensory displays and motion-tracking systems to simulate effectively the structure of perceptual engagement with a real environment (Cummings and Bailenson 2016). Embodiment of avatars is in part a function of the VR system's adherence to the spatiotemporal principles of multisensory integration between visual, auditory, tactile, proprioceptive and vestibular processes (Menzer et al. 2010; Ionta et al. 2011; Kilteni et al. 2015). Accordingly, avatars are often presented from a 'first-person' point of view, visually presenting the virtual body from a location and direction congruent with the view a participant would have of their real body (Petkova et al. 2011). Besides facilitating bottom-up multisensory processing, including a virtual avatar in the virtual scene in this way also serves to increase the adherence of the content to the users' expectations (Slater 2009; Gonzalez-Franco and Lanier 2017; Skarbez et al. 2017). In addition, virtual mirror exploration can serve to both enhance embodiment effects and provide further visual information about body part size and other characteristics (Gonzalez-Franco et al. 2010).

3.2. Virtual embodiment and self-ascription

Besides behavioural and physiological measures of embodiment, researchers have employed questionnaires to gain some sense of participants' experience of the objects they are induced to embody. This methodology also stems largely from the rubber hand paradigm. In their original study, Botvinick and Cohen (1998) presented their participants with a series of eight statements, to which they were asked to rate their agreement or disagreement on a seven-point scale (from -3 for strong disagreement, to +3 for strong agreement). One of these statements, 'I felt as if the rubber hand were my hand', intended to measure participants' sense of ownership for the rubber hand, was given an average rating of 2.5. Results such as these, together with reports of participants making statements like 'I found myself looking at the dummy hand thinking it was actually my own', served to establish the paradigm's relevance for experimentally manipulating the experience of bodily ownership, that is, the experience of one's body as one's own (de Vignemont 2013).

In a large-scale study, Longo and colleagues (2008) used *principal components analysis* to investigate patterns of co-variation in responses to 27 statements about the participant's experience of the rubber hand illusion. Their results indicated that the experience of the illusion could be decomposed into distinct components. In particular, they distinguished three components: the sense of body ownership (in statements such as 'It seemed like the rubber hand was my hand'), self-location (in statements such as 'It seemed like my hand was in the location where the rubber hand was') and the sense of agency (in statements such as 'It seemed like I was in control of the rubber hand'). We will briefly illustrate how each of these components has been investigated in questionnaires concerning virtual objects.

3.2.1. Ownership

Body ownership is the typical subjective measure of participants' experience of virtual bodies and virtual body parts. Concerning the latter, Slater et al. (2008) presented participants with a questionnaire which included the statement, 'During the experiment there were moments in which I felt as if the virtual arm was my own arm', to which they responded with a median score of +2. Concerning full virtual bodies, Lenggenhager et al. (2007) presented their participants with a 'self-attribution' questionnaire which included the statement, 'It felt as if the virtual body was my body', to which they responded with a mean score of 2.3. Consistent results have been reported in a number of subsequent studies (Kilteni et al. 2015).

3.2.2. Agency

In the VR literature, the sense of agency is a broad term encompassing 'the subjective experience of action, control, motor selection and the conscious experience of will' (Blanke and Metzinger 2009, 7). Naturally, users experience agency over virtual avatars they control through motion-tracking (Kong et al. 2017), analogous to that experienced when participants control a hand in situations analogous to the rubber hand illusion (Tsakiris et al. 2005; Longo and Haggard, 2009; Tsakiris et al. 2010), even despite major incongruencies between motor activity and visual feedback of the kind described above (Kannape et al. 2010). However, in a recent study, Kokkinara et al. (2016) studied users' agency for the actions of virtual avatars they observed whilst remaining passive. Their participants wore a head-mounted display, and, whilst seated, were presented with images of an avatar walking. Despite not engaging in the relevant movements, participants responded to the questionnaire items 'During the experiment I felt that the leg movements of the virtual body were caused by my movements' and 'I felt that I was walking' positively (with median scores of 1 and 2, respectively).

3.2.3. Self-location

Research on mental representations of the body employs the term 'self-location' to refer to the experience of occupying a determinate spatial location that may or may not be coincident with one's body (Lenggenhager et al. 2009). This general sense of the term has been adopted by VR researchers, though it is often with a slightly more specific meaning. For instance, the notion is sometimes used as a means to distinguish the feeling of presence in a virtual environment from the feeling of being within a virtual body (Kilteni et al. 2012, 375–76). By contrast, questionnaire items purporting to measure the experience of self-location often concern the distinction between the subject's actual location and their experienced location: for example, 'I experienced that I was located at some distance behind the image of myself, almost as if I was looking at someone else' (rated 1.5 on a -3 to +3 scale) (Ehrsson 2007); 'It felt as if I were lying in the corner of the room, looking at the MR scanner from this perspective' (rated 8 on a 0–10 scale) (Guterstam et al. 2015). Indeed, some researchers have used VR as a means to induce experiences analogous to out-of-body experiences (Ehrsson 2007; Lenggenhager et al. 2007; Ionta et al. 2011).

3.3. Virtual embodiment and flexibility

Although virtual avatars are typically humanoid, VR researchers have begun to explore the boundaries of the range of virtual avatars that participants can embody, by presenting individuals with avatars that are very different in physical characteristics to the participants' actual bodies.[7] Here is a representative list of manipulations of this kind, each of which used motion-

---

[7] It is also noteworthy that while, for instance, the rubber hand illusion typically fails for 'non-corporeal' objects, such as a block of wood (Tsakiris et al. 2010), it does not seem to itself depend upon similarity between the rubber hand and the participant's actual hand (Longo et al. 2009). Moreover, recent work has demonstrated that similar paradigms can be used to illicit the illusion of having a sixth finger (Newport et al. 2016) or as many as four hands (Chen et al. 2018). We welcome the trend towards probing the boundaries of illusions such as these, not least because they might reveal the extent to which our self-ascriptions can be flexible.

tracked virtual avatars, and each of which found comparable results on appropriately modified versions of the questionnaire items described in the previous subsection:

- Piryankova et al. (2014) presented participants with avatars of drastically altered hip, waist and shoulder width corresponding to the shape of overweight or underweight individuals.
- Banakou et al. (2013) presented adult participants with an avatar the size and shape of a typical four-year old.
- Slater et al. (2010) presented male participants with an avatar shaped like a female.
- Banakou et al. (2016) presented light-skinned participants with dark-skinned avatars.
- Steptoe et al. (2013) presented participants with 'extended' humanoid avatars, which were humanoid in shape but with the addition of a tail.

The accumulation of data of this kind might be taken as *prima facie* evidence for flexibility. In sum, individuals' self-ascriptions may range not only over real bodily properties and processes but also virtual properties and processes. Moreover, VR seems to not only afford demonstration of the flexibility of the self-concept, it also seems well suited for systematically studying it. Thus, by appropriately manipulating the presentation of various kinds of avatar, one might systematically study subjects' categorization of various kinds of object as themselves.

4. Some methodological issues

Despite its obvious prospects, there are two significant methodological issues to be faced in using VR to study the flexibility of the self-concept. Both concern the degree of similarity in context between patterns of self-ascription that do and not involve the use of VR technology. The first issue is that it is possible that VR engages users' imagination, placing them in an imaginative context in which they engage in patterns of self-ascription that are sheared off from their actual conception of themselves. The second issue is that VR alters users' perception to such a degree that their thoughts about the relationship between a virtual body and themselves and their actual body and themselves have a very different structure.

4.1. Imagination and VR

The questionnaire items commonly used in VR clearly require subjects to employ the self-concept in identifying properties of avatars as properties of themselves. But they do so by asking subjects about their experience of properties in VR, using qualifying conjunctions such as 'like' or 'as if'. This reflects a similar practice in the construction of questionnaire items used to test the rubber hand illusion and items used to test the experience of virtual embodiment (cf. Longo et al. 2008; Dobricki and de la Rosa 2013). However, the phrasing is equally compatible with the hypothesis that subjects are using the self-concept in a form of pretence (Nichols and Stich 2000), reflecting on a previously unnoticed imaginings in a manner similar to the experience of fiction (Walton 1978).

To illustrate, consider how such an account would proceed for the experience of body ownership reported in the rubber hand illusion (Alsmith 2015). The explanation is that these participants are imaginatively perceiving a rubber hand as their own. In imaginative perception, objects of perception are experienced in accordance with what one imagines at the time. Often when we imaginatively perceive, we imagine a perceived object to be something it is not, such as seeing Laurence Olivier as Hamlet (Currie and Ravenscroft 2002, 29), a triangle as a bully (Heider and Simmel 1944, 247) or a glob of mud as a pie, in a game of make-believe (Walton 1978, 11). Along these lines, Alsmith (2015) suggests that, in the rubber hand illusion, as participants see a rubber hand being stroked whilst they are feeling a hand being stroked, they imagine that the rubber hand that they see is their own hand. This raises the possibility that, in VR studies, participants are seeing a virtual body and imagining that the body they see is their own. It is imagining that the object that they see is their own, that leads participants to report experiencing it as their own, by responding affirmatively to the relevant questionnaire items.

A fuller account on similar lines would need to clarify exactly how the imagining is itself facilitated by the content of perception. But, roughly, the thought would be that the imaginative experience is facilitated by the consistency between the content of perception and the proposition that they have a virtual body (Walton 1978). As a consequence, the experimental setup fosters the participant's imagining that the virtual body is her own, without her needing to engage her imagination actively, or indeed attend to the fact that she is imagining (Walton 1990). For a comparison: when watching a play, I need not intend to imagine that the actor I see onstage is the character she plays, nor will I necessary notice if I do so imagine that to be the case. Yet, if her performance is compelling, I may naturally be caught up in the imaginative project. Similarly, a compelling experimental setup might facilitate a participant in engaging a fictional scenario in which she imagines that she has a virtual body.

We do not mean to suggest that this interpretation of user experience in VR is in any way exclusively supported by the available data. Rather, we present it as a possible interpretation because of its implications for the use of VR to study flexibility. Imagination is notoriously unconstrained by reality. For this reason, imaginative contexts can constitute a significant departure from contexts in which our thoughts *are* more constrained by reality. This feature of the imagination is well captured by Nichols and Stich's (2003, 2000) model of pretence. According to their model, what we believe on the one hand and what imagine on the other is a consequence of mental operations on propositional contents stored in different 'boxes'. Whereas believing involves operations on contents stored in the 'Belief Box' – contents which serve to represent states of affairs in the actual world, imagining involves operations on contents stored in the 'Possible World Box' – contents which, perhaps unsurprisingly, represent states of affairs in possible worlds.

If it is indeed the case that VR users' self-ascriptions reflect their imaginative experiences, then it ought to significantly affect how we model the flexibility of their self-conception in this regard. For whilst it might be the case that there is a great degree of difference in the kinds of entities involved in their pattern of self-ascription (placing their self-conception high on dimension 1), there is also a great degree of difference between the imaginative contexts of their self-ascriptions in VR and the non-imaginative contexts of their self-ascriptions outside VR (placing their self-conception low on dimension 2).

4.2. Thinking about virtual bodies and real bodies

Determining where an individual's self-conception falls on dimension 2 requires a clear understanding of each context in which subjects can apply the self-concept, in order that they might be compared for similarity. One way in which contexts might be similar or different is in the structure of an individual's self-ascriptions *within* each context. The example we will focus on is the structure of an individual's thinking about the relationship between their actual bodies and themselves.

The human body is an integrated structure of parts forming a whole. Since at least the 1970s, researchers have used a variety of methods to determine whether and how individuals locate themselves within specific parts of their body. Some of this work has relied upon manipulating subjects' attention towards parts of their bodies to elicit judgements of their location in relation to those parts. For instance, in an early study by Dixon, participants were asked to manually stimulate various body parts to and answer various questions about the relative location of these parts to the 'vantage point of "I"' on three spatial dimensions and thereby provide a 'three-dimensional fix on locus of self' (Dixon 1972, 104). In more recent work, Bertossa et al. (2008) used a structured interview which also consisted of a series of questions about the relative spatial location of body parts to the 'I', which they used to elicit gradually narrowing self-location judgements to a single bodily location.

Other studies have had participants make projective judgements about the location of the self in a depiction of humanoid or non-humanoid figures. For instance, Limanowski and Hecht (2011) provided participants with a description of the notion of the self and asked participants to mark the location of the self within an outline of a human body and within a rectangular shape containing depictions of a human heart and brain. Starmans and Bloom (2012) used a slightly less direct method in which participants (children and adults) were asked to judge the relative distance of an object from a humanoid a character with a humanoid or alien body.

These approaches have produced somewhat mixed results concerning individuals judgements of the location of the self within the body. Whilst Starmans and Bloom's (2012) results clearly suggested that 'children and adults intuitively think of the self as occupying a physical location within the body, close to the eyes' (Starmans and Bloom 2012, 317), Limanowski and Hecht's (2011) participants' responses clustered around both the head and torso of the outline of the human body. And whereas 51 of Bertossa et al.'s (2008) 59 interviewees judged themselves to be centred within their heads, this was true of only 40 of Dixon's (1972) 80 respondents.

Modifying a paradigm used to study binocular vision, Alsmith and Longo (2014) developed a method for eliciting precise self-location judgements concerning one's own body, rather than a depiction of a body, which also allowed specification of multiple bodily locations across trials.[8] We found a clear bimodal distribution of judgements between the upper-face and upper-torso. Taken

---

[8] We adapted a version of a task developed by Howard and Templeton (1966), originally designed for locating the point of projection of binocular vision. The task required subjects to manually align a visually presented rod along the horizontal plane such that the near end pointed 'directly at himself'. We adapted this task, requiring subjects to align a rod along a sagittal plane, with individual trials split equally between two directions of rotation (upwards or downwards). See Alsmith and Longo (2014) for further details.

together with previous results, this suggests that when someone thinks about her body as herself, she does not necessarily identify with the body *simpliciter*, nor does she necessarily identify with a single part of her body exclusively. Rather, her thinking might be structured around particular body parts.

Identification with particular body parts might vary across judgements according to a range of contextual factors. For instance, in our study, the pointer that was used to pick out bodily locations rotated in the sagittal plane, moving from either an upwards starting direction or a downwards starting direction. Our participants were clearly affected by the starting direction, effectively resolving the choice between two likely locations (upper-face vs upper-torso) by which one came first. We suspect that other contextual factors might have affected the results of the previous studies described above. Indeed, we think that this serves to illustrate the generally thorny issue of determining precisely which contextual factors might influence an individual's self-ascriptions, and thus determining similarity or difference of context.

Moreover, we suspect that this problem is particularly acute in VR studies. For the use of VR technology might introduce a range of distinctive contextual factors due to the fact that it remains a predominantly visual medium. Here it is noteworthy that the vast majority of refinements of sensory displays in recent years has focussed on visual displays (Cummings and Bailenson 2016). In a recent study, we adapted our 2014 method to VR, by presenting a virtual pointer to participants using a head-mounted display (Van der Veer et al. forthcoming). We found a very strong preference for upper face responses, with the majority of participants (16 of 23) additionally reporting that they intended to point to their heads. These initial results suggest that simply wearing a head-mounted display might affect how individuals think about the relationship between their bodies and themselves. But further work is required, particularly incorporating avatars, and using a variety of display technologies. More generally: without parallel studies using comparable methods, the similarity or difference between VR and non-VR contexts of self-ascription remains unclear.

5. Conclusion

Contemporary experimental philosophical work on the self has principally been focused on essentialist notions of selfhood, and especially their relations to moral judgements (see e.g. Newman et al. 2015) and intentional action (see e.g. Sripada and Konrath 2011). In this chapter, we have offered a complementary approach, focusing not on what might remain constant in an individual's conception of the self, but rather on the degree to which it might involve properties attributable to various kinds of entities.

We have suggested that flexibility can be captured by placing pairs of self-ascriptions on two dimensions of higher-order difference/similarity. The first dimension concerns the degree of difference in the range of entities to which self-ascriptions are made, where greater difference yields greater flexibility. The second dimension concerns the degree of similarity between the contexts in which self-ascriptions are made, where greater similarity yields greater flexibility.

The methodological focus of the chapter has been on the use of VR technology as a means to investigate flexibility. Under certain conditions, this technology is sufficient to induce subjects into various forms of illusion in which they might categorize various kinds of virtual objects as themselves, thereby serving to demonstrate flexibility on the first dimension. However, we have highlighted two significant issues to be faced in determining the degree to which these self-ascriptions are flexible on the second dimension: the possible role of the imagination in users' experience of VR; and the perceptual differences resulting from the use of contemporary display technologies.

If self-conceptions are indeed flexible, the implications that this has for the nature of the self-concept, and thus the nature of the self, are surely significant, even if not entirely obvious. In any case, it would seem strange for a comprehensive theory of selfhood to ignore the range of entities encompassed by an individual's pattern of use of the first-person concept. For this would be a theory of selfhood that ignored how individuals conceive of themselves. However, the ease with which we may be prone to flit back and forth, happily equivocating on that which we conceive ourselves to be, ought to give pause on the extent to which any robust metaphysical theory of the self can avoid being revisionary.

Suggested Readings

Kilteni, K., Maselli, A., Kording, K. P., and Slater, M. (2015). Over my fake body: Body ownership illusions for studying the multisensory basis of own-body perception. *Frontiers in Human Neuroscience*, 9(141), 1–20.

Martin, M. G. F. (1997). Self–Observation. *European Journal of Philosophy*, 5(2), 119–140.

Slater, M., and Sanchez-Vives, M. V. (2016). Enhancing our lives with immersive virtual reality. *Frontiers in Robotics and AI*, 3(74), 1–47.

Starmans, C., and Bloom, P. (2012). Windows to the soul: Children and adults see the eyes as the location of the self. *Cognition, 123*(2), 313–318.

References

Alsmith, A. J. T. (2015). Mental activity & the sense of ownership. *Review of Philosophy and Psychology, 6*, 881–896.

Alsmith, A. J. T., and de Vignemont, F. (2012). Embodying the mind and representing the body. *Review of Philosophy and Psychology, 3*, 1–13.

Alsmith, A. J. T., and Longo, Matthew (2014). Where exactly am I? Self-location judgements distribute between head and torso. *Consciousness and Cognition, 24*, 70–74.

Armel, K. C., and Ramachandran, V. S. (2003). Projecting sensations to external objects: Evidence from skin conductance response. *Proceedings of the Royal Society B: Biological Sciences, 270*: 1499–1506.

Banakou, Domna, Groten, Raphaela, and Slater, Mel (2013). Illusory ownership of a virtual child body causes overestimation of object sizes and implicit attitude changes. *Proceedings of the National Academy of Sciences, 110*, 12846–12851.

Banakou, Domna, Hanumanthu, Parasuram D., and Slater, Mel (2016). Virtual embodiment of white people in a black virtual body leads to a sustained reduction in their implicit racial bias. *Frontiers in Human Neuroscience, 10*, 601.

Berger, Christopher C., Gonzalez-Franco, Mar, Tajadura-Jiménez, Ana, Florencio, Dinei, and Zhang, Zhengyou (2018). Generic HRTFs may be good enough in virtual reality. Improving source localization through cross-modal plasticity. *Frontiers in Neuroscience, 12*, 1–9.

Bergström, Ilias, Azevedo, Sérgio, Papiotis, Panos, Saldanha, Nuno, and Slater, Mel (2017). The plausibility of a string quartet performance in virtual reality. *IEEE Transactions on Visualization and Computer Graphics, 23*, 1352–1359.

Bertossa, F., Besa, M., Ferrari, R., and Ferri, F. (2008). Point zero: A phenomenological inquiry into the subjective physical location of consciousness. *Perceptual and Motor Skills, 107*, 323–335.

Blanke, Olaf, and Metzinger, Thomas (2009). Full-body illusions and minimal phenomenal selfhood. *Trends in Cognitive Sciences, 13*, 7–13.

Bloom, Paul (2003). *Descartes' Baby: How the Science of Child Development Explains What Makes Us Human*. New York: Basic Books.

Botvinick, M., and Cohen, J. (1998). Rubber hands 'feel' touch that eyes see. *Nature, 391*, 756.

Brewer, Bill (1995). Bodily awareness and the self. In N. Eilan, A. Marcel, and J. L. Bermúdez (eds.), *The Body and the Self*. Cambridge, MA: MIT Press.

Campbell, John (2004). The first person, embodiment, and the certainty that one exists. *The Monist, 87*, 475–488.

Chen, Wen-Yeo, Huang, Hsu-Chia, Lee, Yen-Tung, and Liang, Caleb (2018). Body ownership and the four-hand illusion. *Scientific Reports, 8,* 2153.

Clark, Stephen R. L. (2003). Non-Personal Minds. *Royal Institute of Philosophy Supplement, 53,* 185–209.

Cummings, James J., and Bailenson, Jeremy N. (2016). How immersive is enough? A meta-analysis of the effect of immersive technology on user presence. *Media Psychology, 19,* 272–309.

Currie, Gregory, and Ravenscroft, Ian (2002). *Recreative Minds: Imagination in Philosophy and Psychology.* Oxford: Oxford University Press.

de Vignemont, Frederique. (2011). Embodiment, ownership and disownership. *Consciousness and Cognition, 20,* 82–93.

de Vignemont, Frederique (2013). The mark of bodily ownership. *Analysis, 73,* 643–651.

Descartes, René (1642/1984). Meditations on first philosophy. In *The Philosophical Writings of Descartes.* Cambridge: Cambridge University Press.

Descartes, René (1643/2007). Descartes to Elizabeth: 28 June 1643, Egmond du Hoef. In Lisa Shapiro (ed.), *The Correspondence between Princess Elizabeth of Bohemia and Rene Descartes.* Chicago: University of Chicago Press.

Dixon, J. C. (1972). Do shifts in attention change perceived locus of self? *The Journal of Psychology, 80,* 103–109.

Dobricki, Martin, and de la Rosa, Stephan (2013). The structure of conscious bodily self-perception during full–body illusions. *PLoS ONE, 8,* e83840.

Ehrsson, H. Henrik (2007). The experimental induction of out-of-body experiences. *Science, 317,* 1048.

Ehrsson, H. Henrik, Wiech, Katja, Weiskopf, Nikolaus, Dolan, Raymond J., and Passingham, Richard E. (2007). Threatening a rubber hand that you feel is yours elicits a cortical anxiety response. *Proceedings of the National Academy of Sciences, 104,* 9828–9833.

Evans, Gareth (1982). *The Varieties of Reference.* Oxford: Oxford University Press.

Guterstam, Arvid, Björnsdotter, Malin, Bergouignan, Loretxu, Gentile, Giovanni, Li, Tie–Qiang, and Ehrsson, H. Henrik (2015). Decoding illusory self-location from activity in the human hippocampus. *Frontiers in Human Neuroscience, 9,* 412.

Heider, Fritz, and Simmel, Marianne (1944). An experimental study of apparent behavior. *The American Journal of Psychology, 57,* 243–259.

Howard, I. P., and Templeton, W. B. (1966). *Human Spatial Orientation.* London: John Wiley & Sons.

Ionta, Silvio, Heydrich, Lukas, Lenggenhager, Bigna, Mouthon, Michael, Fornari, Eleonora, Chapuis, Dominique, Gassert, Roger, and Blanke, Olaf (2011). Multisensory mechanisms in temporo–parietal cortex support self-location and first-person perspective. *Neuron, 70,* 363–374.

Kannape, O. A., Schwabe, L., Tadi, T., and Blanke, O. (2010). The limits of agency in walking humans. *Neuropsychologia, 48,* 1628–1636.

Keil, F. (1989). *Concepts, Kinds and Cognitive Development*. Cambridge, MA: MIT Press.

Kilteni, Konstantina, Groten, Raphaela, and Slater, Mel (2012). The sense of embodiment in virtual reality. *Presence: Teleoperators and Virtual Environments*, *21*, 373–387.

Kilteni, Konstantina, Maselli, Antonella, Kording, Konrad P., and Slater, Mel (2015). Over my fake body: Body ownership illusions for studying the multisensory basis of own-body perception. *Frontiers of Human Neuroscience*, *9*, 1–20.

Kokkinara, Elena, Kilteni, Konstantina, Blom, Kristopher J., and Slater, Mel (2016). First person perspective of seated participants over a walking virtual body leads to illusory agency over the walking. *Scientific Reports*, *6*, 28879.

Kong, Gaiqing, He, Kang, and Wei, Kunlin (2017). Sensorimotor experience in virtual reality enhances sense of agency associated with an avatar. *Consciousness and Cognition*, *52*, 115–124.

Lenggenhager, Bigna, Blanke, O., and Mouthon, M. (2009). Spatial aspects of bodily self–consciousness. *Consciousness and Cognition*, *18*, 110–117.

Lenggenhager, Bigna, Tadi, Tej, Metzinger, Thomas, and Blanke, Olaf (2007). Video ergo sum: Manipulating bodily self-consciousness. *Science*, *317*, 1096–1099.

Limanowski, Jakub, and Hecht, Heiko (2011). Where do we stand on locating the self? *Psychology*, *2*, 312.

List, Christian, and Pettit, Philip (2011). *Group Agency: The Possibility, Design, and Status of Corporate Agents*. Oxford: Oxford University Press.

Longo, Matthew R., Schüür, Friederike, Kammers, Marjolein, P. M., Tsakiris, Manos, and Haggard, Patrick (2009). Self awareness and the body image. *Acta Psychologica*, *132*, 166–172.

Longo, Matthew R., Schüür, Friederike, Kammers, Marjolein, P. M., Tsakiris, Manos, and Haggard, Patrick (2008). What is embodiment? A psychometric approach. *Cognition*, *107*, 978–998.

Lotze, Hermann (1888). *Microcosmus: An Essay concerning Man and His Relation to the World*. New York: Edinburgh, T. and T. Clark.

Martin, M. G. F. (1997). Self–observation. *European Journal of Philosophy*, *5*, 119–140.

Menzer, Fritz, Brooks, Anna, Halje, Pär, Faller, Christof, Vetterli, Martin, and Blanke, Olaf (2010). Feeling in control of your footsteps: Conscious gait monitoring and the auditory consequences of footsteps. *Cognitive Neuroscience*, *1*, 184–92.

Moseley, G. Lorimer, Olthof, Nick, Venema, Annemeike, Don, Sanneke, Wijers, Marijke, Gallace, Alberto, and Spence, Charles (2008). Psychologically induced cooling of a specific body part caused by the illusory ownership of an artificial counterpart. *Proceedings of the National Academy of Sciences*, *105*, 13169–13173.

Newman, George E, De Freitas, Julian, and Knobe, Joshua (2015). Beliefs about the true self explain asymmetries based on moral judgment. *Cognitive Science*, *39*, 96–125.

Newport, Roger, Wong, Dominic Y., Howard, Ellen M., and Silver, Eden (2016). The Anne Boleyn illusion is a six-fingered salute to sensory remapping. *i-Perception*, *7*: 2041669516669732.

Nichols, Shaun and Stich, Stephen. (2000). A cognitive theory of pretense. *Cognition*, *74*, 115–147.

Nichols, Shaun, and Stich, Stephen (2003). *Mindreading: An Integrated Account of Pretence, Self-awareness, and Understanding Other Minds*. Oxford: Oxford University Press.

Papineau, David (2002). *Thinking about Consciousness*. Oxford: Oxford University Press.

Peacocke, Christopher (2014). *The Mirror of the World: Subjects, Consciousness, and Self-consciousness*. New York: Oxford University Press.

Perry, John (1978). *A dialogue on personal identity and immortality*. Indianapolis. Hackett Publishing.

Petkova, Valeria I., Khoshnevis, Mehrnoush, and Ehrsson, H. Henrik (2011). The perspective matters! Multisensory integration in ego-centric reference frames determines full body ownership. *Frontiers in Psychology*, *2*, 1–7.

Piryankova, Ivelina V, Wong, Hong Yu, Linkenauger, Sally A, Stinson, Catherine, Longo, Matthew R., Bülthoff, Heinrich H., and Mohler, Betty J. (2014). Owning an overweight or underweight body: Distinguishing the physical, experienced and virtual body. *PLoS ONE*, *9*, e103428.

Putnam, Hilary (1975). The meaning of 'meaning'. *Minnesota Studies in the Philosophy of Science*, *7*, 131–193.

Rawls, J. (1971). *A Theory of Justice*. Cambridge, MA: Harvard University Press.

Shoemaker, Sydney S. (1968). Self-reference and Self-awareness. *The Journal of Philosophy*, *65*, 555–567.

Skarbez, R., Neyret, S., Brooks, F. P., Slater, M., and Whitton, M. C. (2017). A psychophysical experiment regarding components of the plausibility illusion. *IEEE Transactions on Visualization and Computer Graphics*, *23*, 1369–1378.

Slater, M., Perez-Marcos, Ehrsson, D., Henrik, H., and Sanchez-Vives, M. V. (2008). Towards a digital body: the virtual arm illusion. *Frontiers in Human Neuroscience*, *2*, 1–8.

Slater, Mel (2009). Place illusion and plausibility can lead to realistic behaviour in immersive virtual environments. *Philosophical Transactions of the Royal Society of London B: Biological Sciences*, *364*, 3549–3557.

Slater, Mel, and Sanchez–Vives, Maria V. (2016). Enhancing our lives with immersive virtual reality. *Frontiers in Robotics and AI*, *3*, 1–47.

Slater, Mel, Spanlang, Bernhard, Sanchez–Vives, Maria V., and Blanke, Olaf (2010). First person experience of body transfer in virtual reality. *PLoS ONE*, *5*, e10564.

Spanlang, Bernhard, Normand, Jean–Marie, Borland, David, Kilteni, Konstantina, Giannopoulos, Elias, Pomés, Ausiàs, González-Franco, Mar, Perez–Marcos, Daniel, Arroyo–Palacios, Jorge, Navarro Muncunill, Xavi and Slater, Mel (2014). How to build an embodiment lab: Achieving body representation illusions in virtual reality. *Frontiers in Robotics and AI*, *1*, 1–22.

Sripada, Chandra Sekhar, and Konrath, Sara (2011). Telling more than we can know about intentional action. *Mind & Language*, *26*, 353–380.

Starmans, Christina, and Bloom, Paul (2012). Windows to the soul: Children and adults see the eyes as the location of the self. *Cognition*, *123*, 313–318.

Steptoe, W., Steed, A., and Slater, M. (2013). Human tails: Ownership and control of extended humanoid avatars. *IEEE Transactions on Visualization and Computer Graphics*, *19*, 583–590.

Strawson, Peter F. (1959/2003). *Individuals: An Essay in Descriptive Metaphysics*. London: Routledge.

Tsakiris, Manos, Carpenter, Lewis, James, Dafydd, and Fotopoulou, Aikaterini (2010). Hands only illusion: Multisensory integration elicits sense of ownership for body parts but not for non–corporeal objects. *Experimental Brain Research*, *204*, 343–352.

Tsakiris, Manos, and Haggard, Patrick (2005). The rubber hand illusion revisited: Visuotactile integration and self-attribution. *Journal of Experimental Psychology: Human Perception and Performance*, *31*: 80–91.

Van der Veer, A., Alsmith, A., Longo, M., Wong, H.Y., and Mohler, B. (forthcoming). Where am I in virtual reality? *PLoS ONE*.

Walton, K. L. (1978). Fearing fictions. *The Journal of Philosophy*, *75*, 5–27.

Walton, K. L. (1990). *Mimesis and Make-believe*. Cambridge, MA: Harvard University Press.

Wilson, Margaret Dauler (1978). *Descartes*. London: Routledge.

6

Experimental Economics for Philosophers

Hannah Rubin, Cailin O'Connor and Justin Bruner

1. Introduction

Over the last 20 years or so, game theory and evolutionary game theory – mathematical frameworks from economics and biology designed to model and explain interactive behaviour – have proved fruitful tools for philosophers. Ethics, philosophy of language, philosophy of cognition and mind, social epistemology, philosophy of biology, social science and social and political philosophy, for example, all focus on questions related to human interaction, meaning that game theory and evolutionary game theory have been useful in illuminating problems of traditional interest in these fields.

This methodological osmosis is part of a larger trend where philosophers have blurred disciplinary lines in order to use the best epistemic tools available when tackling the questions that interest them. In this vein, experimental philosophers have drawn on practices from the social sciences, and especially from psychology, to expand philosophy's grasp on issues from morality to epistemology to consciousness.

In this chapter, we argue that the recent prevalence of formal work on human interaction in philosophy opens the door for new methods in experimental philosophy. In particular, we discuss methods from experimental economics, focusing on studies of strategic behaviour, to show how these methods can supplement, extend and deepen philosophical inquiry. This branch of experimentation emphasizes induced valuation – the idea that if we want to understand strategic behaviour in humans, we have to create a situation which mimics the strategic structure of the world. In other words, we have to allow people to make real choices that will impact actual outcomes that they value, as opposed to, say, reporting what choices they would make in such a scenario. The experimental framework also uses minimal framing, on the assumption that we are looking for general behavioural patterns. This contrasts with some commonly used methods in experimental philosophy that emphasize responses to specific cases and speculation on the counterfactual behaviour of the subject.

We will ground our discussion of these methods in a small literature we have been part of developing that uses experimental economics to investigate signalling, language and communication in humans. In particular, we will describe two studies we have recently completed. The first asks: under what conditions does common interest communication arise in small experimental groups? The second asks: can partially honest patterns of communication emerge between humans? We will also present a novel study on the emergence of communication in humans. We consider how the structure of the world that people encounter impacts the languages that emerge. In particular, we ask: do similarity structures ease the development of conventional terms, especially in complex worlds?

As we will argue, these studies are important complements to the theoretical work that inspired them. They lend credence to evolutionary game theoretic predictions, both in the specific cases, but also as a general tool for predicting human communicatory behaviour. In this way, they play a double epistemic role, telling us something about human behaviour as well as about our other methods for understanding human behaviour. In sum, we argue that these experimental methods have much to offer experimental philosophy, for extending and improving existing game-theoretic explorations in philosophy, but also for any inquiry into the nature of strategic interaction – cooperation, altruism, communication, social coordination, social learning, etc. – in humans.

In Section 2 we will describe the methods we import from experimental economics. In Section 3 we describe our past work using these methods and make clear how they facilitate fruitful work in experimental philosophy. Section 4 contains a novel experiment on the emergence of human communication, including background theoretical work and a detailed presentation of our experimental design and results. In Section 5 we conclude by addressing the philosophical upshots of the experiments presented here, and discussing, more generally, how economic methods can be fruitfully used to explore areas of philosophy involving strategic interactions.

2. Experimental economics

The methods we present here are derived from experimental economics, a discipline which dates back to the middle of the last century (e.g. see Allais 1953).

As in the other social sciences, experimentation in economics has allowed scholars to investigate human behaviour in a highly regulated environment that controls for confounding factors, and thus to test and update theoretical predictions in the social sciences. The large body of work that has emerged addresses a range of phenomena spanning almost all imaginable strategic behaviour in humans: cooperation, bargaining, altruism, coordination, price setting, norms, communication, market behaviour, etc. (Kagel and Roth 2016 give an overview).

For example, Guth et al. (1982), in a seminal paper, have subjects play what is called an ultimatum game. One subject is given a set amount of money (say, $10) and told that they may offer some of this to a second subject. The second subject has the option to accept what they are offered, or to reject it, in which case neither subject gets any money. The surprising result, which has since been widely replicated, is that subjects do not behave according to what the best models of rationality would predict. These models suggest that smart subjects should offer as little as possible, and that their partners should accept these offers since otherwise they get nothing. Instead, subjects make substantial offers and reject ones that are too small. This experiment thus supplemented economists' understanding of bargaining behaviour, challenged theoretical assumptions about human rationality and prompted further theoretical work explaining why humans would not behave rationally in an ultimatum game.

The most important cornerstone of this branch of experimentation is *induced valuation*, the practice of prompting subjects to make choices in strategic situations where their choices have real consequences that the subjects care about (Smith 1976). This is usually done by making the payments rendered to subjects dependent on their performance in a trial. In the Guth et al. (1982) experiment on ultimatum bargaining, for example, subjects actually received the amount of money they earned. If one subject offered $5, and the other subject accepted this offer, they both went home with $5. If a subject offered $2, and the other subject rejected it, they went home with nothing.

Why induced valuation? Economic theory says nothing about what subjects say they would do in some strategic scenario, it only makes predictions about what subjects would, in fact, do (Croson 2005). To test such predictions, then, subjects must be induced to make real economic choices. A study that merely asked subjects what they would offer in an ultimatum game, and what they would accept would not test the behaviour that economic theory is about. There are many areas of theoretical philosophy where, likewise, predictions address subject behaviour, rather than self-reports about predicted behaviour. While one might think these collapse, empirical evidence suggests that humans are often quite bad at accessing and reporting their own cognitive states and predicting their own actions (Wilson and Dunn 2004; Poon et al. 2014). In such cases, philosophers would do well do focus on experiments using methods of induced valuation.

Another key aspect of this methodology is that experiments tend to be largely context free. Experimenters often present subjects with just enough structure and information to capture the strategic situation they wish to induce. The goal is to abstract away from framing and structural features that might systematically bias the behaviour of subjects. Consider again the ultimatum bargaining study. One way to present this study is simply to tell the subjects what the strategic structure of the game is. Another way might be to concoct a story about partners with two skill sets necessary to complete a project, where one offers some portion of the proceeds to the other. This latter framing, though, might bias participants to give more since their cultural norms about fairness in the workplace now become relevant in determining their behaviour. Likewise, in the studies we will present here, we avoid describing the study as about language or communication. Since all participants are language users, this could influence their behaviour by, for example, prompting them to behave in communicatively helpful ways, rather than in their own best interests (Bruner et al. 2018). This is not to say that framing effects are unimportant, and, indeed, experimenters regularly add framing to their studies to see how this influences subject behaviour. For example, Leliveld et al. (2008) find that subjects act differently in the ultimatum game depending on which subject is described as 'owning' the money. See also Fagley and Miller (1997). The point is that these additions should be deliberate so that experimenters gain control as to how various additions to their paradigm impact subjects. Again, this practice is relevant to experimental philosophy, which often depends on highly specific vignettes or cases to test the intuitions of subjects.

One last standard practice in experimental economics is to avoid deceiving subjects. This is so that experimenters maintain control over subject expectations and motivations. If subjects have previously been deceived, or know that deception is possible in such experiments, they may not trust the experimental set-up they are presented with. For example, subjects may believe experimenters will secretly rig outcomes so as to pay out the least possible amount (Cooper 2014). In the ultimatum bargaining game, subjects making offers might believe, for example, that there is no second subject, or that experimenters are pretending that the second subject rejected the offer. The data gathered from such subjects will fail to track what experimenters are trying to test. For experimental philosophers who adopt the practice of induced valuation, adhering to the no-deception rule, and making sure subjects are aware of this, will help ensure that subjects are motivated by payoffs in the right ways.

The detailed examples we will present here are specifically within the realm of game-theoretic and evolutionary game-theoretic experimentation. Game theory is the study of *games* – simplified models of strategic interactions between humans. A game is specified by four elements: who interacts *(players)*, what they may do *(strategies)*, what players get for various combinations of strategies they might play *(payoffs)*, and what players know about the game *(information)*. In the ultimatum game the two players are the subject who makes the offer and the one who decides whether to accept or reject it. The strategies are different for the two players – possible monetary offers for the first, and the choice to either accept or reject given various offers for the second. The payoffs are the amounts of money they get for some combination of these strategies – $5 each for an offer of $5 and an acceptance, for example. And the information in this case is that they know who the players are, what the strategies are and what the payoffs are, that is, they have full knowledge of the game.

Classic game theory uses these models, plus assumptions about human rationality, to predict and explain strategic behaviour. These assumptions are, basically, that each player will try to maximize the amount of money they take home given what the other players are doing. Experiments are often useful in showing where such predictions do or do not hold, as with the ultimatum game.[1]

Evolutionary game theory, on the other hand, as applied to human culture, is the study of how humans learn and culturally evolve to deal with strategic scenarios as modelled by games. These models typically take a group of actors playing a game and add *dynamics* – rules for how their strategic behaviour will change over time as a result of learning and cultural evolution. For example, such a model might involve a group of humans all playing the ultimatum game again and again. The dynamics might assume that each individual has a

---

[1] For another example, work on the famous prisoner's dilemma game has consistently shown that humans have a preference for altruistic behaviour that does not accord with the predictions of rational choice (Sally 1995).

preferred strategy (maybe offer \$4, and accept any offer over \$3), but that each time they play there is some chance that they switch strategies and copy the most successful individual in their group. The model then sees what happens over time if everybody in the group learns via this rule. As such, evolutionary game theory makes predictions and provides explanations about how groups of humans will come to behave in strategic learning scenarios.

Predictions from these cultural evolutionary models are often different from those derived in classic game theory. For example, with respect to the ultimatum game, Skyrms (2014) shows how in an evolutionary model, people will sometimes learn to make high offers and reject ones that are too low, providing an explanation for laboratory behaviour that previous models did not. Such cultural evolutionary predictions can themselves be tested by having groups of individuals play a game repeatedly in the lab and seeing what behaviour emerges. In Sections 3 and 4 we will give several examples.

Wherever philosophy uses game theory and evolutionary game theory, and wherever it makes predictions, or offers explanations of, strategic human interactions including communication, coordination, altruism, cooperation, social dilemmas, social norms and resource distribution, the experimental methods we outline here can be of use. To date, there is a small literature in philosophy demonstrating just this point. In political philosophy experimental methods from economics have tested claims in the social contract tradition (Powell and Wilson 2008; Smith et al. 2012; Bruner 2018). These methods are particularly apt as they allow one to explore behaviour in the many hypothetical scenarios contract theorists have utilized to justify a variety of social arrangements.[2] In social philosophy, Devetag et al. (2013) as well as Guala (2013) have used economic methods to probe issues relating to conventions and common knowledge, while Bicchieri and Lev-On (2007), Bicchieri and Xiao (2009), and Bicchieri and Chavez (2013) have used economic experiments to help develop and defend Bicchieri's influential account of social norms. Within epistemology Koppl et al. (2008) and Jonsson et al. (2015) have designed experiments to explore the ways in which group structure makes for better or worse epistemic groups.

In addition, techniques from experimental economics have been used to reinforce previous findings from more traditional experimental philosophy.

---

[2] For more on the relationship between ethics and experimental economics, see Ernst (2007) and Güth and Kliemt (2017). Also, see Guala (2005) for a discussion on what philosophers of science can learn from experimental economics.

Utikal and Fischbacher (2014), for instance, identify a version of the side-effect effect in an economics-style experiment. And Gold et al. (2014) conducted a 'real-life' version of the trolley problem (involving financial losses) and found reactions in this economics experiment were similar to reactions to the more hypothetical cases common in the philosophical literature.

3. Previous results and theoretical grounding: The evolution of meaning

Recently, philosophers and social scientists have developed a huge empirical, experimental and theoretical literature on the evolution of communication and language.[3] This social-scientific exploration dovetails with more traditional philosophical work regarding the meaning of linguistic terms, as well as debates over the conventionality of meaning. Quine, for instance, famously argued that conventionalist accounts of meaning are circular, as conventions themselves appear to presuppose language and meaning. In response, David Lewis developed a game-theoretic account of convention that was then used to demonstrate how, sans explicit agreement, linguistic terms can acquire meaning (Lewis 1969). In particular, Lewis develops a novel communicative game, a *signalling game,* and argues that messages in this strategic scenario can acquire conventional meaning.

Lewis's signalling game has proved an extraordinarily useful framework for explorations of language and communication in philosophy. Brian Skyrms, for instance, has used signalling games to develop a notion of 'informational content,' a generalization of the more familiar propositional content (Skyrms 2010a). Signalling games have also been used by philosophers to work out accounts of deception appropriate for non-intentional organisms (Martinez 2015; Fallis and Lewis 2017). Moreover, signalling games have been used to frame a variety of issues in the philosophy of biology relating to the evolution of language and proto-language (Sterelny 2012a, b). Outside of the philosophy of language, signalling games have been used in social epistemology because they allow philosophers to better understand the conditions that allow for

---

[3] Biologists, too, have for some time pondered the evolution of language. John Maynard Smith and Eors Szathmary, for instance, went as far as to view the evolution of language as one of the major evolutionary transitions (Maynard Smith and Szathmary 1995).

the transfer of information among peers (Skyrms 2010b; Martinez and Godfrey-Smith 2016).

We now describe the communicative game introduced by Lewis. Lewis considers a simple strategic setting involving two players, a *sender* and a *receiver*. The sender is able to observe what *state* the world is in. These states might be, for example, that it is either sunny or raining. The sender can then select a *signal* (or *message*) from some available set to relay to the receiver. This set might be the words 'sunny' and 'raining', or the sounds 'bleh' and 'schmorg', or lighting either one or two lanterns in a belfry. Upon receipt of this signal, the receiver then picks an act to perform. It is assumed that certain acts performed by the receiver match particular states of the world and that both sender and receiver prefer that the receiver perform the act that best matches the underlying state of the world. In this example, the acts might be to get out either an umbrella or sunscreen. Since both actors would like the receiver to grab sunscreen if it is sunny, and an umbrella if it is raining, their interests in this example completely align.

Lewis considered the simplest possible version of this game – one with two possible states, two possible signals (or messages) and two possible actions. His key observation was that such models have two *signalling systems*. These are strategies where the sender always sends one signal in state 1 ('bleh' when it is raining, for example) and a different signal in state 2 ('schmorg' when it is sunny), and the receiver uses this regularity to coordinate action with the world perfectly (bring umbrellas in the rain and sunscreen in the sun). These systems are conventional, since either will do equally well as a communication system. In other words, it does not matter to the actors whether they use 'bleh' to mean rain or sun, as long as they coordinate. These systems are also stable in the sense that once actors have settled on one they will have no incentive to change their behaviour. In this way, a signalling system is what game theorists call an *equilibrium* – a social arrangement where none of the parties can switch behaviour and get a better payoff. Moreover, this particular equilibrium is optimal in the sense that it allows the players to get the highest payoffs possible, by always coordinating their behaviours with the state of the world. Figure 6.1 shows one of these signalling systems – where upon observing state one the sender sends message one, which induces act one. The other signalling system would match message 1 to state 2 and message 2 to state 1.

Brian Skyrms (1996, 2010) was one of the first to explore these common-interest signalling games in an evolutionary context. This program was in part motivated by the fact that while Lewis' game-theoretic approach could explain

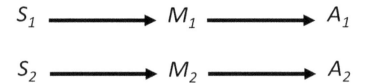

Figure 6.1 One of two signalling systems in David Lewis's signalling game with two states (S1, S2), two messages (M1, M2) and two acts (A1, A2).

the persistence of linguistic conventions, the account developed by Lewis did not provide a satisfactory explanation of the origin of these conventions. Using evolutionary game theory, on the other hand, Huttegger (2007) and Pawlowitsch (2008) show that in Lewis's version of the game, assuming the states are equally likely to occur, signalling systems are guaranteed to emerge under reasonable assumptions about how actors learn or evolve. In the words of Skyrms (1996: 93), in these simple evolutionary models, 'The emergence of meaning is a moral certainty'.

If the underlying signalling interaction is more complex – including, for example, more states of the world – it is possible for an evolutionary process to result in suboptimal outcomes where the sender sends the same signal for multiple states of the world. For instance, if the simple signalling game considered above is modified so that state 1 occurs most of the time (i.e. it is almost always rainy) the following arrangement is now stable: the sender sends one signal regardless of the state of the world and the receiver always performs the act appropriate for the more likely state (always brings umbrellas). This is an instance of what is called a *pooling equilibrium*. The sender's behaviour is the same across both states of the world, and as a result the receiver is unable to glean information regarding the underlying state of the world by attending to the signal. As a result, the receiver essentially ignores the behaviour of her counterpart and opts to take the act which is more likely to match the state of the world (see Figure 6.2).

When there are more states of the world, actors can also learn or evolve to send the same signal in several states of the world (but not all of them). These *partial pooling equilibria* emerge despite the fact that they, like pooling equilibria, are inefficient in that actors can do better by learning signalling systems. It may seem strange that these inefficient outcomes can culturally evolve. They are possible because despite their suboptimality, given what the other player is doing, no one can do any better by trying another strategy.

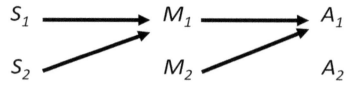

Figure 6.2 An example of a pooling equilibrium in a David Lewis signalling game.

Another complication emerges when the interests of sender and receiver diverge, such that they do not always prefer the same receiver action be performed in a given state. Imagine, for example that the sender is a job candidate, and the state is either that they are capable or lazy. Suppose the receiver is an employer trying to glean information about the candidate. If the candidate is capable, the employer would like to hire them, otherwise not. The candidate, on the other hand, would always like to be hired. In other words, their interests line up only if the state is that the candidate is capable.

In such cases, signalling systems can often emerge, but only when messages are costly to send and only if the cost of a message depends in part on the underlying state of the world or the type of sender (Spence 1973).[4] In the example just given, the costs of earning a college degree might be much lower for a capable candidate than a lazy one, so if the employer receives a 'signal' of a college transcript, they can deduce that the sender will be a capable employee.

Wagner (2013) showed that if this incentive structure is slightly modified to allow for slightly less costly messages, then a partially informative signalling system is possible (often referred to as the *hybrid equilibrium*).[5] In this case, capable candidates might always go to college, but lazy ones sometimes do too, because the costs are a bit lower. As a result, the employer is not able to identify, perfectly, the quality of the candidate upon receiving a signal, although they do considerably better than chance. As a result, they sometimes, but not always, hire the candidate upon seeing a college

---

[4] This is a bit of a simplification since there are other ways to ensure communication in partial-conflict of interest settings. For alternative ways of ensuring communication, see Crawford and Sobel (1982), Akerlof (1970), Viscusi (1978), Grossman (1981), Milgrom (1981) and Jovanovic (1982).

[5] See also Huttegger and Zollman (2010) as well as Zollman (2013). For other experiments investigating conflict of interest signalling games (but not the hybrid equilibrium) see Cai and Wang (2006), Dickhaut, McCabe and Mukherji (1995), Forsythe, Lundholm and Rietz (1999) and Blume, Dejong, Kim and Sprinkle (2001).

transcript. Together, these insights form the basis of what is often referred to as *costly signalling theory*, which has been employed throughout the social and biological sciences in order to explain a variety of initially puzzling signalling behaviours such as signalling by potential mates in the biological world, people on first dates, teenagers and their parents, students and their teachers, etc.

Much attention has been devoted to better understanding when and under what circumstances signalling systems will be likely to emerge in these games.[6] In Sections 3.1 and 3.2 we describe laboratory experiments designed to test predictions which originate from this rich theoretical literature. See Blume et al. (2017) for a survey of related experimental literature.

3.1. David Lewis in the lab

How likely is it that actual subjects will learn to assign meaning to initially meaningless symbols, successfully using these symbols to communicate with each other? And how does making the task more complex affect the likelihood of the emergence of meaning? Bruner et al. (2018) report the results of a laboratory experiment designed to explore these questions when the interests of sender and receiver coincide. Each run of the experiment proceeded as follows. A total of twelve subjects were recruited to the Experimental Social Science Lab at UC Irvine where they interacted anonymously via individual computer terminals. Six of these subjects were randomly assigned to be 'senders,' while the remaining six were 'receivers.' In order to avoid context effects, along the lines of the experimental methods described in Section 2, the labels 'sender' and 'receiver' were replaced by the neutral labels 'role 1' and 'role 2'.

The experiment consisted of 60 rounds. During each round, each sender was randomly matched with a receiver. The sender was then randomly shown one of two symbols (# and * for example), intended to represent the state of the world. Upon observing the state symbol, senders then selected one of two different signal symbols (@ and ^ for example) to relay to the receiver. These

---

[6]   See, for instance, Huttegger et al. (2010), Wagner (2009), Barrett and Zollman (2009), Skyrms (2012), and Brusse and Bruner (2017) for common interest signalling games and Wagner (2011, 2013, 2014), Zollman, Bergstrom and Huttegger (2013), Huttegger and Zollman (2010), Bruner, Brusse and Kalkman (2017), Bruner and Rubin (forthcoming), Bruner (2015), Huttegger, Bruner and Zollman (2015), Kane and Zollman (2016) and Martinez and Godfrey-Smith (2016) for work on conflict of interest signalling games.

random symbols were intended to prevent actors from using salience clues to choose which signals matched each state.[7] In addition, notice again the completely context-free set up. The receiver, upon observing only the signal, would then guess which state symbol the sender saw. At the end of each round both sender and receiver were told what symbol was initially presented to the sender as well as the receiver's guess. Subjects received $1 USD for each out of four randomly chosen rounds where receivers guessed correctly, as well as a show-up fee of $7. This randomization helps prevent wealth effects from influencing later rounds of experimentation (David and Holt 1993). Subjects were made aware of the payment structure and the structure of the signalling game they played at the beginning of the experiment. Since we were testing evolutionary predictions, we did break from standard economic practice by providing subjects with less information about population structure and play of their peers than is typical.[8]

Notice that this set-up embodied induced valuation in that actors were incentivized to signal in hopes of earning payoffs for coordination. It used minimal framing; presenting the signalling game without even using the language of signalling. And it avoided deception by making subjects aware of the strategic scenario they would face, and their potential payoffs.

In the game involving just two states of the world, two signals and two possible acts Bruner et al. (2018) find that, consistent with theoretical predictions, small groups tend to learn signalling systems when the states are equally likely. Subjects converged on a signalling system rather rapidly, usually taking less than 20 or 30 rounds (out of a total of 60). We also find that pooling behaviour becomes increasingly likely as one state becomes more probable than the other. This, again, is consistent with theoretical predictions. We also considered a signalling game involving three states, three signals and three possible receiver responses. In the laboratory setting subjects often developed behaviour that mimicked the expected partial-pooling outcomes, although observed play frequently resulted in a signalling system as well. In sum, the behaviours of lab subjects showed just how easy it is to develop common interest signalling in a lab group (extending results from Blume et al. 1998), and also that evolutionary game-theoretic

---

[7] Mehta et al. (1994) find that saliency can impact coordination behaviour in game-theoretic experiments, so we made every effort to reduce the saliency of any signal for any state.

[8] Blume et al. (1998), for example, in a similar experiment, gave all subjects a running history of play of all subjects (not just their interactive partners) in each round, which influenced the way subjects could learn. In order to more closely fit day-to-day learning environments, in Bruner et al. (2018) we did not provide such information.

predictions are, indeed, reflected in the behaviours of humans learning to signal in the lab.[9]

3.2. Communication without the cooperative principle

Rubin et al. (*manuscript*) use similar methods to investigate the emergence of communication when the interests of the sender and receiver are not perfectly aligned, that is, in situations like the employer-job candidate interaction described earlier. In particular, receivers tried to guess the 'type' of the sender (high or low). As described, senders always want the receiver to guess they are a 'high' type, while receivers want to guess correctly what type the sender is. One standard question here is whether different costs for high and low types can facilitate the evolution of honest signalling. Remember that theoretical work predicts that when these costs are small a hybrid equilibrium, with partially honest communication, will emerge. Rubin et al. (*manuscript*) test whether subjects learn this sort of partial communication transfer in the lab.

In this experiment, senders could choose to either pay a cost to send a signal or pay nothing and not send the signal. Senders were divided into high and low types, where the high type paid less to signal. To keep the language neutral, these types were referred to as blue and red, respectively. There were two treatments: a control treatment, where, as a result of signalling costs, the hybrid equilibrium did not exist, and an experimental treatment where it did. See Rubin et al. (*manuscript*) for details of the payoff structure and more details of the experimental set-up.

Rubin et al. (*manuscript*) found that the experimental results were largely consistent with theoretical predictions. This was done by first comparing the control treatment to the experimental treatment, to see whether there was less information transfer in the experimental treatment as expected. Then, we checked whether there was still some information transfer of the type expected in the hybrid equilibrium: in particular, that the signal increased the likelihood that the sender was a high type. Again, these results confirmed the success of evolutionary game-theoretic predictions as applied to human communication.

---

[9] Of course, learning to communicate with only two or three signals is just a simple starting place. To get anywhere close to explaining human communication, we need to also talk about things like learning new signals (see e.g. Alexander, Skyrms, and Zabell (2012)) and syntax (see e.g. Barrett (2009)).

In both of the studies just described, of course, subjects learn to communicate in situations that are quite removed from the everyday communication of humans. Subjects interact in small groups, over computers, with highly restrictive rules of interaction and artificial incentives. One might wonder whether such studies can, indeed, tell us much about the emergence of real language. Notice, first, that the restrictive lab environment allows for a kind of conclusion that is not possible in more ecologically valid explorations. In particular, we are able to assess much more accurately whether humans learn the kind of behaviour that game-theoretic and evolutionary game-theoretic models predict, and thus validate these models. As for the real world, these experiments can still tell us plenty despite their artificial features. In the first case, contra Quine, conventional meaning can emerge, with no meta-level discussion, very easily. In the second case, as we argue in Rubin et al. (*manuscript*), human communication is often assumed to be fully cooperative, but we show that communication can still occur despite persistent conflicts of interest.

4. Sim-max games and the evolution of categories

The new experiment we present here looks at the emergence of communication in a variation of the Lewis signalling game called the 'sim-max' (similarity maximizing) game. This model was introduced by Jäger (2007) and has been used by him and others to study the evolution of categories, both linguistic (Jäger 2011) and cognitive (Jäger 2007; O'Connor 2014b), the evolution of vague terms (Franke et al. 2010; O'Connor 2014a; Franke & Correia 2016), the emergence of linguistic ambiguity (O'Connor 2015a) and natural kind terms (O'Connor, forthcoming).[10]

The sim-max model adds structure to the basic signalling game by assuming similarity relationships between states of the world. In particular, states are arrayed in a space where distance in the space tracks similarity. For instance, Figure 6.3 shows two possible state spaces for such a game, with five states on a line and nine in a plane. In Figure 6.3a, state 1 is more similar to state 2 than it is to state 4, say, because they are closer in the state-space. This similarity is instantiated in the payoffs of the game. It is assumed that each state has some ideal action that will yield the highest payoff. Actions are also successful,

---

[10] This exploration connects to previous work in economics both on common interest communication and on vagueness and ambiguity. See Blume et al. (1998, 2001, 2017) and Lipman (2009).

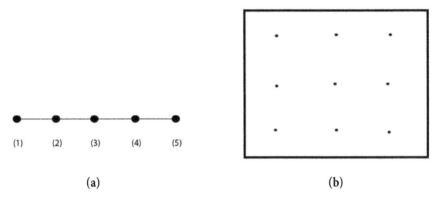

Figure 6.3 Two state spaces of sim-max games with (a) five states arrayed on a line and (b) nine states arrayed in a plane.

though less so, in states similar to the ideal one. Suppose that the five states in 3a represent five levels of rain from completely sunny to a downpour. In state 5, the downpour, the ideal action might be to wear galoshes and a raincoat. In state 4, the heavy rain, this would also be a perfectly fine action because the states are similar, even though it might be strictly better to bring an umbrella rather than a heavy coat. In particular, these games usually assume that the payoff is strictly decreasing as actors take actions that are less appropriate to the state of the world.

Play of the game is otherwise just like the Lewis signalling game. A sender observes the state of the world and sends a message. The receiver gets the message and chooses an action conditional on it. There is complete common interest between the actors, meaning that they always get the same payoff.[11] One typical assumption in these games is that the sender and receiver have access to fewer messages than there are states of the world. This means that they must use the same message for multiple states, that is, develop communicative categories. If we take the states to be different levels of rain, and the actions to be appropriate responses, this would correspond to a situation where actors use just the terms 'rainy' and 'sunny' to communicate about a world that is in fact much more complex. This extends the Lewis model, where states are pre-specified, by inducing a strategic situation where agents must develop their own sets of states to which terms can be applied.

---

[11] If interests diverge, the game is more like the well-studied Crawford-Sobel signalling game in economics (Crawford and Sobel 1982).

Previous results have shown that the categories we should expect to evolve in these games are (more or less) the optimal categories. Jäger et al. (2011) call these categories *Voronoi languages*, after Voronoi tessellations in mathematics. An optimal categorization will minimize, on average, the distance between the state of the world and the act taken, since this maximizes payoff to the actors. To do this, senders should use categories that are about equally sized, and receivers should respond with an action appropriate to a central, prototypical state in the category. Figure 6.4 shows examples of Voronoi languages for two state spaces: (a) a line and (b) a plane. Cells represent categories and open dots represent the action taken by the receiver in response to each category (not states as in Figure 6.3). Returning to the rain example, an optimal strategy of this sort might be to use the term 'rainy' for any state from a downpour to a mild drizzle and the term 'sunny' for the rest of the states. Such a communicative strategy allows actors to achieve a high level of successful action, without having to develop a term for every possible state.

Jäger (2007) argues that these are the types of languages we should expect to evolve, though Elliott Wagner (personal correspondence) has shown that this may not always be the case and O'Connor (2014a, 2017) has found that under some learning rules categories that are similar to, but do not exactly correspond to Voronoi languages emerge. In particular, the categories may not exactly equally divide the state space, but nearly do. The slightly 'handwavy' take-away is that in general we should expect actors in these games to develop categories

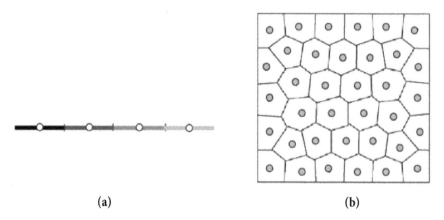

(a)   (b)

Figure 6.4 Two Voronoi languages. Cells represent categories and open dots the receiver response to a category. (a) shows four categories in a linear state space and (b) 37 categories in a plane.

that look more or less like the ones in Figure 6.4, that is, to use terms to represent sets of similar states.[12]

In addition, previous work has shown that actors in sim-max games have an advantage when it comes to learning to signal, in that they can generalize lessons they learn over multiple states. Consider, for example, a standard signalling game (with no similarity structure) with 100 states. Actors must develop conventions for how to signal in every state anew, which is difficult and time consuming. In sim-max games, on the other hand, actors who learn a lesson in state 5 (bring raincoats) can apply that lesson to similar states that they have never encountered, speeding learning and improving payoffs (O'Connor 2014a; O'Connor 2015b). In addition, they may be able to avoid suboptimal equilibria through this sort of generalization (Franke and Correia 2016). This may help explain how real actors manage to signal successfully about so many real-world states: the structure of the world helps, by providing natural similarity classes over which to generalize learned linguistic conventions.

In the following, we use experimental work to ask: do actors playing real sim-max games develop convention categories that facilitate information transfer? Are these categories optimal or near-optimal? And do they generalize learning so as to improve their communicative success and speed of learning?

4.1. Experimental design

The subjects consisted of undergraduate and graduate students from the University of California, Irvine recruited from the Experimental Social Science Laboratory subject pool. The experiment was programmed and conducted with the software z-Tree (Fischbacher 2007). As in Bruner et al. (2018) and Rubin et al. (*manuscript*), subjects interacted over 60 rounds. However, while subjects in Bruner et al. (2018) and Rubin et al. (*manuscript*) were matched randomly within a group of 12 every round, subjects in this experiment interacted with the same partner throughout. This made it easier for subjects to learn signalling behaviour quickly, allowing us to use data from earlier rounds of the experiment.[13] This early data was crucial since we looked at games with

---

[12] O'Connor (2015a) gives a much more detailed overview of evolutionary predictions in these games and the work of Jäger (2007) and Jäger et al. (2011).
[13] We use data from rounds 20–60, compared with Bruner et al. (2018) and Rubin et al. (*manuscript*) who used data from rounds 50–60 to test convergence to equilibria.

many states, and thus needed a large number of data points to detect signalling patterns over these states.

At the start of each session, experimental subjects were asked to sit at a randomly assigned computer terminal. As in Bruner et al. (2018) and Rubin et al. (*manuscript*), subjects were given information about the strategic situation in a manner that was as context-free as possible. After every round, subjects were shown the state of the world, the signal sent, the receiver's action and their own payoff for that round of the experiment. Each run of the experiment consisted of two treatments (so as to gather more data points given limits on time and resources).[14]

There were eight different treatments, summarized in Table 6.1.[15] For each treatment, the first number is the number of states, and the second is the number of available signals. For the games we consider, the number of actions is the same as the number of states. First, in order to investigate how structured state spaces might influence signalling behaviour, we tested some standard Lewis signalling games, that is, without structure to the state space, for comparison. The 2×2 and 3×3 (read 'two-by-two' and 'three-by-three') treatments involved the same Lewis-style signalling games explored by Bruner et al. (2018), except that subjects were now interacting in pairs rather than groups, and successful coordination was rewarded with 100 points (to be translated into money at the end of the trial). In the 100×2 and 100×3 treatments, senders were shown a

Table 6.1 Summary of the different treatments.

	2 signals	3 signals
Unstructured	2×2	3×3
	100×2	100×3
Structured	100×2 structured numbers	100×3 structured numbers
	100×2 structured colours	100×3 structured colours

---

[14] Because we had to reuse signals between the different treatments, we paired treatments that did not have an overlap in any of the signals to prevent using a signal in one treatment that had already gained meaning in a previous treatment. In each particular run, participants were given the same treatments in the same order. The order of these pairs was reversed across runs, for example, one run had the 2×2 treatment followed by the 100×3 structured numbers treatment while another had the 100×3 structured numbers treatment followed by the 2×2 treatment.

[15] In the 2×2 treatment there were 8 subjects (meaning there were 4 interacting pairs), the 3×3 had 10 subjects, the 100×2 had 12, the 100×3 had 10, the 100×2 structured numbers had 16, the 100×3 structured numbers had 30, the 100×2 structured colours had 22, and the 100×3 structured colours had 16.

number from 1 to 100 (representing the state) and had only two or three signals available to communicate the state of the world to the receiver. Upon receipt of a signal, receivers had to guess the state of the world by typing in a number from 1 to 100. If the receiver guessed the correct state of the world, both subjects received 100 points, otherwise they received zero points. While this game sets subjects up for failure, it provides an important comparison to similar sim-max games, as we will see.

The rest of the treatments involved subjects playing sim-max games. The 100 × 2 and 100 × 3 structured numbers treatments were the same as the 100 × 2 and 100 × 3 treatments – senders encounter states 1–100, and have either two or three signals available to coordinate action. But the payoffs were such that close guesses still paid off. In particular, subjects lost two points for each number away from the actual state that their guess was. For example, if the actual state was 20 and the receiver guessed 33, both sender and receiver would receive 74 points that round $(100 - |20–33| \times 2 = 74)$.

The 100 × 2 and 100 × 3 structured colours treatments were also sim-max games, formally identical to the numbers treatments, but where the state stimuli were colours instead of numbers. We displayed a line that faded from very light to very dark green and subjects were told that it was divided into 100 evenly sized parts. The senders were then randomly presented each round with a state, in the form of this colour line with an arrow pointing to one spot, as shown in Figure 6.5. They then chose one of their available signals. Upon receipt of the signal, the receiver was presented with the same colour line, and clicked on state they thought had occurred. Again, they received 100 points for guessing exactly right and lost two points for each unit away from the real state. The goal with these treatments was to test whether a different presentation of the state space would influence communicative behaviour.

As in Bruner et al. (2018) and Rubin et al. (*manuscript*), the signals available to senders were chosen so as to minimize the chance of subjects importing any pre-established meaning (e.g. we did not use the '>' sign as a possible signal for the structured numbers treatments, as subjects might already associate this with

Figure 6.5 Sample of the state chosen in the 100 × 3 structured colours treatment.

larger numbers).[16] For the 2 × 2 and 3 × 3 treatments, we also chose meaningless symbols to represent the states of the world.[17] These symbols were presented in a random order each round to as to prevent ordering from allowing subjects to coordinate.

Subjects received a show-up fee of $7 for participating in the experiment. In addition, subjects were paid for 10 rounds of the experiment: 5 rounds were randomly selected from each treatment for payment, excluding the first 10 rounds of each treatment to allow time for learning. Each subject's score, in terms of experimental points, for these 10 rounds was summed, and subjects were paid $1 for every 100 points they earned (rounded up to the nearest dollar). Subjects were made aware of this payment scheme, and which rounds could be chosen for payment, in the instructions.[18]

4.2. Results

In what follows, we collapse data for the structured colours and numbers treatments and talk just about 'structured' treatments, where subjects played sim-max games.[19] First, we compare the structured treatments to the unstructured treatments, to see whether adding structure to the state space improves subject learning. Then, we look within the structured treatments in order to see whether the subjects could be said to use categories, and how close they were to equilibrium predictions in sim-max models.

4.2.1. Comparison to unstructured treatments

Does the fact that the world has underlying structure make it easier for people to communicate? Below, we test whether adding structure to the state space can

---

[16] The available signals were as follows: ∧ and + for the 2 × 2 treatment; %, * and # for the 3 × 3 treatment; " and \ for the 100 × 2 treatment; ", \ and : for the 100 × 3 treatment; ` and ~ for the 100 × 2 structured treatments; and ?, / and [] for the 100 × 3 structured treatments.

[17] These were as follows: $ and @ for the 2 × 2 treatment and !, @ and > for the 3 × 3 treatment.

[18] Subjects participating in the 100 × 2 and 100 × 3 treatments received a 'bonus' payment of $3. It is not a common practice in experimental economics to hand out bonus payments that are independent of subjects' performance, but this ensured that subjects participating in these treatments were paid fairly for their time compared with other subjects. They were not told about this bonus payment until after the experiment was completed, so it will not have affected their behaviour in the experiment.

[19] These treatments were collapsed because the experiment was designed to test the effect of adding structure to the state-space. Behaviour for the structured numbers and structured-colours treatments was qualitatively similar and there were few significant differences between the two types of treatments. In the three signal treatments, subject's success rate was significantly higher for the structured colours ($p=.0498$), likely because there was significantly less variation in receivers' guesses ($p=.02$, see Section 4.2.2 for a discussion). We will note places where this might influence our analysis and show that it does not affect the conclusions we draw.

improve subjects' communicative success. Based on O'Connor (2014a), we expect that success of subjects in the structured treatments will not be significantly different from the 2×2 or 3×3 treatments but will be significantly different from 100×2 and 100×3 treatments. The idea is that receiving payoffs from being approximately correct can help subjects to reinforce categorization strategies and learn optimal signalling behaviour. We follow O'Connor (2014a) in using the following measure of success to compare how well learners are signalling across games where the base success rate is different:

$$\text{Success rate} = \frac{\text{average payoff in experiment}}{\text{expected payoff at equilibrium}}$$

Average payoff in the experiment was calculated using the points subjects earned in each round of the experiment.

Comparing the 2×2 with the 100×2 structured treatments, and the 3×3 with the 100×3 structured treatments, we used this measure and asked whether there was significantly less success in the structured treatments than in the standard Lewis signalling games. A one-tailed t-test revealed no significant difference in success rate ($p=0.32$ and $p=0.50$, respectively). As shown in Figure 6.6, subjects reached their highest level of success very quickly, and there was little qualitative difference between the compared treatments.

Comparing the success rates of the 100×2 and 100×3 treatments to the structured state-space treatments is less straightforward. This is because the success rates in the 100×2 and 100×3 treatments could vary wildly if a receiver

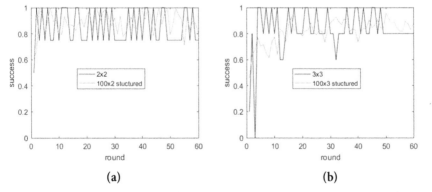

(a)   (b)

Figure 6.6 Success rates over time for (a) the 2×2 versus the 100×2 structured treatments and (b) the 3×3 versus the 100×3 structured treatments. Success rates are averaged over all subjects.

managed to guess the correct state a few times. That is, if the signal meant 'states 1–33' someone could by chance manage to guess exactly right twice in ten rounds, giving them a much higher than expected payoff. Again, using a one-tailed *t*-test, we found that for the two-signal treatments, there was significantly less success for the unstructured versus the structured treatments for rounds 20–60 ($p = .03$). This confirmed our prediction that adding structure would improve learning for the subjects. However, for the three-signal treatments, if we look at rounds 20–60, there is no significant difference ($p = .41$), but if we look at rounds 30–60, then there is ($p = .001$).[20] This was likely the result of a few chance correct guesses that happened in rounds 20–29 of the unstructured treatments.

4.2.2. Categorization

When people have a limited number of signals to describe many possible states of the world, do they use categories in their communication? Do those categories divide up the world such that communication is as effective as possible? Jäger et al. (2011) and O'Connor (2014a) predict that subjects will learn to use (approximate) Voronoi languages, where senders use categories that are about equally sized, and receivers respond with an action appropriate to a central, prototypical state in the category. We test this prediction in three parts below. First, we analyse sender behaviour, then receiver behaviour. Finally, we check that there was information transferred using the signals. For this analysis, we use data from rounds 20 to 60, because, as Figure 6.6 shows, subjects had reached their maximum success rate by round 20 for both the two- and three-signal treatments. In other words, they had learned stable strategies by this point.

Are the categories approximately equally sized? Is each signal sent for approximately $1/n$ of the state-space (where n is the number of signals available)? This would involve, for example, categorizing states 1–50 and 51–100 in the treatments with two signals, and categorizing 1–33, 34–66, and 67–100 in the treatments with three signals. As Figure 6.7 shows, we observe a qualitative match with the prediction. To test this more rigorously, we used the following procedure: take the grouping implied by the senders' strategies (e.g. generally sending signal one for low states and signal two for high, etc.) to get an idea of what they take each signal to mean. Then recode so that signal one corresponds to the signal most frequently sent in the first $1/n$ of the state-space, etc. Then,

---

[20] Since subject's success rate was significantly higher for the 100 × 3 structured colours (versus numbers) treatment, we also tested whether those had a significantly higher success rate than in the 100 × 3 treatment and found that they did not ($p = .36$).

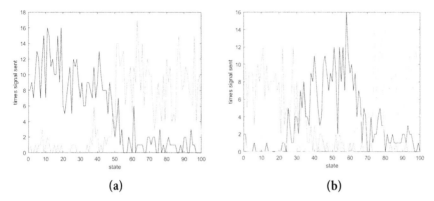

(a)   (b)

Figure 6.7 Senders' signalling behaviour, averaged over all subjects, for (a) the 2 × 2 versus the 100 × 2 structured treatments, and (b) the 3 × 3 versus the 100 × 3 structured treatments.

assume they have divided the state-space into categories of size $1/n$ and look at proportion of time subjects sent the right signal in each category. Finally, check whether this is significantly less than 100% (as would be expected).

Using a one-tailed t-test, we find that subjects' strategies are significantly different from equilibrium strategies of dividing the state space into categories of size $1/n$ ($p = .015$ when there are two signals and $p < .01$ when there are three signals). However, this is mostly due to subjects dividing the state space into categories of non-optimal size, rather than improperly signalling within the categories they have divided the state space into. In Figure 6.7a, for instance, if you look at when the signal meaning 'low' states is sent versus the one meaning 'high' states you can see that most 'mistakes' occur close to the boundary. This is mostly because different subjects drew the boundary between 'high' and 'low' at different places: for one subject high states might be from 45–100, for another 60–100. In fact, learning and evolutionary models often predict some conventionality as to where boundaries between categories are drawn, which accords with the behaviour just described. This is, in part, because languages that are 'close' to Voronoi languages in that they draw the boundaries between categories near to the optimal spot are also very successful (O'Connor, forthcoming). If we look at sender behaviour away from the boundary between categories in the two-signal case, the difference from 100% is no longer significant ($p = .067$), lending credence to the argument that conventionality of boundary position is causing deviation from expected behaviour.[21]

---

[21] We did this by ignoring the middle states (from 36 to 64) and looking at behaviour in roughly the top and bottom thirds of the state space.

Do the receivers take an action appropriate for a central, prototypical state in the category? That is, are the receiver's guesses in the middle of the $1/n$ sized categories? As a first check on whether this was the case, we measured the distance from the equilibrium strategy, assuming the sender uses categories of size $1/n$. The receivers' guesses are significantly off from the equilibrium predictions ($p <<.001$ in both cases).[22] After some initial learning, receivers' guesses are on average about six units (either in number or units of the colour spectrum) away from the equilibrium prediction for both the two- and three-signal treatments. This is because there was high variance in receivers' guesses. So, for instance, if the equilibrium prediction was to guess 25, a receiver may have guessed 19 in one round, then 31 in the next, etc.

We can see the various receiver strategies in Figure 6.8, which shows what each receiver guessed after receiving each signal, averaged over rounds 20–40. As we can see, most subjects' strategies were on average close to the equilibrium prediction, but, as mentioned, their actual guesses tended to vary quite a lot.

Figure 6.9 shows some of the receiver strategies from the 100×2 structured numbers treatment. Each of the three receivers guessed, on average, close to the equilibrium prediction, though only the receiver strategy in 6.9c represents an equilibrium strategy. More common were strategies like that shown in 6.9b, where receivers centred around the equilibrium strategy but often made guesses which were somewhat higher or lower. Receivers also occasionally employed a strategy where they divided the state-space into approximately $1/n$ sized categories, then

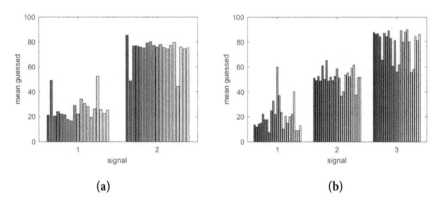

(a)   (b)

Figure 6.8 Receivers' mean guesses for (a) the 100×2 structured treatments and (b) the 100×3 structured treatments. Each bar represents one subject's guesses.

[22] Though in the 3 signal structured colours treatment there was less variance in receiver guesses, these guesses were still significantly off from the equilibrium prediction ($p<<.001$).

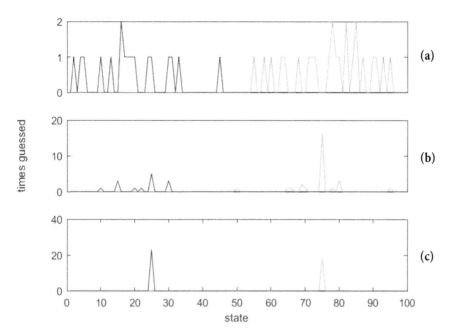

Figure 6.9 Sample receiver behaviour from the 100 × 2 structured numbers treatment.

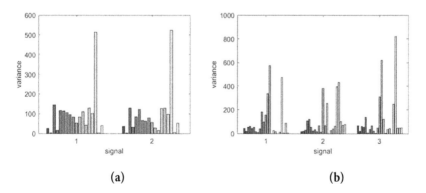

(a)   (b)

Figure 6.10 Variance in receivers' guesses for (a) the 100 × 2 structured treatments and (b) the 100 × 3 structured treatments. Each bar represents the variance in one subject's guesses.

guessed random states within each category, as shown in 6.4a. A summary of the variance in receiver strategies is shown in Figure 6.10.

The variance in receiver guesses can help explain the ambiguous evidence found in Section 4.2.1 for the fact that structure can aid in learning categories:

even though subjects in the structured treatments learn to categorize, their success rate is lower than expected because they do not optimally respond to learning that the state of the world is in a certain category.

The variance in receiver guesses might be explained by a phenomenon similar to probability matching, which is a well-known phenomenon in experimental economics. When subjects employ a probability matching strategy, the frequency of their predictions of a state of the world matches the state's probability of occurring. For instance, if there are two states of the world and state one occurs 70% of the time, then seven out of ten times the subject is asked to predict the state of the world they will guess state one and the other three times they will guess state two. This happens despite the fact that the optimal strategy is to guess the more likely state every time. One explanation of this phenomenon is that subjects try to look for patterns, even when there are none, and predict the next state based on these patterns (Vulkan 2000). For instance, a subject may guess state two when they think it is 'due' to come up. Our subjects may have employed similar reasoning, making their guesses based on an anticipation of a particular state (within the range of states associated with a particular signal) fitting some pattern, rather than based on utility maximization, despite being told that states of the world were randomly determined by the computer.

Was there information transfer? Did the signals contain information about the state of the world? In order to test whether this was the case, we compared subject behaviour to a null hypothesis of no information transfer. More specifically, we compared the average payoff of subjects in the experiment to the expected payoff subjects would receive if they were to ignore the signals, or if there were no information about the state of the world contained in the signal. If there was no information transfer, the best possible strategy would be to guess in the centre of the state space (e.g. either 50 or 51 for the structured numbers treatments). This minimizes the expected difference between one's guess and the state of the world, and it yields an expected payoff of 50.[23] Using a one-tailed *t*-test, we found that subjects in the structured state space treatments did utilize the information content present in the signals to perform significantly better than if there were no information transfer ($p < 0.05$ in both the two-signal and three-signal cases).

---

[23] Subjects employing this strategy would be on average 25 units away, meaning their average payoff would be $100 - 25 \times 2 = 50$.

5. Conclusion: Experimental economics and philosophy

The experiments described in this paper have implications for traditional questions in philosophy. Simple lab experiments completely invalidate theoretical arguments that symbolic meaning cannot emerge naturally (Quine 1936) – on the contrary, it can emerge in a matter of minutes. Furthermore, these experiments help validate the evolutionary models that philosophers have used to argue this point previously. Experiments on partial conflict of interest signalling show how groups of individuals can develop partially communicative conventions. This is relevant to claims in philosophy of language, especially in pragmatics, that human language is usually maximally helpful and informative (Grice 1991). Game-theoretic models of sim-max games suggest the emergence of a categorization scheme that allows for informative communication. In line with the theoretical predictions, we show that near-optimal categories are in fact utilized by subjects. These findings have implications for understanding prototypically vague terms, which look much like those that emerge in the experiments here. Despite the imprecision that has worried logicians and philosophers, these terms do a good job of helping humans communicate when the world is complex (O'Connor 2014; Franke and Correia 2016).

Economics experiments of this kind have broad applicability across philosophy. Wherever philosophers examine the strategic interactions of people – be these cooperative, communicative, antagonistic, political, etc. – economics style experiments, and game-theoretic experiments in particular can be illuminating. In particular, whenever experimental philosophers wish to learn about the actual behaviour of subjects, rather than self-report, these methods can come into play.

Suggested readings

Blume, A., Lai, E. K., and Lim, W. (2017). *Strategic Information Transmission: A Survey of Experiments and Theoretical Foundations.* Working paper. http://wooyoung. people.ust.hk/Info-Transmission-Survey-04-30-2017.pdf

Croson, R. (2005). The method of experimental economics. *International Negotiation,* *10*(1), 131–148.

Davis, D. D., and Holt, C. A. (1993). *Experimental Economics.* Princeton NJ: Princeton University Press.

Friedman, D., and Sunder, S. (1994). *Experimental Methods: A Primer for Economists.* Cambridge: Cambridge University Press.

Smith, V. L. (1976). Experimental economics: Induced value theory. *The American Economic Review, 66*(2), 274–279.

References

Akerlof, G. (1970). The market for 'lemons': Quality uncertainty and the market mechanism. *The Quarterly Journal of Economics, 84*(3), 488–500.

Allais, P. M. (1953). Le comportement de l'homme rationnel devant le risque: Critique des postulats etaxiomes de l'ecole americane. *Econometrica, 21*, 503–546.

Alexander, J. M., Skyrms, B., and Zabell, S. L. (2012). Inventing new signals. *Dynamic Games and Applications, 2*(1), 129–145.

Barrett, J. A. (2009). The evolution of coding in signalling games. *Theory and Decision, 67*(2), 223–237.

Barrett, J. and Zollman, K. (2009). The role of forgetting in the evolution and learning of language. *Journal of Experimental and Theoretical Artificial Intelligence, 21*(4), 292–309.

Bicchieri, C., and Chavez, A. K. (2013). Norm manipulation, norm evasion: Experimental evidence. *Economics and Philosophy, 29*(2), 175–198.

Bicchieri, C., and Lev-On, A. (2007). Computer-mediated communication and cooperation in social dilemmas: An experimental analysis. *Politics, Philosophy and Economics, 6*(2), 139–168.

Bicchieri, C., and Xiao, E. (2009). Do the right thing: But only if others do so. *Journal of Behavioral Decision Making, 22*(2), 191–208.

Blume, A., DeJong, D. V., Kim, Y. G., and Sprinkle, G. B. (1998). Experimental evidence on the evolution of meaning of messages in sender–receiver games. *The American Economic Review, 88*(5), 1323–1340.

Blume, A., DeJong, D., Kim, Y., and Sprinkle, G. (2001). Evolution of communication with partial common interest. *Games and Economic Behavior, 37*, 79–120.

Blume, A., Lai, E. K., and Lim, W. (2017). *Strategic Information Transmission: A Survey of Experiments and Theoretical Foundations.* Working paper.

Bruner, J. (2018). 'Decision making behind the veil: An experimental approach'. In T. Lombrozo, S. Nichols and J. Knobe (eds.), *Oxford Studies in Experimental Philosophy.* Oxford: Oxford University Press.

Bruner, J., O'Connor, C., Rubin, H., and Huttegger, S. (2018). David Lewis in the Lab. *Synthese, 195*(2), 603–621.

Bruner, J., Brusse, C., and Kalkman, D. (2017). Cost, expenditure and vulnerability. *Biology and Philosophy, 32*(3), 357–375.

Bruner, J., and Rubin, H. (forthcoming). Inclusive fitness and the problem of honest communication. *British Journal for the Philosophy of Science.*

Bruner, J. (2015). Disclosure and information transfer in signalling games. *Philosophy of Science, 82*(4), 649–666.

Brusse, C, and Bruner, J. (2017). Responsiveness and robustness in David Lewis signalling games. *Philosophy of Science, 84*(5), 1068–1079.

Cai, H., and Wang, J. (2006). Over–communication in strategic information transmission games. *Games and Economic Behavior, 56*, 7–36.

Cooper, D. J. (2014). A Note on Deception in Economic Experiments. *Journal of Wine Economics, 9*(2), 111.

Crawford, V. P., and Sobel, J. (1982). Strategic information transmission. *Econometrica: Journal of the Econometric Society, 50*(6), 1431–1451.

Croson, R. (2005). The method of experimental economics. *International Negotiation, 10*(1), 131–148.

Davis, D. D., and Holt, C.A. (1993). *Experimental Economics.* Princeton NJ: Princeton University Press.

Devetag, G., Hosni, H., and Sillari, G. (2013). You better play 7: Mutual versus common knowledge of advice in a weak-link experiment. *Synthese, 190*(8), 1351–1381.

Dickhaut, J., McCabe, K., and Mukherji, A. (1995). An experimental study of strategic information transmission. *Economic Theory, 6*, 389–403.

Ernst, Z. (2007). Philosophical issues arising from experimental economics. *Philosophy Compass, 2*(3), 497–507.

Fallis, D. and P. Lewis (2017). Toward a formal analysis of deceptive signalling. *Synthese* (2017) 1–25.

Fagley, N. S., and Miller, P. M. (1997). Framing effects and arenas of choice: Your money or your life? *Organizational Behavior and Human Decision Processes, 71*(3), 355–373.

Fischbacher, Urs. (2007). z-Tree: Zurich toolbox for ready-made economic experiments. *HYPERLINK "https://econpapers.repec.org/article/kapexpeco/"* *Experimental Economics, 10*(2), 171–178.

Forsythe, R., Lundholm, R., and Rietz, T. (1999). Cheap talk, fraud and adverse selection in financial markets: Some experimental evidence. *Review of Financial Studies, 12*, 481–518.

Franke, M., Jäger, G., and Van Rooij, R. (2010). Vagueness, signalling and bounded rationality. In *JSAI International Symposium on Artificial Intelligence* (pp. 45–59).

Franke, M., and Correia, J.P. (2016). Vagueness and imprecise imitation in signalling games. *The British Journal for the Philosophy of Science* axx002. Doi: 10.1093/bjps/axx002.

Gold, Natalie, Briony D. Pulford, and Andrew M. Colman (2014). The outlandish, the realistic, and the real: Contextual manipulation and agent role effects in trolley problems. *Frontiers in Psychology, 5*, 35.

Grice, H. Paul (1991). *Studies in the Way of Words.* Cambridge MA: Harvard University Press.

Grossman, S. (1981). The informational role of warranties and private disclosure about product quality. *The Journal of Law and Economics, 24*(3), 461–483.

Guala, F. (2005). *The Methodology of Experimental Economics*. Cambridge: Cambridge University Press.

Guala, F. (2013). The normativity of Lewis conventions. *Synthese, 190*(15), 3107–3122.

Guth, W., Schmittberger, R., and Schwarze, B. (1982). An experimental analysis of ultimatum bargaining. *Journal of Economic Behavior and Organization, 3*(4), 367–388.

Güth, W., and Kliemt, H. (2017). *Experimental Economics–A Philosophical Perspective*. Oxford: Oxford Handbooks Online.

Huttegger, S. (2007). Evolution and the explanation of meaning. *Philosophy of Science, 74*(1), 1–27.

Huttegger, S., Skyrms, B., Smead, R., and Zollman, K. (2010). Evolutionary dynamics of Lewis signalling games: Signalling systems vs partial pooling. *Synthese, 172*(1), 177–191.

Huttegger, S., and Zollman, K. (2010). Dynamic stability and basins of attraction in the Sir Philip Sidney game. *Proceedings of the Royal Society B, 94*, 1–8.

Huttegger, S., Bruner, J., and Zollman, K. (2015). The handicap principle is an artefact. *Philosophy of Science, 82*(5), 997–1009.

Jäger, G. (2007). The evolution of convex categories. *Linguistics and Philosophy, 30*(5), 551–564.

Jäger, G., Metzger, L. P., and Riedel, F. (2011). Voronoi languages: Equilibria in cheap-talk games with high-dimensional types and few signals. *Games and Economic Behavior, 73*(2), 517–537.

Jonsson, M., Hahn, U., and Olsson, E. (2015). The kind of group you want to belong to: Effects of group structure on group accuracy. *Cognition, 142*, 191–204.

Jovanovic, B. (1982). Truthful disclosure of information. *Bell Journal of Economics, 13*, 36–44.

Kagel, John H. and Alvin E. Roth (eds.). (2016). *The Handbook of Experimental Economics, Vol. 2* Princeton, NJ: Princeton University Press.

Kane, P., and Zollman, K. (2016). An evolutionary comparison of the handicap principle and hybrid equilibrium theories of signalling. *PLoS ONE, 10*(9), e0137271. doi:10.1371/journal.pone.0137271.

Koppl, R., Kurzban, R., and Kobilinsky, L. (2008). Epistemics for forensics. *Episteme, 5*(2), 141–159.

Leliveld, M. C., van Dijk, E., and Van Beest, I. (2008). Initial ownership in bargaining: Introducing the giving, splitting, and taking ultimatum bargaining game. *Personality and Social Psychology Bulletin, 34*(9), 1214–1225.

Lewis, David. (1969). *Convention: A Philosophical Study*. Cambridge, MA: Harvard University Press.

Lipman, Barton L. (2009). "*Why is Language Vague?*" Boston University: Unpublished manuscript.

Martinez, M. (2015). Deception in sender–receiver games. *Erkenntnis, 80*, 215–227.

Martinez, M., and Godfrey-Smith, P. (2016). Common interest and signalling games: A dynamic analysis. *Philosophy of Science*, *83*(3), 371–392.

Mehta, J., Starmer, C., and Sugden, R. (1994). The nature of salience: An experimental investigation of pure coordination games. *The American Economic Review*, *84*(3), 658–673.

Milgrom, P. (1981). Good news and bad news: Representation theorems and applications. *Bell Journal of Economics*, *12*, 380–391.

O'Connor, C. (2014a). The evolution of vagueness. *Erkenntnis*, *79*(4), 707–727.

O'Connor, C. (2014b). Evolving perceptual categories. *Philosophy of Science*, *81*(5), 840–851.

O'Connor, C. (2015a). Ambiguity is kinda good sometimes. *Philosophy of Science*, *82*(1), 110–121.

O'Connor, C. (2015b). Evolving to generalize: Trading precision for speed. *The British Journal for the Philosophy of Science*, *68*(2), 389–410.

O'Connor, C. (forthcoming). Games and Kinds. *The British Journal for the Philosophy of Science*.

Pawlowitsch, C. (2008). Why evolution does not always lead to optimal signalling systems. *Games and Economic Behavior*, *63*, 203–226.

Poon, C. S., Koehler, D. J., and Buehler, R. (2014). On the psychology of self-prediction: Consideration of situational barriers to intended actions. *Judgment and Decision Making*, *9*(3), 207.

Powell, B., and Wilson, B. (2008). An experimental investigation of Hobbesian jungles. *Journal of Economic Behavior and Organization*, *66*, 669–686.

Quine: W. V. O. (1936). Truth by convention. In *Philosophical Essays for Alfred North Whitehead* (pp. 90–124). New York: Longman Green and co.

Rubin, H., Bruner, J., O'Connor, C. and Huttegger, S. (manuscript). *Communication without the Cooperative Principle*.

Sally, D. (1995). Conversation and cooperation in social dilemmas: A meta-analysis of experiments from 1958 to 1992. *Rationality and society*, *7*(1), 58–92.

Skyrms, B. (1996). *Evolution of the Social Contract*. Cambridge: Cambridge University Press.

Skyrms, B. (2010a). *Signals: Evolution, Learning and Information*. Oxford: Oxford University Press.

Skyrms, B. (2010b). The flow of information in signalling games. *Philosophical Studies*, *147*, 155. https://doi.org/10.1007/s11098-009-9452-0

Skyrms, B. (2012). Learning to signal with probe and adjust. *Episteme*, *9*, 139–50.

Skyrms, B. (2014). *Evolution of the Social Contract*. Cambridge: Cambridge University Press.

Smith, A., Skarbek, D., and Wilson, B. (2012). Anarchy, groups and conflict: An experiment on the emergence of protective associations. *Social Choice and Welfare*, *38*(2), 325–353.

Smith, J. M., and Szathmary, E. (1995). *The Major Transitions in Evolution*. Oxford: Oxford University Press.

Smith, V. L. (1976). Experimental economics: Induced value theory. *The American Economic Review, 66*(2), 274–279.

Spence, M. (1973). Job Market Signalling. *The Quarterly Journal of Economics, 87*(3), 355–374.

Sterelny, K. (2012a). *The Evolved Apprentice.* Cambridge, MA: MIT Press.

Sterelny, K. (2012b). A glass half-full: Brian Skyrms' signals. *Economics and Philosophy, 28,* 73–86.

Utikal, V., and Fischbacher, U. (2014). Attribution of externalities: An economic approach to the Knobe effect. *Economics and Philosophy, 30,* 215–240.

Viscusi, W. (1978). A note on 'lemons' markets with quality certification. *Bell Journal of Economics, 9*(1), 277–279.

Vulkan, N. (2000). An economist's perspective on probability matching. *Journal of Economic Surveys, 14* (1), 101–118.

Wagner, E. (2009). Communication and structured correlation. *Erkenntnis, 71,* 377–393.

Wagner, E. (2011). Deterministic chaos and the evolution of meaning. *British Journal for the Philosophy of Science, 63*(3), 547–575.

Wagner, E. (2013). The dynamics of costly signalling. *Games, 4*(2), 163–181.

Wagner, E. (2014). Conventional semantic meaning in signalling games with conflicting interests. *British Journal for the Philosophy of Science, 66*(4), 751–773.

Wilson, T. D., and Dunn, E. W. (2004). Self–knowledge: Its limits, value, and potential for improvement. *Annual review of psychology, 55,* 493–518.

Zollman, K. J. S., C. T. Bergstrom, and S. M. Huttegger (2013). Between Cheap and Costly Signals: The Evolution of Partially Honest Communication. *Proceedings of the Royal Society London, B,* 280, 20121878.

Part Two

Digital X-Phi: Introducing Digital and Computational Methods

7

Causal Attributions and Corpus Analysis

Justin Sytsma, Roland Bluhm, Pascale Willemsen and
Kevin Reuter

1. Introduction

Most studies in experimental philosophy have used questionnaires involving vignettes. There are good reasons for the prevalence of questionnaire methods in experimental philosophy, including that these methods are fairly easy to use and are well-suited to investigating many of the philosophical questions that have been asked. As the present volume amply illustrates, however, questionnaire methods are not the only methods available to experimental philosophers, nor are they the only ones that experimental philosophers have used. In this chapter we will offer a brief introduction to a powerful set of non-questionnaire methods that can aid experimental philosophers in investigating a wide range of questions – methods of corpus linguistics.[1]

Our primary goal in this chapter is to introduce experimental philosophers to working with corpora, to survey some of the tools available and to demonstrate how these tools can complement the use of questionnaire-based methods. Toward this, we will put some of these tools to use in an area of research that has seen a flurry of interest in recent years – investigations of the effect of norms on ordinary causal attributions. Specifically, we focus on four questions:

Acknowledgements

We would like to thank an anonymous reviewer for the many insightful comments, as well as Eugen Fischer for his philosophical and administrative support.

[1]   Although a handful of philosophers have used different tools of corpus linguistics (for a brief overview, cf. Bluhm 2016), and although the interest in the approach seems to have increased recently (e.g. Reuter 2011; Bluhm 2012; 2013; Hahn et al. 2017; Fischer and Engelhardt 2017; Sytsma and Reuter 2017 and the contribution by Mejia et al. in this volume – Ch.8), the use of corpora is still marginal to philosophy.

(a) Can corpus analysis provide independent support for the thesis that ordinary causal attributions are sensitive to normative information?
(b) Does the evidence coming from corpus analysis support the contention that outcome valence matters for ordinary causal attributions?
(c) Are ordinary causal attributions similar to responsibility attributions?
(d) Are causal attributions of philosophers different from causal attributions we find in corpora of more ordinary language?

We argue that the results of our analysis provide evidence for a positive answer to each of these questions.

Here is how we will proceed. In Section 2, we will briefly discuss recent work in experimental philosophy on ordinary causal attributions, highlighting our four questions. In Section 3, we introduce corpus linguistics. In Section 4, we bring corpus analysis methods to bear on our target questions. In Section 5, we use methods of distributional semantics to support the previous analyses. We conclude with some general methodological advice concerning the integration of corpus analysis techniques into experimental philosophy and philosophy as a whole.

2. Ordinary causal attributions and injunctive norms

Philosophical discussions of causation are often concerned with what has been termed *actual causation*. Actual causation is usually contrasted with *general causation*. A general causation statement describes a law-like relation between two types of events that stand in a causal relation, such as 'smoking causes cancer' or 'throwing rocks at windows causes them to break'. An actual causal statement, in contrast, describes the relation between two event tokens, such as 'Peter's smoking caused his lung cancer' or 'Jenny's throwing the rock caused the window to break'. For both general and actual causation, most philosophers assume that the concept of causation is a purely descriptive notion, referring to a relation in the world. As a consequence, a *causal attribution* such as 'A caused B' is true if and only if the relation of causation holds between A and B. Such an understanding of causation, however, means that normative considerations are irrelevant to causal attributions. The basic idea here is that whether or not an action is permitted by morality or convention simply does not matter for purposes of assessing whether that action, or the entity carrying it out, *caused* the outcome. Similarly, whether an action causes a morally good or bad outcome is irrelevant for causal considerations. Call this the *standard view* on causation.

Against the standard view, a growing body of empirical findings indicates that ordinary causal attributions are sensitive to normative information, prominently including *injunctive norms* (Hilton and Slugoski 1986; Alicke 1992; Knobe and Fraser 2008; Hitchcock and Knobe 2009; Sytsma et al. 2012; Reuter et al. 2014; Kominsky et al. 2015; Livengood et al. 2017a).[2] Injunctive norms include both *prescriptive norms*, which tell people what they should do, and *proscriptive norms*, telling people what they should not do.[3] Moral norms, such as the impermissibility of killing or hurting others, are prime examples of injunctive norms, but there is also a variety of non-moral norms that have similar action-guiding functions, such as social rules and regulations, etiquette norms and so on.[4]

Here are a couple of the empirical findings that have received much attention in the literature. Knobe and Fraser (2008) presented people with a story in which a secretary keeps her desk stocked with pens and both administrative assistants and faculty members help themselves from this stock. However, faculty members are not supposed to take pens. One day, both Professor Smith and the administrative assistant take a pen. Later that day, the secretary has no pen left to take an important message. Who caused the problem? In this case, Professor Smith and the administrative assistant performed symmetric actions (each took a pen), jointly leading to a bad outcome. The key difference between them is that while Professor Smith violated an injunctive norm (faculty members are not supposed to take pens), the administrative assistant did not (administrative assistants are allowed to take pens). Despite the two agents performing symmetric actions, participants were significantly more likely to agree that Professor Smith, the norm-violating agent, *caused* the problem than that the administrative assistant did.

To make matters more interesting, in a follow-up study, Sytsma et al. (2012) tested what happens if you remove the injunctive norm from the Pen Case, such

---

[2] These results do not directly challenge the standard view. Rather they put pressure on it insofar as philosophers are committed to what Livengood et al. (2017a) call the *'folk attribution desideratum'*. The folk attribution desideratum asserts that a key measure of the acceptability of an account of actual causation is that the verdicts it issues about specific cases line up with ordinary causal attributions about those cases. And there is reason to think that many, perhaps most, philosophers working on causation are committed to this desideratum.

[3] It should be noted that in the recent literature in experimental philosophy of causation, 'prescriptive norm' is often used indiscriminately to refer to both prescriptive norms and proscriptive norms as we understand them.

[4] Injunctive norms can be distinguished from descriptive norms (or 'statistical norms'). While there is an ongoing debate among experimentalists about whether descriptive norms have an independent effect on ordinary causal attributions (e.g. Knobe and Fraser 2008; Sytsma et al. 2012; Livengood et al. 2017a), we will focus on injunctive norms in this chapter.

that both Professor Smith and the administrative assistant are allowed to take pens. They found that participants now tended to *disagree* that Professor Smith caused the problem, while continuing to deny that the administrative assistant caused the problem.

Livengood et al. (2017a) used a computer case scenario, for which they found the same effects. More specifically, their studies revealed that participants were significantly more likely to agree that an agent who violated a norm caused a bad outcome, compared to the norm-conforming agent. Agreement that the norm-conforming agent caused the bad outcome was significantly below the neutral point, while agreement for the norm-violating agent was significantly above the neutral point.

While the Pen Case and the Computer Case are probably two of the most prominent examples in the literature, similar effects have repeatedly been found in subsequent research, and they seem to be robust for different causal structures (Kominsky et al. 2015; Sytsma et al. ms; Livengood and Sytsma under review), for both actions and omissions (Henne et al. 2015; Willemsen 2016; Willemsen and Reuter 2016) and across multiple test queries (Livengood et al. 2017a; Livengood and Sytsma under review).

Several explanations of the relevance of injunctive norms have been proposed in the literature. The most fundamental dispute regards the question of whether the observed effects reveal a real effect of injunctive norms on *causal attributions*. In two recent papers, Samland and Waldmann (2014, 2016) have denied this. According to their alternative explanation, when participants answer the question in studies like those noted above, they do not read the questions as being about causation, but about a related notion such as accountability or responsibility.

Those researchers who are convinced that norms do affect causal attributions have offered a variety of explanations of why and how this effect occurs. In the following, we will focus on two specific explanations, the *norm violation* and the *responsibility account*, as they make empirical predictions that we believe can be effectively tested with help of corpus analyses.[5]

The *norm violation account* put forward by Hitchcock and Knobe (2009) holds that the effects of norms in cases like we saw above are best explained in terms of the *cognitive processes* that lead to causal attributions being sensitive

---

[5] While we will focus on these two accounts here, it should be noted that these are not the only two plausible explanations in the literature nor are they the only two worth discussing. Just a few notable examples are the work of Alicke 1992, Cushman 2013, Malle et al. 2014, and Reuter et al. 2014.

to norms. According to Hitchcock and Knobe's account, causal judgements serve to identify suitable points for intervention in a system, and norms come into play because they affect which counterfactuals are most salient for determining the suitability of different intervention points. The basic idea is that in considering a situation, people think about how the outcome could have been prevented. But they do not consider every way in which the outcome might have been prevented; rather, they focus on those aspects of the situation in which something abnormal (i.e. counter-normative) has occurred. As such, while the norm violation account holds that the evaluation of norms is a crucial component in causal cognition, this does not mean that the concept of causation at play in ordinary causal attributions is a normative concept. Instead, norms come into play when people attempt to identify suitable intervention points in normatively laden situations.

The *responsibility account* put forward by Sytsma et al. (2012) contends that the concept of causation at play in ordinary causal attributions is itself normative. Thus, the cognitive process of making causal attributions starts off from a normatively laden concept. Instead of making a purely descriptive causal judgement that is later tainted by norms, the causal judgements are already normative. Sytsma and Livengood (2018, 7–8) express the idea this way:

> Saying that an agent caused an outcome […] typically serves to indicate something more than that the agent brought about that outcome or that the agent's action produced that outcome. Rather, it serves to express a normative evaluation that can be roughly captured by saying that the agent is responsible for that outcome or that the agent is accountable for that outcome, whether for good or for ill.

The norm violation account and the responsibility account make a number of diverging predictions. One such prediction is that while Hitchcock and Knobe hold that normative considerations have the same effect on causal attributions regardless of the outcome valence (regardless of whether the outcome is good or bad), the responsibility account allows that outcome valence might often make a difference.

As we noted above, Hitchcock and Knobe explain the influence of norms on causal attributions exclusively in terms of shaping the counterfactuals that are considered. And whether the outcome is good or bad would not seem to be directly relevant to assessing whether a counterfactual in which a candidate cause did not occur was more normal than what actually happened. Thus, Hitchcock and Knobe argue that to assess Alicke's (1992) competing account, which they read as explaining the influence of norms in terms of the desire to assign blame for the outcome, what is needed are cases in which a norm is violated and yet where no one is assigned blame because the outcome is good. Hitchcock and Knobe write that for such cases their account 'suggests that the impact of normative considerations *should remain unchanged* (because people still see that a norm has been violated)' (2009, 603; emphasis added).

The responsibility account, by contrast, does not make a direct prediction about the role of outcome valence in ordinary causal attributions; rather, it makes a prediction when coupled with a plausible prediction about responsibility attributions – that people are more likely to assert that an agent is responsible for a bad outcome than a good outcome.[6] If this is correct, and if causal attributions are relevantly akin to responsibility attributions, then we would expect that outcome valence will often make a difference.

One way to make progress on the issues concerning the role of norms on ordinary causal attributions that we have surveyed in this section would be to run still more questionnaire studies. However, we want to suggest that there is also benefit in approaching the problem from another angle. What we aim to demonstrate in this chapter is that another source of evidence can be brought to bear on these questions, namely corpus linguistics, and that its methods can both complement and enhance the use of questionnaire-based studies. After offering a brief introduction to corpus linguistics and applying some of its methods to the domain of ordinary causal attributions, we will return to the use of questionnaire studies and discuss potential shortcomings of such studies and how they can be alleviated through the use of corpus analysis.

3. The basics of corpus analysis

Corpus linguistics is a branch of linguistics that is defined by its use of *corpus analysis*. In its most basic sense, the term 'corpus' simply refers to 'a collection of texts' (Kilgarriff and Greffenstette 2003, 334) and 'analysis' to the process of

---

[6]   We find this plausible because we expect that people are generally more concerned with assigning blame than praise, as a number of philosophers have noted. For instance, Prinz (2007, 79) writes: 'We blame someone for stealing, but we don't issue a medal when he refrains from stealing. We don't lavish the non-pedophile with praise for good conduct. In other words, we tend to *expect* people to behave morally.' While we will not argue for the veracity of this prediction here, it does find some support in the corpus analyses detailed below.

looking at the linguistic data that the corpus contains and assessing it for some research purpose.[7]

Briefly summarized, there are three sources of this method (*cf.* McCarthy and O'Keeffe 2010a). Its historical roots can be traced back as far as the Middle Ages, when concordances of Biblical words and phrases in context were compiled to assist exegesis. One of the basic functions of present-day corpora still is – particularly important for qualitative assessment of corpus data – to provide listings of queried linguistic expressions in context, in much the same mode of presentation as the one that mediaeval concordances used. The second important source of corpus analysis is the recognition of the importance of data representing actual use of a language, as opposed to data generated by the linguists themselves.[8] The third factor driving the development of corpus linguistics – of particular importance for quantitative analyses of corpus data – is the fast-paced development and spread of computer technology and the Internet in the late 20th century. Thus, while all corpora used in corpus linguistics are basically 'collections of texts', present-day corpora are typically collections of digitized texts that are accessed with computers.

Paradigmatic examples of well-known, large and freely accessible English-language corpora are the British National Corpus (BNC) and the Corpus of Contemporary American English (COCA).[9] The World Wide Web also offers a rich repository of digital texts, and while it is somewhat problematic to simply use the web *as* a corpus (termed WaC in the literature), there are some ways to tap into the Web's wealth of data by using extracts of it *for* building a corpus (termed WfC in the literature). We will make use both of COCA and a WfC approach below.[10]

One of the aspects that make a corpus out of a mere collection of texts is the decision to look within it for evidence of some use of particular linguistic expressions. The access to the data granted by the search engine is therefore not at all marginal to corpus analysis. Every run-of-the-mill search engine can do a full text search and handle wildcards, that is, it is possible to execute a query with a search string (a sequence of alphanumeric characters) in which some letter or

---

[7] Helpful overviews of the discipline are given by Biber et al. 1998, McEnery and Wilson 2001, and McCarthy and O'Keeffe 2010b. For a quite comprehensive collection of articles on many aspects of corpus linguistics, see Lüdeling and Kytö 2008–2009.

[8] See Leech 1992.

[9] The number and variety of corpora compiled by linguists is ever-growing. Xiao 2008 and Lee 2010 give an overview of extant corpora, usefully sorted by type.

[10] The useful terms 'WaC' and 'WfC' go back to de Schryver 2002. For a brief introduction to the rapidly growing field of 'web linguistics', you may turn to Bergh and Zanchetta 2008.

letters are substituted with a variable. For example, 'cause*' will not only find all instances of the use of 'cause' as a verb and a noun in the corpus, but also tokens of 'causes', 'caused' and more unexpected words such as 'causeway' and 'causer'. A more sophisticated corpus is pre-analysed and annotated with linguistic information, allowing, for example, to specifically search for the *lemma* 'cause' (all instances of the root word 'cause' regardless of inflection). It further allows to find tokens by grammatical category, for example, only instances of the verb 'causes' in its third person singular form, instead of the noun in its plural form; or co-occurrences of expressions, for example, the lemma 'cause' together with some noun (within a specified distance). Such queries are obviously much more powerful than mere full-text search. They are indispensable if a pertinent linguistic expression cannot be specified by a definite search string, or if, as in our case, the relevant linguistic phenomena include phrases such as 'responsible for' or 'caused' followed by some noun.

4. Exploring causal attributions with corpus analysis

We have already hinted at some very basic search options offered by common corpus search engines. Generally speaking, there are two approaches to using corpora (cf. Biber 2010): Corpus-*driven* research uses corpora to generate theories on linguistic phenomena from bottom up. Accordingly, the corpus is approached with minimal hypotheses as to the linguistic forms relevant to a given research question, for example, searching for tokens of 'cause' as a starting point to develop an understanding of causal attribution language. Corpus-*based* research, on the other hand, uses corpora to verify or falsify extant hypotheses about the use of language based on available theories about linguistic forms (for example, trying to confirm that 'cause' has some specific collocations). In practice, these approaches overlap to some extent, with researchers switching back and forth between them in their research process.

Similarly, corpora can be approached in a qualitative manner, that is, focusing on interpreting corpus findings with respect to meaning, or a quantitative manner, focusing on analyses based on countable objects and statistical facts. Both the methods used in this section and the next would count as quantitative on this definition. However, despite relying on frequency counts, the methods employed in this section have a somewhat more qualitative aspect to it, while the methods employed in the next are decidedly more quantitative, as will be apparent.

All the approaches to corpora we have hinted at depend on the existence of linguistic phenomena that can be traced with the help of available search engines. But in studying the language of causation, broadly construed, it is not immediately clear which linguistic expressions to look for. Linguists have identified an astonishing number and variety of ways that (arguably) are used to express causal relations.[11] Apart from the verb 'to cause' and the noun 'cause', as well as (partially) synonymous expressions, there are conjunctions for the subordination of clauses like 'because', 'since', 'as' (cf. Altenberg 1984; Diessel and Hetterle 2011), but also causative verbs, adverbs, adjectives and prepositions (cf. Khoo et al. 2002), and no exhaustive list of such means to express causal connection is available. There are also linguistic means to express a causal relation without lexical means, for example, the coordination of sentences and text organization (cf. Altenberg 1984; Achugar and Schleppegrell 2005). In consequence, it is only possible to find *some*, but not *all*, instances of causal language in a corpus with the help of a search engine. Moreover, most of the expressions mentioned above serve not only to express causal relations, but may also be used differently. To give but one example, 'cause' may also refer to a concern or purpose, as in 'her cause was just'.

With these caveats in mind, it is, of course, possible to access *some* of the causal language contained in a corpus. In our present context, we are interested in similarities and differences of the language of causal attributions and the language of responsibility attributions, and it seems plausible to approach our linguistic study with a focus on 'cause' as a relational verb and the phrase 'is responsible for', matching the type of phrases used in questionnaire studies to elicit causal attributions (e.g. 'Lauren caused the system to crash', 'Marcy is responsible for the death of the bystander').[12]

In the second section of this chapter, we surveyed some recent studies on ordinary causal attributions. These studies suggest that injunctive norms play a substantial role. Various theories have been proposed to account for this effect. We focused on two of these – the norm-violation account and the responsibility account. According to the norm-violation account, while ordinary causal attributions are influenced by norms, the underlying concept of causation

---

[11] We need not pass judgement at this point on whether such utterances really are intended to express or really do refer to a *causal* relation of some kind, let alone whether they express the specific relation at issue for causal attributions as we have defined them.

[12] The phrase 'is responsible for' is used to cue responsibility attributions in studies in Sytsma and Livengood (2018).

is descriptive and, thus, diverges notably from the concept of responsibility. Importantly for our research purposes, only norms, but not the valence and the severity of the outcome are said to affect causal attributions. From that we can infer the empirically testable prediction that the language of causal attributions and the language of responsibility attributions should be clearly distinguishable. In contrast, the responsibility account holds that the language of causal attributions and the language of responsibility attributions are quite similar. Moreover, the responsibility account predicts that the valence of the outcome will often have a notable effect on causal attributions.

If the ordinary concept of causation was a purely descriptive concept, then we would have no *a priori* reason to expect it to be used disproportionately in contexts with any particular valence. Rather, we would expect 'cause' to be used indiscriminately in the contexts where the outcome is good, in contexts where the outcome is bad, and contexts where the outcome is neutral. And if this was the case then we should see a good mix of positively and negatively connotated causal expressions without one of these types of expression dominating people's use of 'cause'. This is not what we find, however.

In order to investigate the nature of the terms used most commonly when expressing a causal statement, we looked at the most frequent nouns appearing after the causal phrase 'caused the' using the Corpus of Contemporary American English (COCA). The ten most frequently used nouns (numbers in brackets indicate the number of hits) are 'death' (103), 'accident' (87), 'crash' (80), 'problem' (79), 'explosion' (47), 'fire' (46), 'collapse' (27), 'injury' (26), 'damage' (24) and 'loss' (23). Independent judges classified all of these terms as negative.[13] Of the top 50 nouns, 30 were classified as negative, 19 neutral and only 1 positive. These data support the results from questionnaire studies indicating the relevance of norms to ordinary causal attributions. Furthermore, the commanding presence of negative terms strongly suggests that 'cause' is not only partly normative, but also primarily directed at negative outcomes. In other words, the results of

---

[13] Three independent raters were given a prompt – 'Please classify each of the following items based on whether you think instances of this type are most often positive, negative or neutral?' – followed by 780 items for classification. Items were the top 50 hits for 'caused the', the top 50 hits for 'responsible for the', and the top 20 hits for each of the eight synonymous expressions used below. This produced a list of 260 items that were then presented to each rater in three randomized orders. For the ten most frequently used nouns just reported, there was 99.4% agreement across the classifications with 169 out of 170 occurrences of these items being classified as negative. Overall, inter-rater agreement was high with a Kendall's tau of 0.751 and Spearman's rho of 0.803 averaging across the values for each possible pair of raters. For subsequent classifications we treated a term as negative (positive) if it was classified as negative (positive) at least two-thirds of the time.

our analysis strongly indicate that outcome valence has a substantial effect on ordinary causal attributions.

A key objection to our interpretation of these corpus data might be raised at this point. First, one important source of the corpus we have used is newspaper articles. And newspaper articles are notorious for focusing on negative events. Accordingly, it would not be surprising to find statements about causal relations for which the outcome is often negative. However, when we limited our search to sources of other types, like fiction, as is possible in COCA, no differences were found. For example, the top ten list of nouns following the phrase 'caused the' for spoken language only were 'death' (41), 'accident' (40), 'crash' (38), 'fire' (30), 'problem' (30), 'explosion' (26), 'plane' (11), 'damage' (9), 'recession' (9) and 'collapse' (8).[14] And a similar list was obtained when the corpus was restricted to academic texts.

It might be offered in rejoinder that a focus on the negative is simply a part of the human condition. As such, it might be suggested that terms that are arguably synonymous with 'cause' will also tend to be used in negative contexts. If that were true, then we could not conclude that we have discovered a specific characteristic of the language of causal attributions, but rather a more general phenomenon, for which an entirely different explanation would seem to be required. In order to investigate this objection, we posited the following null hypothesis:

> *Synonymy Effect*: There is no significant difference in the normative use between 'cause' and synonymous expressions.

If 'cause' is indeed specific in being used in a predominantly normative way, then we should be able to falsify Synonymy Effect. To test the hypothesis, we executed a corpus search with the eight terms that are listed by the English Thesaurus as being often used synonymously with the verb 'to cause': 'create', 'generate', 'induce', 'lead to', 'make', 'precipitate', 'produce' and 'provoke'. We inserted the phrase 'Φed the' and noted the 20 most frequent nouns that appear after that phrase for all eight synonymous terms. (Table 7.1 lists those terms that were rated negatively for 'caused the' as well as the synonymous phrases.)

Considering only the 20 most frequent nouns, we calculated whether there was any significant difference in the use of 'caused the' compared to synonymous

---

[14] Only 'plane' was classified as neutral, with 160 out of 170 occurrences of these 10 terms being classified as negative by our raters.

Table 7.1 Most frequent negatively connotated nouns after phrases synonymous to 'caused the'.

Phrase	Number of negative terms (out of 20)	Negative terms
caused the	16	death, accident, crash, problem, collapse, injury, damage, loss, crisis, destruction, decline, extinction, harm, demise, explosion, fire
created the	3	problem, need, illusion
generated the	3	waste, war, killing
induced the	4	coma, panic, defendant, opposition
led to the	7	death, arrest, collapse, demise, loss, firing, end
made the	2	mistake, cut
precipitated the	13	crisis, war, attack, decline, conflict, collapse, downfall, fight, invasion, violence, demise, end, split
produced the	1	plutonium
provoked the	9	anger, fight, violence, murder, resignation, rebellion, strike, crisis, evacuation

expressions. Pearson's chi-square tests revealed that only 'precipitated the' was not significantly different ($\chi^2(1.13, 1) = 0.288$). All other comparisons were highly significant: 'provoked the' ($\chi^2(5.23, 1) = 0.022$), 'led to the' ($\chi^2(8.29, 1) = 0.004$); $p < 0.001$ for all other phrases.

The results are noteworthy in a couple of respects. (a) The searches demonstrate that the COCA corpus is not – at least not strongly – tilted towards texts that feature negative events. (b) The Synonymy Effect is likely to be false. Seven out of eight synonymous expressions of the form '*Φed the*' are not only significantly different in their most frequent uses compared to 'caused the', most of the synonymous terms seem to be used mainly in a neutral fashion. This means that the effect we recorded for 'cause' seems to be rather specific.

We have seen that Sytsma and Livengood propose that causal attributions are inherently normative, being used to assign responsibility. If this is correct, then we would expect responsibility attributions to be similar to causal attributions, including that they should also tend to occur more often in negative contexts. To test this expectation, we ran the same corpus search as before, this time entering the phrase 'responsible for the' and recorded the most frequent nouns that occur after that phrase. The ten most frequent nouns are 'death' (130), 'attack' (46), 'murder' (44), 'actions' (43), 'safety' (42), 'development' (42), 'loss' (35), 'design' (34), 'decline' (31) and 'violence' (31).[15] Of the 50 most frequent nouns occurring after 'responsible for the', 19 terms were rated negative, 17 neutral and 14 positive. It should be further noted, however, that many of the positive terms seem to belong to an alternative sense of 'responsible' from the responsibility attributions we are after – a sense indicating the duties involved in a role (e.g. 'content', 'creation', 'design', 'implementation', 'safety', 'security').

Nonetheless, the results show that 'responsible for the' has a similar environment to 'caused the' in being normatively laden and more often directed at negative events than positive. And by looking in greater detail at the respective numbers of hits for various terms, we found more striking similarities. For many terms like 'death', 'decline', 'damage', 'destruction', 'crisis', the use of responsibility language is roughly as frequent as causal language, suggesting that at least for some terms, both phrases might be used interchangeably (Table 7.2 lists the number of hits for these terms as well as the frequency ratio). This provides further support for the hypothesis that the causal attributions and responsibility attributions are often used to express the same state of affairs.

However, other comparative results between 'responsible for the' and 'caused the' do not quite match, which might suggest that we cherry-picked the data that fits our hypothesis. Table 7.2 lists two terms ('attack', 'murder') which are far more commonly used with responsibility language than causal language; for example, people seem to be far more likely to say 'she is responsible for the attack' than 'she caused the attack'. An explanation is easy to give: when we want to express a causal relation between a person and an attack or a murder, we would usually just rely on the causative aspect of the verbs and say that 's/he attacked' or 's/he murdered'. In other words, the English language has a simpler means to express causal language when it comes to attack and murder. The opposite result was found for the terms 'problem' and 'accident'. The corpus analysis revealed that 'caused the problem' is far more frequent than 'responsible for the problem'. If both concepts are akin, should we not expect that their uses are similarly frequent? Here, a closer look at the search hits is helpful.

What we find is that speakers often raise questions like 'what caused the problem?' leaving it open that it was not an agent but rather an event that

---

[15] Every occurrence of six of these terms was classified as negative by our raters. Every occurrence of 'safety' and 'design' was classified as positive. Classifications for 'actions' and 'development' were split, although they were positive overall. In total, 154 out of 198 occurrences of these 10 terms were classified as negative.

Table 7.2 Hits for some of the most frequent nouns after the phrases 'responsible for the' and 'caused the' and the ratio between them.

Word	**responsible for the**	**caused the**	**ratio**
death	130	103	1.26:1
decline	31	15	2.07:1
damage	24	24	1:1
destruction	18	16	1.13:1
crisis	11	17	0.65:1
attack	46	3	15.33:1
murder	44	2	21:1
problem	13	79	0.16:1
accident	11	87	0.13:1

caused the problem. In contrast, responsibility language is mostly used in relation to agents.[16] Thus, it is relatively rare that people make claims such as 'the malfunctioning brakes are responsible for the accident' but rather speak of malfunctioning brakes causing accidents. This in turn suggests that the semantic similarity between the language of causal attributions and responsibility attributions might be most pronounced for agent causation. A further investigation into this possibility is, however, beyond the scope of this paper.

It might be objected that the similar frequency in use of 'responsible for the' and 'caused the' for many nouns are merely coincidental and do not reveal any semantic similarity between these phrases; other phrases may be just as frequent. To counter this objection, we further examined which verbal phrases occur most frequently before nouns such as 'death', 'decline' and 'destruction', for which we have observed the same frequencies. The word 'death' was most frequently preceded by the verbal phrases 'caused the' and 'responsible for the'.[17] For the term 'decline', only 'contributed to' was a more common phrase than 'caused the' and 'responsible for the'. And for the term 'destruction' only 'stopped the' and 'prevented the' were more common than 'caused the' and 'responsible for the'. These data suggest that the similarity in use between the language of

---

[16] There is some quite strong evidence coming from further corpus analyses that support such a view. When entering the phrase 'what caused', COCA delivers 1,189 search hits compared to only 250 hits for 'who caused'. The situation is reversed for responsibility language. A search on COCA lists 412 hits for 'who is responsible for' but only 16 hits for 'what is responsible for'.

[17] In fact, 'seek the', 'face the' and 'get the' were even more common, but occurred not together with 'death', but mostly with the fixed expression 'death penalty'.

responsibility attributions and the language of causal attributions is unlikely to be a mere matter of coincidence. Rather, corpus analysis indicates that these languages are highly similar in meaning. This also puts pressure on Samland and Waldmann's (2014, 2016) alternative account of the effect of normative information on ordinary causal attributions: it does not seem to be the case that participants read questions in vignette studies to be about a related notion such as responsibility. Instead, the data suggests that the notion of causation is in itself inherently normative.

5. Corroborating the findings with distributional semantics

In addition to the somewhat qualitative approach to corpus analysis taken in the previous section, there is also a more mathematical way of exploiting corpus data by using an array of computational methods for investigating word meaning. One prominent approach is based on the 'distributional hypothesis', which follows Firth's dictum that 'you shall know a word by the company it keeps' (Firth 1957, p. 11; see also, Harris 1954). Accepting this, word meaning can be explored by using computational methods to look at the distribution of words across a corpus. One set of such tools are distributional semantic models (DSMs). The typical DSM represents terms as geometric vectors in a high-dimensional space that can be compared to give a quantitative measure of similarity. This is typically done by taking the cosine of two vectors, with a value of 1 indicating identical meanings while a cosine of 0 would indicate completely dissimilar meanings., [18] [19]

There are a number of different ways of carrying out distributional analyses. Unfortunately, the details of these different approaches can get quite daunting, especially for researchers who are new to the area. That said, we believe that even the more accessible techniques for distributional analyses are of value. As such, we encourage experimental philosophers to begin employing these tools, and to tackle their more complex aspects and sophisticated varieties in due course. We will begin with some tools that any experimental philosopher could employ, then expand the analysis to tools that require greater familiarity with programming.

---

[18] The cosine can also take on a negative value. It is at best unclear how negative values should be interpreted, however, and these are generally treated as being 0.

[19] For more detailed discussions of DSMs see Baroni et al. 2014a, Erk 2012, and Turney and Pantel 2010.

Perhaps the most prominent type of distributional analysis is Latent Semantic Analysis (LSA; Deerwester et al. 1990), and this is the method we will begin with in this section. LSA starts with the texts of a corpus being broken down into pre-defined documents, such as paragraphs of text. The frequency of each term in the corpus is then counted for each document to produce a term-by-document matrix. It should be noted that this matrix does *not* include information about the relative location of terms in a document. Because of this, LSA is sometimes referred to as a 'bag-of-words' approach. And in this, LSA is perhaps most markedly different from the approach used in the previous section, which specifically looked at the relative position of terms in a sentence.

While LSA has had a good deal of empirical success, one should be mindful of the limitations of the bag-of-words approach and recognize that other approaches are available. In LSA, the context for a target word is the rest of the document. Alternatively, window-based methods use the terms surrounding the target word as context (while this can be thought of as a bag-of-words, it is a relatively *small* bag of words). For instance, a window of size 5 would take the two terms to either side of the target word as context. Another option is to use the words that stand in a particular syntactic relation to the target word as the context.[20] In contrast to these approaches, the 'new kids on the distributional semantics block' are what Baroni et al. (2014b) term *context-predicting models*. Instead of counting the terms occurring in a given context around a target word, these models use artificial neural networks to set vector weights that 'optimally predict the contexts in which the corresponding words tend to appear' (Baroni et al. 2014b, 238).[21]

The easiest way to begin using LSA is to query a premade semantic space. One option is the LSA website from the University of Colorado Boulder.[22] This website allows users to run a number of different types of queries for a range of semantic spaces. To illustrate, we used the pairwise comparison tool for the General Reading up to 1st Year College space[23] to look at cosine values for 'cause' and 'caused' as compared to four terms relevant to assessing causal attributions and their relation to outcome valence – 'responsibility', 'blame', 'fault', and 'praise'.

---

[20] For a comparison of these approaches, as well as a number of other parameters involved in constructing vector-based DSMs, see Kiela and Clark 2014.

[21] Baroni et al. (2014b) conduct an extensive comparison between context-predicting and count-based models. To their surprise, they 'found that the predict models are so good that [...] there are very good reasons to switch to the new architecture' (245).

[22] See http://lsa.colorado.edu/

[23] This space is built from a corpus of 37,651 documents and covers 92,409 unique terms.

As predicted on the basis of our previous analyses, both terms show a notable similarity to 'responsible'. Further, in line with our previous analyses we found that both terms show a notable similarity with the negative terms 'blame' and 'fault', but essentially no similarity with the positive term 'praise' (see Table 7.3).

It would be nice to be able to say something absolute about the degree of similarity indicated by a given cosine value. Unfortunately, this is complicated by differences in the sizes of LSA spaces. As a result, the values should be thought of as relative measures. One option for getting a sense of the relative values for a space is to test some comparison terms. For instance, the value we found for 'cause' and 'responsible' is slightly higher than the value we find for 'dog' and 'wolf' (0.30), while the value for 'dog' and 'animal' (0.15) is half that, and the value for 'dog' and 'sandwich' is slightly higher than we found for 'cause' and 'praise'. While such comparisons can help you get an initial sense of the space, it is dependent on the terms that you select and might be misleading. A more systematic approach is to look at a predefined list of comparisons. One option is to use a list that is part of a benchmark, such as MEN (Bruni et al. 2013). MEN includes a test set of 1,000 comparisons whose relatedness has been assessed by a large sample of human judges. We can run each of these comparisons, then look at the pairs of terms that have a similar cosine value to the pairs we want to assess.

Another tool available through the LSA website of the University of Colorado Boulder is to search for the nearest neighbours of a given term. This provides the terms closest to the target term *in the semantic space*. When we did this for 'cause' and 'caused', we found that the nearest neighbours, excluding terms sharing the same word stem, generally have a negative cast (e.g. 'damage' (0.66), 'symptoms' (0.61), 'disease' (0.69), 'infections' (0.67)). Again this is in keeping with our previous findings. When we turned to 'responsible', however, we found that many of the nearest neighbours for this term are of a different sort. For instance, we found that 'duties' (0.57) is the nearest neighbour, followed directly by 'supervision' (0.56) and 'personnel' (0.55). This suggests that the responsibility

Table 7.3 Cosine values for term comparisons for the General Reading up to 1st Year College space on the LSA website from the University of Colorado Boulder.

	responsible	blame	fault	praise
cause	0.32	0.28	0.30	0.05
caused	0.32	0.31	0.26	0.01

attributions we are after are getting drowned out by the alternative usage of 'responsibility' noted in the previous section – that of the duties associated with a role.

In addition to getting a sense of degree of similarity indicated by a cosine value in a given space, we also assessed whether it is doing a good job in capturing the semantic relatedness of terms. To do this we employed the MEN list, mentioned above, and used the list of cosine values for the test set to analyse how well this correlates with the scores from the human judges. When we did this for the General Reading space, we found that it does a relatively good job: we got a Spearman's rho of 0.67. For comparison, Kiela and Clark (2014) report values of 0.66 to 0.71 for the spaces they compared in their Table 6, while Baroni et al. (2014b) report in Table 7.2 a top value of 0.72 for the best count-based model tested and a value of 0.80 for the best context-predicting model tested.

It is also possible to build an LSA space oneself. While the details that go into the construction of an LSA space are complicated, a number of tools are available to facilitate the process. We will begin with tools that can be used through the statistical software package R. While there are several benefits to using R in the present context,[24] it also suffers from some limitations, as we will see.

The easiest way to get started with LSA in R is to use the LSAfun package to import a premade semantic space (Günther et al. 2015). To illustrate, we used the EN_100k_lsa space to further explore the relationship between 'cause'/'caused' and 'responsible'. This space was created from a corpus of some two billion words combining the British National Corpus, the ukWaC corpus and a 2009 Wikipedia dump. We began by looking at the same set of comparisons that we performed above. Again we see that 'cause'/'caused' show a notable similarity to 'responsible', and that both terms are much more similar to 'blame' and 'fault' than to 'praise' (see Table 7.4). Next, we looked at the nearest neighbours for 'cause', 'caused' and 'responsible'. The results were in line with what we saw previously, with 'cause' and 'caused' being close to a number of negative terms (e.g. 'excessive' (0.78), 'suffer' (0.78), 'damage' (0.82), 'fatal' (0.71)), while 'responsible' was close to a range of terms related to duties associated with a role (e.g. 'overseeing' (0.76),

---

[24] One is that R supports a large range of statistical analyses of use to experimental philosophers. Another is that the only book-length guide to the practice of experimental philosophy currently available (Sytsma and Livengood 2016) uses R as its preferred statistical program and Chapter 10 of that work provides a general introduction to the use of R.

Table 7.4 Cosine values for term comparisons for the EN_100k_lsa space.

	responsible	blame	fault	praise
cause	0.32	0.46	0.50	0.23
caused	0.34	0.47	0.57	0.18

'supervising' (0.63)). Like the previous space, the EN_100k_lsa space performs well on the MEN benchmark with a Spearman's rho of 0.67.[25]

R also offers tools for creating corpora and building LSA spaces. To illustrate, we used the RCurl package to scrape the text for all entries in the Stanford Encyclopedia of Philosophy and the Internet Encyclopedia of Philosophy from their websites. The result was a corpus including 2,378 entries split into 136,946 paragraphs (the documents for our analyses) and composed of over 149 million words and 115,644 unique terms. We then used the koRpus package, to annotate the documents with lemma information. The tm package was used to convert this into a corpus, which was fed into the lsa package to generate the term-by-document matrix and create the semantic space. It is in this final step that we ran into the limitations of R noted above. Specifically, in R the data for analysis is held in RAM, which places severe limits on the size of the matrix it can process on a typical home computer. One option is to reduce the size of the term-by-document matrix by removing infrequently occurring terms before creating the semantic space. For the philosophy corpus, we needed to reduce the matrix to the 8,821 most frequently occurring terms.

Given that philosophers have often treated the ordinary concept of causation as being a purely descriptive concept and that many have expressed either surprise or outright scepticism towards the results surveyed above we would not expect to find the valence effect for the philosophy corpus that we saw in our previous investigations. With regard to the relation between 'cause' and 'responsible', one might predict that these terms would also be largely unrelated. Alternatively, one might note that many philosophers hold that causing an outcome is a prerequisite for being responsible for that outcome. As such, one

---

[25] There is also a window-based space available from the same corpus – the EN_100k space – that performs slightly better on the MEN benchmark (Spearman's rho of 0.71). Another option are the spaces available from the COMPOSES Semantic Vectors website, which provides the best models from Baroni et al. (2014b) as text files. Their context-predicting model performs especially well with a Spearman's rho of 0.80 on the MEN benchmark. Both spaces paint a similar picture to what we saw for the EN_100k_lsa space, both in terms of the comparisons in Table 7.4 and the nearest neighbours of 'cause'/'caused' and 'responsible'.

might expect to see a notable similarity between these terms in the philosophy corpus. What we found is that 'cause' showed virtually no relation to 'blame', 'fault' or 'praise', and that it showed virtually no relation to 'responsible' (see Table 7.5).[26] The space performed better than expected on the MEN benchmark, with a Spearman's rho of 0.48. Because the term-by-document matrix was significantly reduced, however, the correlation was only calculated on 269 comparisons.

Given the degree to which the term-document-matrix was reduced, the results for the philosophy corpus LSA space should be taken with a hefty grain of salt. To further test these results, we switched to the Gensim toolkit implemented in Python, which is designed to handle large corpora efficiently and is able to carry out a wide variety of distributional analyses, including LSA and the context-predicting models using word2vec that performed best in Baroni et al.'s (2014b) comparisons. We exported the processed philosophy corpus from R into Gensim, then analysed it using word2vec with recommended parameters, including a five-word context window. The results were quite different from what we found for the LSA space. Most notably, we found a much higher cosine value for 'cause' and 'responsible' (see Table 7.6). We also saw a moderate relation between 'cause' and 'blame' or 'fault', but no relation between 'cause' and 'praise'. Further, the nearest neighbours of 'cause' and 'responsible' were quite distinct.[27] While these results are more like what we've seen for the general corpora, they continue to

Table 7.5 Cosine values for term comparisons for the philosophy corpus LSA space.

	responsible	blame	fault	praise
cause	−0.0003	0.0004	−0.0011	−0.0051

Table 7.6 Cosine values for term comparisons for the philosophy corpus word2vec space.

	responsible	blame	Fault	praise
cause	0.41	0.18	0.17	−0.04

[26] We ran the comparisons only for 'cause' since we lemmatized the text and 'cause' and 'caused' belong to the same lemma.

[27] Five nearest neighbours for 'cause': 'proximate' (0.64), 'efficient' (0.60), 'effect' (0.59), 'volition' (0.51) and 'necessitate' (0.50); five nearest neighbours for 'responsible': 'accountable' (0.69), 'blameworthy' (0.64), 'attributable' (0.58), 'culpable' (0.56) and 'negligent' (0.55).

suggest that the philosophical usage diverges from the ordinary usage, as will be spelled out below. The space performed comparably on the MEN benchmark, with a Spearman's rho of 0.48 on a much higher number of comparisons.[28]

The same tools can be applied to other corpora that are available for download, including COCA. To facilitate comparison to philosophical usage, we excluded academic texts. Since COCA comes with lemma information, we did not need to annotate the documents. Other than this we followed the same procedure detailed for the philosophy corpus to generate a word2vec space. The results were in keeping with what we saw for the other general corpora above, with there being a notable similarity between 'cause' and 'responsible', between 'cause' and the negative lemmas 'blame' and 'fault', and no similarity between 'cause' and the positive lemma 'praise' (see Table 7.7). As expected, the space performed extremely well on the MEN benchmark with a Spearman's rho of 0.80.

With access to a full corpus it is also possible to target causal attributions and responsibility attributions more carefully by directly comparing multi-word expressions. To do this we replaced the phrases 'caused the' and 'responsible for the' with single tokens before lemmatizing and processing the non-academic COCA corpus. We then analysed it using the same predictive model as above. The cosine value between the causal attribution token and the responsibility token was quite large, and notably larger than for the previous comparison between 'cause' and 'responsible' (see Table 7.8). Further, each token was one of

Table 7.7 Cosine values for term comparisons for the non-academic COCA corpus word2vec space.

	responsible	blame	fault	praise
cause	0.41	0.56	0.48	−0.06

Table 7.8 Cosine values for term comparisons for the non-academic COCA corpus word2vec space with causal attribution and responsibility attribution tokens.

	responsible for the	blame	fault	praise
caused the	0.63	0.55	0.42	−0.10
responsible for the		0.61	0.35	0.16

[28] 944 of the 1,000 comparisons were used (25 terms were missing from the corpus).

the five nearest neighbours of the other. The nearest neighbours for each token included a number of terms with a negative cast – e.g. 'catastrophic' (0.68), 'fatal' (0.63), 'culpable' (0.62), 'complicit' (0.58) – including that 'blame' was one of the five nearest neighbours for the responsibility attribution token. In addition, none of the nearest neighbours for this token indicated the notion of duties associated with a role that marred our previous results.

To better compare philosophical usage with ordinary usage, we tokenized the philosophy corpus and repeated the analysis. We found that the causal attribution token was quite similar to the responsibility attribution token. In line with the alternative prediction noted above, this might reflect that many philosophers hold that causing an outcome is a prerequisite for being responsible for that outcome. Although the cosine values for the two tokens are similar for both the philosophy space and the COCA space, when we look deeper we find evidence that the causal attributions of philosophers are quite different from the causal attributions of more ordinary language. Thus, while there was a strong relation between the two tokens in the philosophy space, the causal attribution token was much less similar to 'blame' and 'fault' (see Table 7.9). This stands in marked contrast to what we saw for the COCA space. Further, the same contrast holds for responsibility attributions. The results suggest that the ordinary usage of both causal attributions and responsibility attributions has a negative cast that the philosophical usage lacks.

Overall the results of our latent semantic analyses nicely line up with the results of the analyses in the previous section, with the two approaches providing a consilience of evidence. Looking across the two sets of analyses, we find compelling evidence for a positive answer to each of the questions we opened this chapter with: our results provide independent support for the thesis that ordinary causal attributions are sensitive to normative information; they provide support for the contention that outcome valence often matters for ordinary causal attributions; they indicate that causal attributions are similar to responsibility attributions[29]; and they suggest that philosophers use the language of causal attribution differently from lay people.

---

[29] It could be objected here that while our results indicate that causal attributions are similar to responsibility attributions, they do not indicate that causal attributions are themselves normative. For instance, it might be suggested that they are close in semantic space because causing an outcome is a prerequisite for being responsible for that outcome. This would not explain the results of our analysis in the previous section, however, or the valence effect observed for causal attributions in the semantic spaces. While this could be investigated further using DSMs to assess the Synonymy Effect hypothesis, space prevents us from doing so here. Alternatively, expanding on Alicke's view, it might be argued that the desire to blame biases both causal attributions *and* responsibility attributions. While we cannot rule this out based on the present results, we hold that the responsibility view offers the simpler explanation.

Table 7.9 Cosine values for term comparisons for the philosophy corpus word2vec space with causal attribution and responsibility attribution tokens.

	responsible for the	blame	fault	praise
caused the	0.55	0.16	0.17	0.03
responsible for the		0.10	0.07	0.00

6. Concluding remarks: Corpus analysis as a method for experimental philosophy

Causation is one of the most contested concepts in philosophy. Recent questionnaire-based studies have produced some rather surprising insights into how we use that concept. Most importantly, they suggest that normative considerations play a central role in ordinary causal attributions. However, it is still an open issue how best to interpret and explain these results.

We believe that to make progress in deciding between the different accounts of the impact of norms on causal attributions, it is fruitful to expand the set of empirical tools used by experimental philosophers working on causation. Specifically, we believe that the study of ordinary causal attributions can benefit from the tools of corpus linguistics. One reason for this belief is that while questionnaire methods are powerful and often well-suited to investigating philosophical questions, they also have limitations. And the questionnaire-based studies on ordinary causal attributions that we have looked at in this chapter do suffer from some of those limitations. To illustrate, we will focus on the first study we discussed in Section 2 Knobe and Fraser's (2008) investigation of the Pen Case.[30]

The basic misgiving one might have about Knobe and Fraser's study is that it relies on an instrument they unintentionally designed in such a way that it would elicit the suspected effect, not because norms actually do have an impact on ordinary causal attributions but because the study leads participants to misread the prompts. For instance, one might object to participants being asked which one of the two agents, the administrative assistant or the professor, caused the problem. While this is certainly a very natural way to ask the question of interest, we believe that asking people who caused 'the problem', as compared

---

[30] We would like to emphasize that even though we focus on Knobe and Fraser here, our worries apply to questionnaire studies more generally, and the authors recognize that their own work is also liable to these methodological concerns.

to 'the outcome' or 'the situation', might trigger an interpretation of the question in normative terms. Alternatively, one might worry that by having participants rate two statements – one about the administrative assistant and one about Professor Smith – and phrasing this in a way that suggests an all-or-nothing state of affairs (as opposed, for example, to asking whether the agent was 'a cause' of the outcome) might prompt participants to feel that they should agree with at most one of the two questions. Since the only distinguishing feature between the two agents' actions is that one violates a norm while the other does not, participants might latch onto this as a relevant cue for fulfilling the task. Or one might note that Knobe and Fraser asked participants whether an *agent* caused an outcome. Typically, when philosophers talk about causation, they talk about *events* as the causal relata, not *people*. Asking about the agent, rather than his or her activities, might therefore create another reason for participants to believe that the researcher is asking about something normative.

All of the potential issues we just noted for Knobe and Fraser's study could be addressed through further questionnaire-based research. And, in fact, a good deal of work has subsequently been done on the Pen Case, or cases like it, that varies these sorts of factors. But follow-up studies addressing one potential confound run the risk of introducing others. This is simply one of the difficulties inherent to this type of research. It does not mean, of course, that questionnaire studies should be abandoned. Rather, the moral we should draw from it is that in the face of these risks we should diversify our set of methods. Turning to corpus linguistics seems natural here, as one strength of corpus analysis is that it is relatively immune from the pragmatic pitfalls we have just highlighted.

One of the motivations of corpus linguistics is the preference for 'real' language data over examples of language use generated by linguists themselves. While a corpus cannot, strictly speaking, be representative for a language in its entirety (because possible utterances of that language are infinite), linguistic corpora aim at balanced sampling from this impressive population. Unless the research interest is focused on a specific genre – such as the usage of a given term in academic texts – a balanced corpus contains a considered choice of texts of various types and from different sources. For example, it does not only contain written, but also (transcribed) spoken language, not only specialized (e.g. academic) language, but also its everyday variety, not only literary texts, but also mundane ones such as operation manuals. In this way, a large corpus does present a meaningful sample of actual language use.

The preference for 'real' language sits nicely with precepts of experimental philosophy, in that it emphasizes the importance of empirical data over that generated by researchers relying on their own intuition or judgement. An important advantage it has over data generated by questionnaire studies is that the linguistic data of a general corpus usually has been generated independently of the researcher and her specific research questions. The data thus is usually uncued, in the sense that the utterances the corpus contains have not been produced in response to some prompt of the researcher. It is then plausible to assume that a corpus is unbiased with respect to the specific research question with which a philosopher approaches it (cf. Schütze 2010).[31] And for the same reason, such a corpus can be considered free of the biases of experimental pragmatics.

Having said that, there are, of course, limits to corpus analysis. The linguistic data of a corpus can be evidence in relation to some philosophical issue only to the extent that the actual use of language is relevant to it. This relevance may be direct or indirect, because the linguistic data in a corpus may well allow us to infer something about deeper structures of language. But, clearly, if the observation of linguistic phenomena in actual use is irrelevant to a philosophical issue, then so is corpus analysis. Moreover, if the pertinent phenomena are of the very particular and subtle kind that is common for philosophical problems, even a large corpus may not yield any, let alone many, examples of their use. By way of contrast, questionnaires can be constructed to elicit informants' responses to precisely worded prompts, and in doing so, the wording can easily be varied to bring out and test subtle differences in language.

Therefore it is clear that corpus analysis is best viewed as a fruitful addition to the methodological toolbox of experimental philosophy. Not only can it be used effectively to explore the actual use of linguistic expressions – something that is called for in philosophy on many occasions – it can, more specifically, be used to complement experimental studies in several helpful ways: to pre-test hypotheses that inform such studies, to help with the general construction of questionnaires and the precise wording of their items, and, most importantly, we believe, to test the findings from questionnaire studies, either giving them independent support from another empirical source or providing evidence against them.

---

[31] It is, of course, possible to come up with examples of corpora that are dependent on the researcher and that contain cued language use. Most simply, for example, in the case that the corpus consists of written responses to a vignette. It is the choice of texts that determines whether the data contained in a corpus is indeed independent and uncued. If a pre-existing general language corpus is used, this objection can be assumed to be moot.

Suggested readings

Bluhm, R. (2016). Corpus analysis in philosophy. In M. Hinton (ed.), *Evidence, Experiment and Argument in Linguistics and the Philosophy of Language* (pp. 91–109). Frankfurt am Main: Peter Lang.

Erk, K. (2012). Vector space models of word meaning and phrase meaning: A survey. *Language and Linguistics Compass*, 6(10), 635–653.

McEnery, T., and Wilson, A. (2001). *Corpus linguistics* (2nd edn.). Edinburgh: Edinburgh University Press.

Turney, Peter, and Pantel, Patrick (2010). From frequency to meaning: Vector space models of semantics. *Journal of Artificial Intelligence Research*, 37, 141–188.

Bibliography

Achugar, M, and Schleppegrell, M. J. (2005). Beyond connectors. *Linguistics and Education*, 16, 298–318.

Alicke, M. D. (1992). Culpable causation. *Journal of Personality and Social Psychology*, 63, 368–378.

Altenberg, B. (1984). Causal linking in spoken and written english. *Studia Linguistica*, 38, 20–69.

Baroni, M., Bernardi, R., and Zamparelli, R. (2014a). Frege in space. *Linguistic Issues in Language Technology*, 9, 241–346.

Baroni, M., Dinu, G., and Kruszewski, G. (2014b). Don't count, predict! A systematic comparison of context-counting vs. context-predicting semantic vectors. *Proceedings of the 52nd Annual Meeting of the Association for Computational Linguistics*, 238–247.

Bergh, G., and Zanchetta, E. (2008). Web linguistics. In A. Lüdeling and M. Kytö (eds.), *Corpus Linguistics*, Vol. 1 (pp. 309–327). Berlin: de Gruyter.

Biber, D. (2010). Corpus-based and corpus–driven analyses of language variation and use. In B. Heine and H. Narrog (eds.), *The Oxford Handbook of Linguistic Analysis* (pp. 159–191). Oxford: Oxford University Press.

Biber, D., Conrad, S. and Reppen, R. (1998). *Corpus Linguistics*, Cambridge: Cambridge University Press.

Bluhm, R. (2012). *Selbsttäuscherische Hoffnung: Eine sprachanalytische Annäherung.* Münster: mentis

Bluhm, R. (2013). Don't ask, look! Linguistic corpora as a tool for conceptual analysis. In M. Hoeltje, T. Spitzley, and W. Spohn (eds.), *Was dürfen wir glauben? Was sollen wir tun? Sektionsbeiträge des achten internationalen Kongresses der Gesellschaft für Analytische Philosophie e.V.* Duisburg: DuEPublico.

Bluhm, R. (2016). Corpus analysis in philosophy. In M. Hinton (ed.), *Evidence, Experiment and Argument in Linguistics and the Philosophy of Language* (pp. 91–109). Frankfurt am Main: Peter Lang.

Bruni, E., Tran, N. K. and Baroni, M. (2013). Multimodal distributional semantics. *Journal of Artificial Intelligence Research 49*, 1–47.

Cushman, F. (2013). Outcome, and value: A dual-system framework for morality. *Personality and Social Psychology Review, 17*(3), pp. 273–92.

De Schryver, G.-M. (2002). Web for/as corpus. *Nordic Journal of African Studies, 11*, 266–282.

Deerwester, S., Dumais, S., Landauer, T., Furnas, G., and Harshman, R. (1990). Indexing by Latent Semantic Analysis. *Journal of the Society for Information Science, 41*(6), 391–407.

Diessel, H., and Hetterle, K. (2011). Causal clauses: A crosslinguistic investigation of their structure, meaning, and use. In P. Siemund (ed.), *Linguistic Universals and Language Variation* (pp. 23–54). Berlin: Mouton de Gruyter.

Erk, K. (2012). Vector space models of word meaning and phrase meaning: A survey. *Language and Linguistics Compass, 6*(10), 635–653.

Firth, J. R. (1957). A synopsis of linguistic theory 1930–55. In *Studies in Linguistic Analysis* (pp. 1–32). Oxford: Blackwell.

Fischer, E., and Engelhardt, P. E. (2017). Diagnostic experimental philosophy. *teorema, 36*(3), 117–137.

Günther, F., Dudschig, C., and Kaup, B. (2015). LSAfun: An R package for computations based on Latent Semantic Analysis. *Behavior Research Methods, 47*, 930–944.

Hahn, U., Zenker, F., and Bluhm, R. (2017). Causal argument. In M. R. Waldmann (ed.), *The Oxford Handbook of Causal Reasoning* (pp. 475–494). New York, NY: Oxford University Press.

Harris, Z. (1954). Distributional structure. *Word, 10*(23), 146–162.

Henne, P., Pinillos, Á., and Brigard, F. De (2015). Cause by omission and norm: Not watering plants. *Australasian Journal of Philosophy, XXX*, 1–14.

Hilton, D., and Slugoski, B. (1986). Knowledge-based causal attribution: The abnormal conditions focus model. *Psychological Review, 93*, 75–88.

Hitchcock, C., and Knobe, J. (2009). Cause and norm. *The Journal of Philosophy, 106*, 587–612.

Khoo, C., Chan, S., and Niu, Y. (2002). The many facets of the cause-effect relation. In R. Green, C. A. Bean, and S. H. Myaeng (eds.), *The Semantics of Relationships* (pp. 51–70). Dordrecht: Springer Netherlands.

Kiela, D., and Clark, S. (2014). A systematic study of semantic vector space model parameters. *Proceedings of the 2nd Workshop on Continuous Vector Space Models and their Compositionality*, 21–30.

Kilgarriff, A., and Greffenstette, G. (2003). 'Introduction' to 'the Web as Corpus'. *Computational Linguistics, 29*(3), 333–347.

Knobe, J., and Fraser, B. (2008). Causal judgments and moral judgment: Two experiments. In W. Sinnott–Armstrong (ed.), *Moral Psychology, Volume 2: The Cognitive Science of Morality* (pp. 441–447). Cambridge: MIT Press.

Kominsky, J., Phillips J., Gerstenberg, T., Lagnado, D., and Knobe, J. (2015). Causal superseding. *Cognition, 137*, 196–209.

Lee, D. Y. W. (2010). What corpora are available? In M. McCarthy and A. O'Keeffe (eds.), *Corpus Linguistics* (pp. 107–121). London, New York: Routledge.

Leech, G. (1992). Corpora and theories of linguistic performance. In J. Svartvik (ed.), *Directions in Corpus Linguistics* (pp. 105–122). Berlin, New York: Mouton de Gruyter.

Livengood, J., Sytsma J., and Rose, D. (2017). Following the FAD: Folk attributions and theories of actual causation. *Review of Philosophy and Psychology, 8*(2), 274–294.

Livengood, J., and Sytsma, J. (under review). *Actual Causation and Compositionality.*

Lüdeling, A., and M. Kytö (eds.) (2008–2009). *Corpus Linguistics* (2 vols.). Berlin: de Gruyter.

Malle, B., Guglielmo S., and Monroe, A. E. (2014). A theory of blame. *Psychological Inquiry 25*(2), 147–86.

McCarthy, M., and O'Keefe, A. (2010a). Historical perspective. In M. McCarthy and A. O'Keefe (eds.), *The Routledge Handbook of Corpus Linguistics* (pp. 3–13). London, New York: Routledge.

McCarthy, M., and O'Keefe, A. (eds.) (2010b). *The Routledge Handbook of Corpus Linguistics.* London, New York: Routledge.

McEnery, T., and Wilson, A. (2001). *Corpus linguistics* (2nd edn.). Edinburgh: Edinburgh University Press.

Prinz, J. (2007). *The Emotional Construction of Morals.* Oxford: Oxford University Press.

Reuter, K. (2011). Distinguishing the appearance from the reality of pain. *Journal of Consciousness Studies, 18*(9–10), 94–109.

Reuter, K., Kirfel, L., Riel, R. van, and Barlassina, L. (2014). The good, the bad, and the timely: How temporal order and moral judgment influence causal selection. *Frontiers in Psychology, 5*, 1336.

Samland, J., and Waldmann, M. R. (2014). Do social norms influence causal inferences? In P. Bello, M. Guarini, M. McShane, and B. Scassellati (eds.), *Proceedings of the 36th Annual Conference of the Cognitive Science Society* (pp. 1359–1364). Austin, TX: Cognitive Science Society.

Samland, J. and Waldmann, M. R. (2016). How prescriptive norms influence causal inferences. *Cognition, 156*, 164–176.

Schütze, C. T. (2010). Data and evidence. In K. Brown, A. Barber, and R. J. Stainton (eds.), *Concise Encyclopedia of Philosophy of Language and Linguistics* (pp. 117–123). Amsterdam et al.: Elsevier.

Sytsma, J., Livengood, J., and Rose, D. (2012). Two types of typicality: Rethinking the role of statistical typicality in ordinary causal attributions. *Studies in History and Philosophy of Biological and Biomedical Sciences, 43*, 814–820.

Sytsma, J., Rose, D., and Livengood, J. (ms). *The Extent of Causal Superseding*.

Sytsma, J., and Livengood, J. (2016). *The Theory and Practice of Experimental Philosophy*. Peterborough: Broadview.

Sytsma, J., and Livengood, J. (2018). *Intervention, Bias, Responsibility … and the Trolley Problem*. http://philsci-archive.pitt.edu/14549/.

Sytsma, J., and Reuter, K. (2017). Experimental philosophy of pain. *Indian Council of Philosophical Research, 34*(3), 611–628.

Turney, Peter and Pantel, Patrick (2010). From frequency to meaning: Vector space models of semantics. *Journal of Artificial Intelligence Research, 37*, 141–188.

Willemsen, P. (2016). Omissions and expectations: A new approach to the things we failed to to do. *Synthese, 19*(4), 1587–1614.

Willemsen, P. and Reuter, K. (2016). Is there really an omission effect? *Philosophical Psychology, 29*(8), 1142–1159.

Xiao, R. (2008). Well-known and influential corpora. In A. Lüdeling and M. Kytö (eds.), *Corpus Linguistics*, Vol. 1 (pp. 383–457). Berlin, New York: de Gruyter.

8

Using Corpus Linguistics to Investigate Mathematical Explanation

Juan Pablo Mejía-Ramos, Lara Alcock, Kristen Lew, Paolo Rago,
Chris Sangwin and Matthew Inglis

1. Corpus linguistics

Corpus linguistics is a methodological approach that involves analysing large collections of naturally occurring texts, known as corpora. Its methods can be used to investigate many types of linguistic questions. Before reporting our study on the notion of explanation in mathematics and physics research papers, we briefly outline the basic concepts of this approach. This outline of corpus linguistics falls into three parts, each focusing on an important stage of conducting a corpus analysis: assembling a corpus, processing raw text to render it suitable for analysis and deciding upon an analytical approach.

1.1. Assembling a corpus

A corpus is simply a large collection of machine-readable texts designed to represent some broader body of natural language. In theory, any text could be considered a corpus, but the term is normally reserved for a set of texts carefully sampled to be representative of a larger body of language. For example, while we might analyse the complete *Diary of Samuel Pepys* with a

Acknowledgements

This project was funded by the British Academy and The Leverhulme Trust. This work was presented at the 20th Annual Conference on Research on Undergraduate Mathematics Education (San Diego 2017) and the 25th International Congress of History of Science and Technology (Rio de Janeiro 2017) and we are grateful to the audiences for their valuable suggestions. We also thank the editor of the present volume and an anonymous referee.

view to understanding linguistic features of Pepys's writing, we would not consider it a corpus representative of the writing of 17th century England: such a generalization would be problematic because we would not know whether a particular linguistic feature was characteristic of the period's writing generally, or only of Pepys's writing.

The desire for generalization arises because corpus linguists are most commonly interested in understanding the properties of some broad body of language, such as broadsheet newspaper articles or political speeches. As it would be difficult to collect the text of every political speech ever made, a first consideration is how to obtain as *representative* a sample as possible (Biber 1993). This gives rise to important issues of sampling that parallel those of traditional empirical research. Just as experimental psychologists ideally seek to sample participants randomly from their population of interest, corpus linguists ideally seek to sample texts randomly from the wider set of texts to which they would like to generalize. The population is referred to as the 'sampling frame', and sometimes genuine random sampling can occur: given access to an appropriate archive, it would be possible to randomly select 10% of all newspaper articles published in a given time period. But corpus linguists are often interested in a less accessible sampling frame, which makes it difficult to randomly sample. For example, in the current investigation we would like to generalize to all research-level mathematical writing. But randomly sampling from this population would be difficult, as some writing in the population is inaccessible. In such situations we must instead appeal to the representativeness of our corpus.

One common approach to ensuring adequately representative sampling is to use a bibliographic index. For instance, researchers might define the sampling frame to be every text included in a particular list of published texts. The Lancaster-Oslo/Bergen (LOB) corpus, designed to be representative of general written British English, took this approach and used the British National Bibliography and Willings' Press Guide as indices (Johansson et al. 1978). Alternatively, it is possible to sample participants rather than texts. For instance, the British National Corpus (a comprehensive collection of 100 million words of spoken and written English, designed to represent a cross-section of current English usage) contains a spoken section where participants – selected using demographic sampling techniques – were asked to record their day-to-day spoken interactions for several days (Crowdy 1993). In either case, representativeness might be further ensured via hierarchical or stratified sampling approaches. A sample might be composed of 10% of texts from one sub-category, 10% from another and so on. The Brown and LOB corpora both adopted this approach in an attempt to be representative of American and British English respectively. Each contains 500 texts sampled from 15 categories (e.g. press reportage, popular lore, general fiction, science fiction, learned and scientific writing).

A second consideration when assembling a corpus is size. The required size depends on the linguistic feature being studied: if the feature is relatively rare, then a much larger corpus will be needed. Biber (1993), for instance, gave the relative frequencies of various linguistic features in a particular corpus, noting that conditional subordination occurred 2.5 times per 1,000 words, whereas prepositions occurred 111 times. Clearly, this means that a larger corpus is required to study conditional subordinations than prepositions. Fortunately, creating extremely large samples of texts has recently become considerably easier, and corpora have been constructed based on webpages (e.g. various corpora based on Wikipedia articles), on television subtitles (e.g. the SUBTLEX-UK corpus; Van Heuven et al. 2014) and on parliamentary proceedings (e.g. the Hansard corpus; The SAMUELS Consortium 2015).

A third consideration concerns dispersion. This refers to how evenly distributed a linguistic feature is across texts in a corpus. If a feature appears in only few texts, perhaps written by only a few authors, then this calls into question the generalizability of any claimed results, even if the corpus is reasonably representative in general (McEnery and Wilson 2001).

In the study we describe below, our decisions concerning the sampling frame meant that we were able to assemble a large corpus of mathematical texts (approximately 31 million words) as well as a control corpus of physics texts (approximately 59 million words). We then addressed considerations of representativeness by following good practice from empirical research in psychology: we assembled two further corpora of mathematics and physics texts of approximately the same size and from the same source, thus allowing us to replicate all our analyses on a new dataset. This should enable the reader to feel confident that we did not conduct a large number of analyses and report only those which gave statistically significant results (*cf.* John et al. 2012 and Simmons et al. 2011, on *p*-hacking).

1.2. Processing the corpus

Assembling a novel corpus, or selecting an existing corpus, is only the first stage of a corpus linguistics research project. Often it is necessary to process texts in some way before proceeding with the analysis. In general contexts it is often important to annotate a corpus with tags relevant to the researchers' questions.

This might involve grammatical tagging (often called part-of-speech, or POS tagging) where each word in the corpus is tagged with a label that categorizes it in some way. For instance, it may be useful to know if a word is an adjective, noun or adverb (and so on). One common way of tagging a corpus is to append the tag after each word (e.g. replacing 'cutting' with 'cutting_NN' where the 'NN' represents the code 'noun, singular or mass'). Leech (2013) pointed out several important features an annotation scheme should have. First, it should always be possible to remove the tags and revert to the raw corpus. Second, the tags should be extricable from the corpus if necessary (e.g. it should be possible to count the number of nouns in a corpus). Third, the annotation scheme should be carefully documented. Leech also emphasized that the quality of the annotation should be documented. If, for example, a computer-based POS tagger is used (e.g. TagAnt), it might be appropriate to manually check a sample of the corpus and record the agreement percentage. Understandably, automated computer-based POS tagging is a complex process (for a review see Garside et al. 2013).

POS tagging has been successfully applied to mathematical corpora. For instance, Dawkins et al. (2018) used a corpus of university-level mathematics textbooks and presented a comparative analysis of the use of 'is' in mathematical and non-mathematical English. Because they were interested in advanced scientific texts that contain complex mathematical notation, this raised particular issues about the processing of LaTeX-encoded mathematical symbols. Our approach to this issue is discussed in the methods section below.

1.3. Analysing the corpus

Once a corpus has been successfully assembled and processed, the next step is to choose an analysis technique that addresses the research question. Of course, there are many possible analytical approaches; here we give a brief overview of only the most common.

Often it is possible to answer research questions by simply studying the frequency with which certain words or phrases occur. Mejía-Ramos and Inglis (2011), for instance, used this approach to analyse 'semantic contamination' from day-to-day English into mathematical language. Semantic contamination refers to the phenomenon in which the meanings of words in natural language 'leak' into a different linguistic register (e.g. Monaghan 1991). Mejía-Ramos and Inglis compared the frequency of the verb and noun forms of the word 'proof' (i.e. 'prove' and 'proof') in the specialist (business, medical, legal proceedings, etc.) and informal (conversations, popular radio, etc.) sections of the British

National Corpus. They found that the verb form was significantly more common in informal than in formal language, and derived the hypothesis that 'proof' was more likely to be associated with the notions of formal validity, whereas 'prove' was more likely to be associated with the less formal notion of conviction. In two subsequent experimental studies, they found that changing a question from 'does the argument prove the claim?' to 'is the argument a proof of the claim?' did indeed change students' responses in the direction predicted.

In the study reported here, our interest in the relative frequency of a particular set of words (which we will refer to as 'explain words') in mathematical writing means that our main analysis too involved defining which words fell into our category and counting their occurrences. By also conducting the same analysis on a different corpus (of physics papers) we were then able to compare the frequencies in two subgenres of research papers. Clearly, when comparing frequencies in corpora of different sizes it is necessary to adopt a frequency rate measure; the number of occurrences per million words is typically used. Counting the (relative) frequencies of a category of words in two corpora generates a two-by-two contingency table, with the two corpora as rows and the hits and non-hits (words in the corpus that are and are not in the category) as columns. This permits use of a chi-squared test or Fisher's exact test to assess statistically whether the relative frequencies differ significantly between the two corpora. Our analysis is reported below.

As well as counting words or categories of words, it is also possible to study the frequencies of more complex linguistic features. For instance, one might be interested in producing a list of the most frequently occurring 'n-grams' or 'lexical bundles' – collections of words of a given length. Or one might be interested in producing a frequency list of 'clusters' – collections of words of a given length that contain a given word. Herbel-Eisenmann, Wagner and Cortes (2010) used the notion of a lexical bundle to study common interaction patterns in mathematics classrooms. Using a corpus formed from transcripts of interactions in secondary mathematics lessons, they found that there are particular types of lexical bundles involved in teacher/student interactions that allow the communication of feelings, attitudes, value judgements and assessments. By comparing their findings with other corpora (of, for instance, university classes) the researchers were able to argue that their findings were particular to the secondary mathematics context.

An alternative way to compare corpora is to identify keywords: words that occur disproportionately often in one corpus compared to another, but that have not been identified a priori by the researcher. These can be identified using chi-squared statistics in a similar manner to the approach described earlier. For instance, we can compare British and American English by identifying the keywords in the LOB corpus compared to the Brown corpus (organized by chi-squared values, these are: London, labour, I, sir, Mr, she, towards, Britain, British and centre). Similarly, we can identify keywords in the Brown corpus compared to the LOB corpus (program, toward, state, states, center, defense, federal, labor, York and American). By identifying the keywords in one corpus with respect to another, researchers can begin to understand differences between the bodies of language represented by each of the corpora.

More qualitative ways to analyse corpora can help researchers to understand how given words are used. For instance, most corpus linguistics software packages allow examination of 'concordances' or 'key words in context' (KWIC). The packages generate lists containing every occurrence of a given word – the search term – with context on either side (perhaps 80 characters to either side of the occurrence). By carefully studying a concordance, or a randomly selected subset of a concordance, researchers can begin to develop categories capturing how the word is used. This can subsequently support further quantitative analysis, especially if multiple equivalent corpora are available (e.g. a concordance analysis can be conducted on one corpus and then a quantitative analysis can be used on the other to triangulate).

Similarly, packages can also permit examination of words that systematically co-occur. For instance, 'back' and 'front' are often found close to each other, and corpus linguists would say that 'back' is a collocate of 'front' (and vice versa). More formally, two words are collocates if there is an above-chance co-occurrence of them within some given span (perhaps plus or minus five words). Collocates can be identified by constructing a word frequency list of all words within a five-word window around the search term, and comparing it to the overall word frequency list of the corpus. Words that disproportionately occur around the search term are its collocates (various statistical criteria can be used to formalize what 'disproportionately' means). Understanding the collocates of a given word can help reveal its meaning, and perhaps uncover implicit associations that it has with other words or ideas (Sinclair 1991; Hunston 2002). In particular, studying the collocates of a word can identify a word's 'semantic prosody', the 'consistent aura of meaning with which a form is imbued by its collocates' (Louw 1993, 157). Baker, Gabrielatos, Khosravinik, Krzyżanowski, McEnery and Wodak (2008) used this method to study representations of refugees and asylum seekers in British newspapers, finding that collocates were often words that may negatively stereotype refugees and asylum seekers. For example, their collocate analysis showed how references to refugees and asylum seekers were often accompanied by quantification via water metaphors (e.g. pour, flood, stream), which 'tend to dehumanize [refugees and asylum seekers], constructing them as an out-of-control, agentless, unwanted natural disaster.' (p. 287) In other words, the words 'refugee' and 'asylum seeker' have negative semantic prosody in British newspapers.

The outline of corpus linguistic research methods given here is necessarily basic. Readers interested in conducting corpus analyses might wish to consult McEnery and Wilson's (2001) excellent textbook for further information. We now turn to our investigation of the notion of explanation in mathematics and physics research papers.

2. Mathematical explanation

Explanations are important, nowhere more so than in education. Teachers routinely offer instructional explanations as part of classroom practice, answering implicit or explicit questions posed by their students or themselves (e.g. Treagust and Harrison 1999; Leinhardt 2001). Also important are self-explanations generated by learners with the aim of increasing their own understanding (Rittle-Johnson et al. 2017). Encouraging students to self-explain can be a highly effective pedagogical strategy: in the context of university level mathematics, Hodds et al. (2014) found that students prompted to explain a mathematical proof to themselves attained comprehension one standard deviation better than peers in a control group (*cf.* Chi et al.1989; Fonseca and Chi 2011). Finally, student-generated explanations can be used in educational assessment, particularly when one wishes to focus on the depth of students' conceptual understanding (e.g. Knuth et al. 2006; Bisson et al. 2016). But what are explanations, especially in mathematics?

Philosophers of science have devoted considerable attention to the question of what it means for A to explain B. Many accounts rely on either statistical associations or causal mechanisms. For instance, Salmon (1971, 1984) suggested that A explains B if B is consistently correlated with A or if there is a causal history that connects B and A (*cf.* Hempel and Oppenheim 1948). So one can say that buying shoes in the wrong size explains why one's feet hurt because there is a causal connection between the two events. But, while causal and statistical accounts work well in scientific contexts, they fail in mathematics (Colyvan 2011; Mancosu 2001). Mathematical concepts are not related causally because there is no temporal order: the fact that the square root of 2 is irrational is not located at a particular time. Nor are they related statistically: mathematical facts take no probabilities other than 0 or 1. Consequently, scientific accounts of explanation do not seem to apply to mathematics.

But if mathematical explanations are not scientific explanations, then what are they? This question has generated significant philosophical interest. A small number of philosophers regard the lack of causal and correlational relations as reason to deny that mathematical explanations exist (Resnik and Kushner 1987; Zelcer 2013), arguing that there is little empirical evidence to suggest that explaining is central to mathematicians' practice. However, others vociferously dispute this (e.g. Colyvan 2011; Weber and Frans 2016). The dispute seems to turn on the extent to which practicing mathematicians use the notion of explanatory value in their own work. For example, Steiner (1978) claimed that 'mathematicians routinely distinguish proofs that merely demonstrate from proofs which explain' (p.135). In contrast, Resnik and Kushner (1987) claimed that mathematicians 'rarely describe themselves as explaining' (p. 151). Hafner and Mancosu (2005) responded by stating that '[c]ontrary to what Resnik and Kushner claim (p. 151), mathematicians *often* describe themselves and other mathematicians as explaining' (p. 223, emphasis in the original). Hafner and Mancosu (2005) supported this claim by presenting several examples of what they called *explanatory talk* in mathematical practice: passages of research mathematics papers in which the authors explicitly describe themselves or some piece of mathematics as *explaining* a given 'mathematical phenomenon'. While this evidence is not sufficient to settle the disagreement, the specific cases discussed by Hafner and Mancosu have been interpreted in significantly different ways:

> I believe that detailed case studies, such as those by Hafner and Mancosu (2005), *decisively* refute Resnik's and Kushner's [claim]. (Lange 2009, 203, our emphasis)

> Though philosophers have lately been pointing out some exceptions, the examples tend to be rather exotic (e.g., in Hafner and Mancosu 2005). There has been no systematic analysis of standard and well-discussed texts illustrating any pattern of mathematical explanations. (Zelcer 2013, 179–180)

Clearly – and contrary to Lange's (2009) suggestion – it is impossible to decisively refute a claim that a given event is rare by identifying one or more instances of the event occurring. Instead, a systematic analysis of the type suggested by Zelcer (2013) is needed. Attempts in this direction have been made, in scientific fields if not in mathematics. Overton (2013), for instance, analysed all regular articles published in the journal Science in a one-year period (a total of 781 papers and approximately 1.6 million words). He searched for all 'explain words' (defined to be: *explain, explains, explained, explaining, explainable, explanation, explanations, unexplained, unexplainable, explicate, explicates, explicated, explicable, inexplicable*) and compared their frequencies to those of words of other types. Overton found that approximately 45% of the 781 papers contained at least one 'explain' word (with an average of 0.96 'explain' words per article), and he concluded that: 'The numbers for "explain" are perhaps surprisingly low if scientific journals are vehicles for explanations. [...] The observed frequencies of "explain" words suggest that explanation is only moderately important in science.' (p. 1387). This low frequency of 'explain words' in articles in science – a field in which explanation is widely regarded as playing a central role – might warrant Zelcer's (2013) scepticism of the predominance of explanatory talk in mathematics – a field in which even the existence of explanation is debated. One goal of this chapter is to shed light on this decades-old dispute among philosophers concerning the frequency with which mathematicians describe themselves or their mathematical work as explaining other mathematics. Like Overton (2013), we do this by analysing large collections of text.

A second goal of this chapter is to explore the types of explanations mathematicians discuss in their explanatory talk. To date, analyses of mathematical explanations tend to differentiate between explanations of other mathematics (mathematics X explains mathematics Y, or X is an explanatory proof of theorem Y), and explanations of physical phenomena (mathematics X explains physical phenomenon Y). Colyvan (2011) referred to these as intra-mathematical and extra-mathematical explanations respectively. Hafner and Mancosu (2005) further differentiated between two uses of intra-mathematical explanations: those that are 'instructions' on how to master the tools of the trade, explaining how to employ mathematical techniques, and those that 'call for an account of the mathematical facts themselves, the reason why' (p. 217). While Hafner and Mancosu considered the latter to be a 'deeper' use of mathematical explanation, others have emphasized the importance of the former in mathematical practice. For instance, Rav (1999) emphasized the mathematical methodologies and problem solving strategies/techniques contained in proofs, and insisted that one of the main reasons mathematicians read proofs is to glean this mathematical know-how:

> Proofs are for the mathematician what experimental procedures are for the experimental scientist: in studying them one learns of new ideas, new concepts, new strategies—devices which can be assimilated for one's own research and be further developed. (p. 20)

This claim is consistent with empirical evidence from both small-scale interview studies and large-scale surveys asking mathematicians about their practice (Weber and Mejía-Ramos 2011; Mejía-Ramos and Weber 2014). But neither this claim nor the distinctions upon which it relies have been examined at scale in written mathematical research papers.

In this chapter, we address this issue, reporting a study that employs methods of corpus linguistics to address the following specific questions:

1. To what extent do mathematicians describe themselves (or their mathematical work) as explaining mathematical phenomena in their research papers, and how does this compare with descriptions of explanation in physics discourse and in general, day-to-day discourse?
2. How does the extent to which mathematicians describe themselves as explaining compare with the extent to which they describe themselves as engaging in related mathematical activities (such as *proving* theorems)?
3. To what extent do mathematicians describe themselves as explaining *why* a certain mathematical statement is true, as compared with explaining *how* to do something in mathematics?

3. Methods

3.1. Collecting the texts

For our study, we needed a large sample of mathematics research papers, together with two comparison corpora: a large sample of research papers from another discipline, and corpora of general, day-to-day English. For our comparison disciplinary corpus, we collected physics research papers, and for our day-to-day English corpus we used both the British National Corpus (BNC) and the larger Corpus of Contemporary American English (COCA).[1] To assemble our

---

[1] COCA contains more than 560 million words of spoken, fiction, popular magazines, newspapers, and academic texts.

corpora of mathematics and physics research papers, we adopted two largely pragmatic criteria:

1) Text should be in LaTeX format to enable consistent processing (discussed below).
2) Text should be published non-commercially and freely available online.

Based on these criteria, we used research papers uploaded to the ArXiv (https:// arxiv.org/). The ArXiv is an online repository of electronic preprints of scientific papers in mathematics, physics, astronomy, computer science, quantitative biology, quantitative finance and statistics; it is one of the main repositories that mathematicians and physicists around the world use to share their work. We downloaded the bulk source files (mostly TeX/LaTeX) containing all papers uploaded to the ArXiv in the first eight months of 2009 (which provided us with a large enough sample of more than 30,000 research papers), then converted the source code to plain text for use with standard corpus analysis software (all analyses reported in this paper were performed using CasualConc, version 2.0.3).

3.2. Processing the texts

Converting mathematical language into a form that can be processed using the standard corpus linguistics software presents a challenge. Most professional mathematics is written using the TeX/LaTeX[2] mark-up language, not plain text. Our first goal was therefore to create a method of converting LaTeX source code to plain text in a way that preserved the natural sentence structure of the language, but removed non-linguistic features of the source code (the code "\textbf{text}" for bold text, for instance). We constructed scripts to achieve this, converting LaTeX to analysis-ready plain text.[3]

Another important question for the would-be creator of a mathematical corpus concerns how to deal with inline mathematical notation. For instance, a typical mathematical sentence might be "Let $f : X \to Y$ be a bijection." How

---

[2] TeX was developed in the 1970s (Knuth 1979) to enable digital typesetting of structured documents containing mathematics. Most professional mathematicians still write using TeX or the subsequent LaTeX markup language. TeX/LaTeX is written as plain text documents that include control codes to structure the document and codes to typeset the special symbolism used in mathematics. The system consciously separated the encoding of the document from the processing and production of the human readable (e.g. printed) output.

[3] These scripts are freely available for research purposes at https://github.com/sangwinc/arXiv-text-extracter.

should "$f: X \rightarrow Y$" appear in a plain text corpus? One approach would be to leave the LaTeX source code intact and analyse the code as if it were natural language. The difficulty with this option is that there are several different ways in which one could encode "$f: X \rightarrow Y$" in LaTeX. For instance, "\$f:X\ rightarrow Y\$" and "\(f:X\rightarrow Y\)" produce identical output, and "\$f\,:\,X\longrightarrow Y\$" differs only stylistically (the spacing is wider and length of the arrow slightly longer). We therefore felt that retaining the LaTeX codes would be unhelpful for the majority of questions a researcher would wish to answer using a mathematical corpus. A second option would be to delete all mathematical code entirely, and record the example above as "Let be a bijection". We rejected this option because failing to preserve the logical structures of sentences would influence certain analyses (those that investigate the collocation of words, for instance). Instead we opted to replace all occurrences of inline mathematics with the string "inline_math" (although this choice of string can be altered by users of our scripts if desired).

3.3. Analysing the corpus

With the source files processed, we sorted the articles using their primary subject classification (mathematics, physics, etc.) to assemble our two disciplinary corpora: one containing the processed text of all mathematics papers and the other containing the processed text of all physics papers. As noted above, one benefit of working with these large datasets is that a researcher can partition a large corpus into smaller samples that remain sufficiently large to conduct statistical analyses. This provides samples for both exploratory and confirmatory analyses: the researcher can perform initial analyses on one sample and then test whether the corresponding findings replicate when the same analyses are performed on a different sample. With this in mind, we split each disciplinary corpus into two smaller samples based on the month in which the papers had been uploaded: for each discipline, the first sample contained the papers uploaded in January–April 2009, and the second sample contained the papers uploaded in May–August 2009.

4. Results

Table 8.1 presents the number of physics and mathematics papers uploaded to the ArXiv in January–April and May–August of 2009, together with the

Table 8.1 Number of papers and words in the physics and mathematics corpora

	January–April 2009		May–August 2009	
	#papers	#words	#papers	#words
Mathematics	5,087	30,892,695	4,970	31,289,569
Physics	11,787	58,859,660	12,370	62,807,075

number of words[4] in each set of papers. We notice that in those eight months, researchers uploaded approximately 2.4 times as many physics papers as mathematics papers. We also notice that, on average, physics papers contained around 5,000 words, whereas mathematics papers contained around 6,200 words.

We used the January–April sample to address each of our research questions and the resulting analyses are presented in Sections 4.1 to 4.3, we briefly present the analyses of the replication using the May–August sample.

4.1. Absolute and relative frequency of 'explanatory talk'

Following Overton (2013), we defined 'explain words' to be 18 words linguistically related to the word *explain*[5]:

'Explain words': explain, explains, explained, explaining, explainable, explanation, explanations, explanatory, unexplained, unexplainable, explicate, explicates, explicated, explicating, explicable, inexplicable, explication, explications.

Table 8.2 shows the frequencies of 'explain words' in our corpus of 5,087 mathematics papers and 11,787 physics papers uploaded between January and April of 2009. 'Explain words' occurred a total of 4,910 times in the mathematics

---

[4] A word here is any string of characters between spaces. As discussed above, for these analyses we opted to replace all occurrences of inline mathematics with the string "inline_math" and count it as one word. For instance, the string "Let $f : X \to Y$ be a bijection" in a paper, coded in LaTeX by the authors as "Let \$f:X\rightarrow Y\$ be a bijection", would have been translated to text as "Let inline_math be a bijection" and coded as having 5 words.

[5] Our aim was to be consistent with Overton's (2013) analysis. However we decided to include the words *explanatory, explication, explications,* and *explicating* given their close relation to some of the 14 words in Overton's (2013) original analysis (e.g. the original list included *explicate,* but not *explication*). However, as Table 8.2 shows, these additional words did not appear frequently in these corpora of research papers. Indeed, only five 'explain words' (*explain, explained, explanation, explains,* and *explaining*) make up more than 95% of all appearances of 'explain words' in each of these two corpora.

papers (around 159 times per million words), an average of 0.97 times per paper, with 1,898 of mathematics papers (approximately 37%) in this sample containing at least one 'explain word'. This provides an existence proof of explicit explanatory talk in this corpus. In order to assess whether this is a large or small frequency, we conducted the same analysis on the physics corpus.

In comparison, 'explain words' showed up 21,345 times in the corresponding set of physics papers (around 363 times per million words), an average of 1.81 times per paper, with 6,499 of these papers (roughly 55%) containing at least one 'explain word'. Thus, the number of 'explain words' per million in the physics papers is around 2.28 times that for the mathematics papers.

Figure 8.1 compares across corpora, displaying frequencies per million words of 'explain words' in the mathematics papers, the physics papers, and the two

Table 8.2 Frequency and frequency per million words of 'explain words' appearing in the January–April mathematics and physics papers

	Mathematics		Physics	
	Frequency	Per million	Frequency	Per million
explain	1,827	59.14	7,768	131.97
explained	1,690	54.71	6,513	110.65
explanation	498	16.12	3,564	60.55
explains	484	15.67	1,601	27.20
explaining	175	5.66	914	15.53
explanations	119	3.85	675	11.47
explanatory	51	1.65	62	1.05
unexplained	22	0.71	177	3.01
explication	13	0.42	4	0.07
explicated	10	0.32	15	0.25
explicate	6	0.19	5	0.08
explicating	5	0.16	0	0.00
unexplainable	4	0.13	8	0.14
explications	4	0.13	2	0.03
explainable	1	0.03	23	0.39
explicates	1	0.03	1	0.02
explicable	0	0.00	9	0.15
inexplicable	0	0.00	4	0.07
Total	4,910	158.92	21,345	362.63

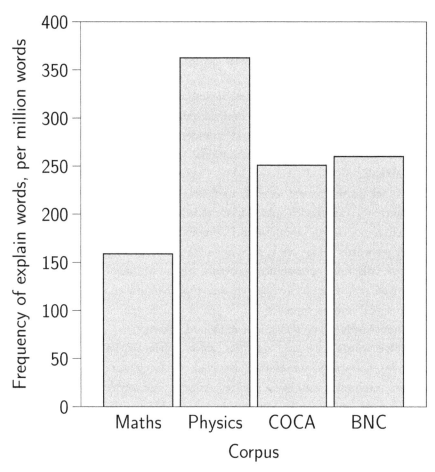

Figure 8.1 Frequency of 'explain words' (per million words) in the mathematics, physics, COCA and BNC corpora.

general English corpora. We note that while the number of 'explain words' per million in the physics papers (362.63) is roughly 1.4 times higher than that in COCA (250.97) or the BNC (260.01), the frequency of 'explain words' in these day-to-day English corpora is still around 1.6 times higher than that for the mathematics papers (158.92).

4.2. Explanation versus related notions

To assess the extent to which the observed frequencies of 'explain words' were high or low *within* mathematical discourse, we compared them against the frequencies of words related to other intuitively important mathematical activities. Table 8.3 presents the frequencies of words linguistically related to the notions of conjecturing, defining, modelling, proving, showing and solving.

'**Conjecture words**':[6] *conjecture, conjectured, conjectures,* conjectural, conjecturally, conjecturing.

'**Define words**': *defined, define, definition, defines, definitions,* defining, definable, undefined, redefine, redefined, definability, redefinition, definably, redefining, welldefined, definedness, interdefinable, predefined, redefinitions, interdefinability, redefines, definitional, definitionally, undefinability, undefinable.

'**Model words**': *model, models,* modeled, modeling, modelled, modelling, countermodel, submodel, submodels, modelized, modelization, modelisation, modelize, modelizing, countermodels, premodel, remodeled.

'**Prove words**': *proof, prove, proved, proves, proofs,* proving, proven, provable, reprove, disprove, provability, provably, reproved, disproved, unprovable, unproven, reproving, disproving, reproves, prover, unproved, subproof, disproof, disproven, disproves, reproven.

'**Show words**': *show, shows, shown, showed,* showing.

'**Solve words**': *solution, solutions, solve, solving, solvable, solved, solves, resolvent,* solvability, subsolution, resolved, resolving, supersolution, resolve, solver, resolvents, unsolved, resolves, solvers, nonsolvable, supersolutions, subsolutions, unresolved, nonsolvable, unsolvable, cosolvable, equisolvable, supersolvable, unsolvability.

Measured against these other frequencies, mathematicians used 'explain words' rather infrequently. For instance, mathematicians used 'explain words' in their papers approximately 12 times less frequently than 'show words' and nearly 23 times less frequently than 'prove words'.

So far, our study of explanatory talk has investigated the use of 'explain words'. This approach has the virtue of focusing on unambiguously explicit discussion of explanation, but could potentially leave unnoticed a significant amount of explanatory talk (i.e. mathematicians describing themselves or their work as explaining some mathematical phenomenon). For instance, the main case of explanatory talk discussed by Hafner and Mancosu (2005) highlights how a mathematician described one of his proofs as providing 'the true reason why' a given mathematical phenomenon was the case. Clearly, this is a case

---

[6]   For each group, words are listed in order of frequency, with the most frequent words in the group listed first. The italicized words in each group make up 95% of all instances of words from that group appearing in the mathematics papers uploaded in the first four months of 2009.

Table 8.3 Frequencies of words related to *explaining, conjecturing, defining, modelling, proving, showing* and *solving* in the January–April mathematics papers. The last two columns provide the number of papers containing at least one word in that group and the percentage of such articles

	Frequency	Per million	Per paper	In #papers	In %papers
define	124,129	4,018.07	24.40	4,838	95
prove	111,838	3,620.21	21.99	4,710	93
show	5,9359	1,921.45	11.67	4,691	92
solve	53,013	1,716.04	10.42	3,073	60
model	23,658	765.81	4.65	2,013	40
conjecture	8,362	270.68	1.64	1,413	28
explain	4,910	158.94	0.97	1,898	37

Table 8.4 Frequencies of alternative expressions related to explanatory talk in the January–April mathematics and physics papers

Alternative expression	Concordance search[7]	Mathematics	Physics
"the deep reasons"	deep* reason*	5	16
"an understanding of the essence"	understand* of the essence	0	0
	understand* the essence	0	5
"a better understanding"	better understand*	161	767
"a satisfying reason"	satisfy* reason	0	0
"the reason why"	reason* why	312	924
"the true reason"	true reason*	3	1
"an account of the fact"	an account of the fact	0	0
"the causes of"	cause* of	16	609
	Total	497	2,322

of explanatory talk that does not use 'explain words'. However, a difficulty of expanding our investigation to expressions that do not use 'explain words' is that these alternative expressions may not actually indicate the presence of explanatory talk. For instance, Overton (2013) argued that the use of words such as 'because' (which he found to be ubiquitous in his corpora) may not

---

[7] In concordance searches an asterisk can be used as a wildcard to find words (or expressions with words) that contain a particular string of characters, but with potentially different beginnings or endings. For instance, a search for "deep* reason*" would find the expressions "deeper reasons" and "deep reasoning".

really indicate the presence of scientific explanations (p. 1387). Fortunately, Hafner and Mancosu (2005) identified eight expressions that they found to be commonly used in the literatures of mathematics and philosophy of mathematics to describe the search for mathematical explanations. Table 8.4 presents these expressions along with the specific concordance search we made to investigate their prevalence in the mathematics and physics papers, and the frequencies with which these alternative expressions appeared. The total number of occurrences of these expressions is only about 10% of the total number of 'explain words' in each set of papers. Furthermore, there were disproportionately more occurrences of these expressions in physics than in mathematics, so this analysis does not materially affect the findings based on 'explain words' only. We suggest, therefore, that Hafner and Mancosu (2005) may have overestimated how common these expressions are in mathematics research papers. We are also left wondering to what extent such *common* alternative expressions exist.

4.3. Explaining why versus explaining how

To compare mathematicians' propensity to describe themselves as *explaining why* a certain mathematical statement is true (Hafner and Mancosu's 'deep' explanation) with their propensity to describe themselves as *explaining how* to do something in mathematics (related to Rav's notion of mathematical know-how), we created a concordance to identify every instance in which an *explain word* was followed immediately by the words *why* or *how*. We did this by searching the concordance for *expla* why and *expla* how, and checking that all results were indeed uses of 'explain words'. We then repeated the process with the corpus of physics papers.

When taken together, the total of *expla*-why and *expla*-how expressions were roughly as common in math papers as they were in physics papers, with approximately 22 of these expressions showing up per million words in each set of papers; they formed a relatively small subset of the wider use of 'explain words' (roughly 14% and 6% of *explain word* usage in mathematics and physics, respectively). However, as shown in Table 8.5, the distributions of these two different types of expressions in the mathematics and physics papers differed significantly (Fisher's exact test, $p < .001$). Mathematicians used nearly twice as many *expla*-how expressions as *expla*-why expressions; physicists used between two and three times as many *expla*-why expressions as *expla*-how expressions. Furthermore, general English is more similar to physics than to mathematics in use of these expressions. Figure 8.2 shows that the frequency of

Table 8.5 Frequencies and frequencies per million words of 'explain words' immediately followed by the words *why* or *how* in the January–April mathematics and physics papers.

	Mathematics		Physics	
	Frequency	Per million	Frequency	Per million
expla why	247	7.99	952	16.17
expla how	458	14.83	353	6.00
Total	705	22.82	1,305	22.17

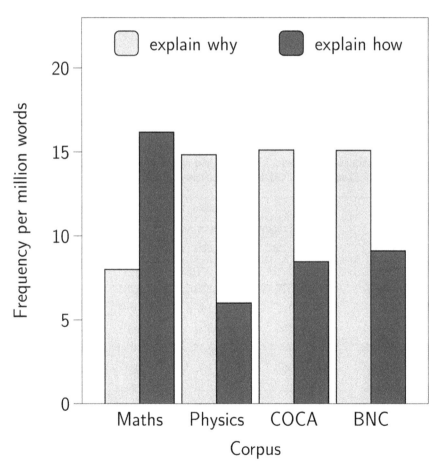

Figure 8.2 Frequencies of 'explain words' immediately followed by the words *why* or *how* in the mathematics, physics, COCA and BNC corpora.

expla-why expressions in general English is around 15 per million words, and that *expla*-how expressions occur roughly half as frequently.

4.4. Replication using the May–August 2009 papers

To replicate our analyses, we used the May–August 2009 papers from the ArXiv. Table 8.6 presents the frequencies of 'explain words' appearing in the May–August set of mathematics and physics papers, while Table 8.7 presents the frequencies of 'explain words' immediately followed by the words *why* or *how* in these sets of papers.

Table 8.6 reveals the same pattern of frequencies as Table 8.2. Indeed, the same five 'explain words' (explain, explained, explanation, explains and explaining) made up 95% of all 'explain words' in each set of papers. Furthermore, the number of 'explain words' per million was very similar in the two sets of papers (158.94 and 163.63 in mathematics; 362.64 and 351.60 in physics), with around 2.15 times more 'explain words' per million in the physics papers than in the mathematics papers.

Table 8.6 Frequency and frequency per million words of 'explain words' appearing in the May–August mathematics and physics papers.

	Mathematics		Physics	
	Frequency	Per million	Frequency	Per million
explain	1,881	60.12	7,974	126.96
explained	1,841	58.84	6,596	105.02
explanation	537	17.16	3,788	60.31
explains	525	16.78	1,694	26.97
explaining	166	5.31	954	15.19
explanations	98	3.13	740	11.78
explanatory	36	1.15	78	1.24
unexplained	19	0.61	159	2.53
explication	1	0.03	5	0.08
explicated	6	0.19	12	0.19
explicate	6	0.19	12	0.19
explicating	0	0.00	5	0.08
unexplainable	0	0.00	1	0.02
explications	2	0.06	5	0.08

Table 8.6 (*Continued*)

	Mathematics		Physics	
	Frequency	Per million	Frequency	Per million
explainable	1	0.03	24	0.38
explicates	0	0.00	0	0.00
explicable	0	0.00	24	0.38
inexplicable	1	0.03	12	0.19
Total	5,120	163.63	22,083	351.60

Table 8.7 Frequency and frequency per million words of 'explain words' immediately followed by the words *why* or *how* in the May–August mathematics and physics papers.

	Mathematics		Physics	
	Frequency	Per million	Frequency	Per million
expla why	277	8.85	970	15.44
expla how	526	16.81	464	7.39
Total	803	25.66	1,434	22.83

Similarly, the frequencies presented in Table 8.7 are consistent with those in Table 8.5: the numbers of *expla*-why and *expla*-how expressions per million were similar in each discipline (25.66 in mathematics; 22.83 in physics), but there were significantly different distributions of these types of expressions in the two sets of papers (Fisher's exact test, $p <.001$). Again, mathematicians used nearly twice as many *expla*-how expressions as *expla*-why expressions, and physicists used a little over twice as many *expla*-why expressions as *expla*-how expressions.

5. Discussion

Our analysis of explanatory language in a large sample of mathematics papers allows us to offer empirically justified contributions to philosophical debates. We relate our findings now to various points raised in our opening sections.

First, our findings do not support the often-made claim in the philosophy of mathematics that explanatory talk is prevalent in mathematical writing. Indeed, mathematics research papers contain less than half the amount of 'explain words' in physics research papers, and less than two-thirds the amount in general

English. Even within the subject, mathematicians discuss explanation less than other practices such as solving problems and proving theorems. Nevertheless, our data are also inconsistent with Zelcer's (2013) claim that mathematicians 'rarely' talk about explanation, whereas this is 'the standard vocabulary' of science: we found around 160 'explain words' per million in mathematics and 360 per million in physics. So, although discussion of explanation is less common in mathematics, it is far from non-existent. Philosophers who appeal to mathematical practice to justify the importance of studying mathematical explanation will find succour in our data.

Second, our data shed light on the types of explanation discussed by mathematicians. We found that when mathematicians engage in explanatory talk, they seem more often interested in explaining how to do something in mathematics than in explaining why things are the way they are. In both physics and general English we found the opposite. This is particularly interesting given the concern philosophers of mathematics devote to intra-mathematical explanations of the form X explains why Y (where X and Y are mathematical assertions), and particularly to the notion of explanatory proofs in which proof X explains why theorem Y is true (Steiner 1978; Colyvan 2011). Perhaps this concern has been inherited from more traditional study of scientific explanation, where scientists wish to answer why-questions about the real world and where, according to our findings about physics, this is reflected in their written discourse. Our findings suggest that a focus on explaining why may be misguided for those interested in explanation in the discourse of professional mathematicians. Indeed, as suggested by Rav (1999), it seems that when it comes to proofs and explanations, mathematicians communicate more in terms of learning how to solve problems than in terms of learning why mathematical results hold true.

Of course, as with any empirical work, one must be careful about several of the inferential jumps made in this kind of analysis. First, while the ArXiv may well be the world's largest, most widely used repository of mathematical preprints and postprints, it nevertheless represents a specific type of mathematical discourse. Our work thus leaves open the possibility that studies of mathematical discourse in settings such as conversational or other digital communications could lead to contrasting findings. Perhaps, for instance, mathematicians are more willing to discuss explanations in general and answers to why questions in particular when communicating in 'live' verbal settings. Second, we have analysed these research papers for a potentially limited type of explanatory talk, requiring the use of 'explain words' or a limited number of alternative, related expressions. While this was an obvious place to start, it is certainly possible that analysing other expressions related to mathematical explanation might alter our results. These limitations indicate clear avenues for future empirical research on mathematical explanation.

Suggested Readings

Baker, P., Gabrielatos, C., Khosravinik, M., Krzyżanowski, M., McEnery, T., and Wodak, R. (2008). A useful methodological synergy? Combining critical discourse analysis and corpus linguistics to examine discourses of refugees and asylum seekers in the UK press. *Discourse & Society, 19*, 273–306.

Garside, R., Leech G., and McEnery, T. (2013). *Corpus Annotation: Linguistic Information from Computer Text Corpora*, 3rd edn. Abingdon, UK: Routledge.

McEnery, A. M., and Wilson, A. (2001). *Corpus Linguistics: An Introduction*. Edinburgh: Edinburgh University Press.

Overton, J. A. (2013). 'Explain' in scientific discourse. *Synthese, 190*, 1383–1405.

References

Baker, P., Gabrielatos, C., Khosravinik, M., Krzyżanowski, M., McEnery, T., and Wodak, R. (2008). A useful methodological synergy? Combining critical discourse analysis and corpus linguistics to examine discourses of refugees and asylum seekers in the UK press. *Discourse and Society, 19*, 273–306.

Biber, D. (1993). Representativeness in corpus design. *Literary and Linguistic Computing, 8*(4), 243–257.

Bisson, M.-J., Gilmore, C., Inglis, M., and Jones, I. (2016). Measuring conceptual understanding using comparative judgement. *International Journal of Research in Undergraduate Mathematics Education, 2*(2), 141–164.

CasualConc (Version 2.0.3) [Computer Software], available from https://sites.google.com/site/casualconc/.

Chi, M. T. H., Bassok, M., Lewis, M. W., Reimann, P., and Glaser, R. (1989). Self-explanations: How students study and use examples in learning to solve problems. *Cognitive Science, 13*, 145–182.

Colyvan, M. (2011). *An Introduction to the Philosophy of Mathematics*. Sydney: University of Sydney.

Crowdy, S. (1993). Spoken corpus design. *Literary and Linguistic Computing, 8*(4), 259–265.

Dawkins, P., Inglis, M., and Wasserman, N. (2018). The use(s) of 'is' in mathematics. *22nd Annual Conference on Research on Undergraduate Mathematics Education*. San Diego, CA.

Fonseca, B. A., and Chi, M. T. (2011). Instruction based on self–explanation. In R. E. Mayer and P. A. Alexander (eds.), *Handbook of Research on Learning and Instruction* (pp. 296–321). New York: Routledge.

Garside, R., Leech, G., and McEnery, T. (2013). *Corpus Annotation: Linguistic Information from Computer Text Corpora*, 3rd edn. Abingdon, UK: Routledge.

Hafner, J. and Mancosu, P. (2005). The varieties of mathematical explanation. In P. Mancosu et al. (eds.), *Visualization, Explanation and Reasoning Styles in Mathematics* (pp. 215–250). Berlin: Springer.

Hempel, C. G., and Oppenheim, P. (1948). Studies in the logic of explanation. *Philosophy of Science*, 15(2), 135–175.

Herbel-Eisenmann, B., Wagner, D., and Cortes, V. (2010). Lexical bundle analysis in mathematics classroom discourse: The significance of stance. *Educational Studies in Mathematics*, 75, 23–42.

Hodds, M, Alcock, L., and Inglis, M. (2014). Self-explanation training improves proof comprehension. *Journal for Research in Mathematics Education*, 45, 98–137.

Hunston, S. (2002). *Corpora in Applied Linguistics*. Cambridge: Cambridge University Press.

Johansson, S., Leech, G. N., and Goodluck, H. (1978). *Manual of Information to Accompany the Lancaster-Oslo/Bergen Corpus of British English, for use with Digital Computer*. Department of English, University of Oslo.

John, L. K., Loewenstein, G. and Prelec, D. (2012). Measuring the prevalence of questionable research practices with incentives for truth telling. *Psychological Science*, 23(5), 524–532.

Knuth, D. E. (1979). *TEX and METAFONT: New Directions in Typesetting*. Bedford, MA: American Mathematical Society and Digital Press.

Knuth, E., Stephens, A. C., McNeil, N. M., and Alibali, M. W. (2006). Does understanding the equal sign matter? Evidence from solving equations. *Journal for Research in Mathematics Education*, 37(4), 297–312.

Lange, M. (2009). Why proofs by mathematical induction are generally not explanatory. *Analysis*, 69(2), 203–211.

Leech, G. (2013). Introducing corpus annotation. In R. Garside, G. Leech and T. McEnery (eds.), *Corpus Annotation: Linguistic Information from Computer Text Corpora*, 3rd edn. (pp. 1–18), Abingdon, UK: Routledge.

Leinhardt, G. (2001). Instructional explanations: A commonplace for teaching and location for contrast. In V. Richardson (eds.), *Handbook of Research on Teaching*, 4th edn. (pp. 333–357). Washington, DC: American Educational Research Association.

Louw, B. (1993). Irony in the Text or Insincerity in the Writer? The Diagnostic Potential of Semantic Prosodies. In M. Baker, G. Francis and E. Tognini-Bonelli (eds.), *Text and Technology: In Honour of John Sinclair* (pp. 157–176). Philadelphia/Amsterdam: John Benjamins.

McEnery, A. M., and Wilson, A. (2001). *Corpus Linguistics: An Introduction*. Edinburgh: Edinburgh University Press.

Mancosu, P. (2001). Mathematical explanation: Problems and prospects. *Topoi, 20,* 97–117.

Mejía-Ramos, J. P., and Inglis, M. (2011). Semantic contamination and mathematical proof: Can a non-proof prove? *Journal of Mathematical Behavior, 30,* 19–29.

Mejía-Ramos, J. P., and Weber, K. (2014). Why and how mathematicians read proofs: Further evidence from a survey study. *Educational Studies in Mathematics, 85*(2), 161–173.

Monaghan, J. (1991). Problems with the language of limits. *For the Learning of Mathematics, 11*(3), 20–24.

Overton, J. A. (2013). 'Explain' in scientific discourse. *Synthese, 190,* 1383–1405.

Rav, Y. (1999). Why do we prove theorems? *Philosophia Mathematica, 7*(3), 5–41.

Resnik, M. D., and Kushner, D. (1987). Explanation, independence and realism in mathematics. *The British Journal for the Philosophy of Science, 38*(2), 141–158.

Rittle-Johnson, B., Loehr, A., and Durkin, K. (2017). Promoting self-explanation to improve mathematics learning: A meta-analysis and instructional design principles. *ZDM Mathematics Education, 49,* 599–611.

Salmon, W. (eds.) (1971). *Statistical Explanation and Statistical Relevance.* Pittsburgh: University of Pittsburgh Press.

Salmon, W. (1984). *Scientific Explanation and the Causal Structure of the World.* Princeton: Princeton University Press.

Simmons, J. P., Nelson, L. D., and Simonsohn, U. (2011). False-positive psychology: Undisclosed flexibility in data collection and analysis allows presenting anything as significant. *Psychological Science, 22*(11), 1359–1366.

Sinclair, J. (1991). *Corpus, Concordance, Collocation.* Oxford: Oxford University Press.

Steiner, M. (1978). Mathematical explanation. *Philosophical Studies, 34,* 135–151.

TagAnt (Version 1.2.0 – 2014) [Computer Software], available from http://www. laurenceanthony.net/.

Treagust, D. F., and Harrison, A. G. (1999). The genesis of effective scientific explanations for the classroom. In J. Loughran (ed.), *Researching Teaching: Methodologies and Practices for Understanding Pedagogy* (pp. 28–43). London: Falmer Press.

The SAMUELS Consortium (2014) (MNOP). The SAMUELS Project. United Kingdom AHRC and ESRC. http://www.glasgow.ac.uk/samuels.

Van Heuven, W. J., Mandera, P., Keuleers, E., and Brysbaert, M. (2014). SUBTLEX-UK: A new and improved word frequency database for British English. *The Quarterly Journal of Experimental Psychology, 67*(6), 1176–1190.

Weber, E., and Frans, J. (2016). Is mathematics a domain for philosophers of explanation? *Journal for General Philosophy of Science, 48*(1), 125–142.

Weber, K., and Mejía-Ramos, J. P. (2011). Why and how mathematicians read proofs: An exploratory study. *Educational Studies in Mathematics, 76,* 329–344.

Zelcer, M. (2013). Against mathematical explanation. *Journal for General Philosophy of Science, 44,* 173–192.

9

Natural Language Processing and Network Visualization for Philosophers

Mark Alfano and Andrew Higgins

1. Introducing big data and semantic networks to philosophy

Progress in philosophy is difficult to achieve because our methods are evidentially and rhetorically weak (Chalmers 2015). In the last two decades, experimental philosophers have begun to employ the methods of the social sciences to address philosophical questions (e.g. Weinberg et al. 2001; Nichols and Knobe 2007; Schwitzgebel and Cushman 2012). However, the adequacy of these methods has been called into question by repeated failures of replication (Open Science Collaboration 2015). In the next decade, experimental philosophers need to incorporate more robust methods to achieve a multi-modal perspective. In this chapter, we explain and showcase exemplars of a promising methodology for experimental and empirical philosophy: network analysis and visualization.

Networks dominate contemporary technologies such as the Internet, social media, so-called 'smart' cities and ports, blockchain cryptocurrencies and many others. In psychological science, constructs as diverse as the neuronal structure of the brain (Raichle 2015), pathologies and disorders (Borsboom and Cramer 2013), personality factors (Epskamp et al. 2012) and belief systems have been successfully modelled as networks (Brandt et al. 2018). In humanistic studies of novels and other literature, researchers employing digital humanities methods have analysed texts and whole corpora (Moretti 2013; Alfano and Stauffer 2015). Yet these methodological tools have been employed sparingly in philosophy. A recent search of www.philpapers.com – the most comprehensive archive of philosophical publications in the world, with over 2 million entries – returned 28 results for 'semantic network', 189 for 'social network' and 0 for 'psychological network'. And most of those results were published in inter- or trans-disciplinary venues such as *MedInfo, Minds and Machines, Frontiers in Psychology, Library Trends, Behavioral and Brain Sciences* and *Journal of Intelligent Systems*.[1] Despite their interest in empirical methods, experimental philosophers have – with few exceptions such as Alfano, Higgins and Levernier (2017), Betti et al. (2014), Chen and Floridi (2013) and Rathkopf (2018) – demonstrated almost no interest in network analytics.

Semantic networks in philosophy of mind and language

This state of affairs is unfortunate as there are strong theoretical reasons for philosophers to take an interest in semantic, social and psychological networks. Using semantic networks, for instance, it should be possible to shed light on word- and sentence-meaning beyond what can be revealed through introspection and conceptual analysis. After all, even for non-Wittgensteinians, it is generally uncontroversial that the meaning of a term is constitutively related to its use in natural language. More specifically, as Alfano (2015) argues, network analytics are presupposed by David Lewis's (1966, 1970, 1972) development of Frank Ramsey's (1931) approach to the implicit definition of theoretical terms. The principle underlying the Ramsey-Lewis approach to implicit definition (often referred to as 'Ramsification') can be illustrated with a well-known story:

> And the Lord sent Nathan unto David. And he came unto him, and said unto him, 'there were two men in one city; the one rich, and the other poor. The rich man had exceeding many flocks and herds: But the poor man had nothing, save one little ewe lamb, which he had bought and nourished up: and it grew up together with him, and with his children; it did eat of his own meat, and drank of his own cup, and lay in his bosom, and was unto him as a daughter. And there came a traveller unto the rich man, and he spared to take of his own flock and of his own herd, to dress for the wayfaring man that was come unto him; but took the poor man's lamb, and dressed it for the man that was come unto him.' And David's anger was greatly kindled against the man; and he said unto Nathan, 'As the Lord liveth, the man that hath done this thing shall surely die: And he shall restore the lamb fourfold, because he did this thing, and because he had no pity.' And Nathan said to David, 'Thou art the man.'

[1] Search conducted 12 January 2018 in The Netherlands.

Nathan uses Ramsification to drive home a point. He tells a story about an ordered triple of objects (two human and one non-human animal) that are interrelated in various ways. Some of the first object's properties (e.g. wealth) are monadic; some of the second object's properties (e.g. poverty) are monadic; some of the first object's properties are relational (e.g. he steals the third object from the second object); some of the second object's properties are relational (e.g. the third object is stolen from him by the first object); and so on. Even though the first object is not explicitly defined as *the X such that ...* , it is nevertheless implicitly defined as *the first element of the ordered triple such that ....* The big reveal happens when Nathan announces that the first element of the ordered triple is the very person he's addressing (the other two, for those unfamiliar with *2 Samuel* 12, are Uriah and Bathsheba).

The story is biblical, but the method has a modern incarnation. To implicitly define a set of theoretical terms (henceforth 'T-terms'), one formulates a theory T in those terms and any other terms (henceforth 'O-terms') one already understands or has an independent theory of. Next, one writes T as a single sentence, such as a long conjunction, in which the T-terms $t1, ... , tn$ occur (henceforth 'the postulate of T'). The T-terms are replaced by unbound variables $x1, ... , xn$, then existentially quantified over to generate the Ramsey sentence of T, $\exists x1, ... , xn\ T[x1, ... , xn]$, which states that T is realized, that is, that there are objects $x1$ through xn that satisfy the Ramsey sentence. An ordered n-tuple that satisfies the Ramsey sentence is then said to be a *realizer* of the theory.

Lewis (1966) famously employed this method to argue for the psychophysical identity theory. He identified the postulate of folk psychology as the conjunction of all folk-psychological platitudes. The Ramsey sentence of folk psychology is formed by replacing all mental state-terms (e.g. 'belief', 'desire', 'pain', etc.) with variables and existentially quantifying over those variables. Finally, one goes on to determine what, in the actual world, satisfies the Ramsey sentence; that is, one investigates what, if anything, is a realizer of the Ramsey sentence. If there is a realizer, then that's what the T-terms refer to; if there is no realizer, then the T-terms do not refer. Using this method ensures that we don't simply change the topic when we try to give a philosophical account of some phenomenon. If your account of the mind is inconsistent with the postulate of folk psychology, then – while you may be giving an account of *something* – you're not doing what you set out to do.

Notably, Lewis never compiled all the platitudes of folk psychology. He just asserted that it was obvious that the realizer of the Ramsey sentence of folk psychology was the brain. But he never wrote down the Ramsey sentence. This was for at least two reasons. First, there are *a lot* of folk psychological platitudes, and they're all (by design) extremely boring. So, the Ramsey sentence of folk psychology would probably fill many stultifying books, making it difficult to assemble and even more difficult to audit. Second, in Lewis's time there was no systematic way to collate the platitudes.

Using contemporary natural language processing tools, we now have the opportunity to implement (a slightly modified form of) the method of Ramsification with semantic networks. To do so, we need access to a large corpus of platitudes. We then parse this corpus using ConText (Diesner 2014) or Python packages such as nltk (Bird et al. 2009) and gensim (Radim and Sojka 2010). After cleaning the raw text, we select the T-terms from the parsed corpus. These are modelled as nodes in a semantic network. Edges (i.e. pairwise connections or binary relations) between the nodes represent semantic relations between the T-terms they represent. The resulting network can then be analysed and visualized to assess whether it has any realizers.

Networks and their visualizations

Networks can be used to map trade, semantics, friendships, heredity, ecosystems and much more. The simplest type of network represents nothing beyond the bare presence of nodes and whether, for each pair of nodes, there exists a connection between them. Network-analytics are valuable for highlighting and analysing the extrinsic, relational properties of objects rather than focusing on intrinsic, atomistic traits.

A network is an abstract structure. In what follows, we will sometimes express such abstract structures using visualizations. A visualization is not the same thing as a network, just as a drawing of a circle or square is not the same thing as a circle or a square. But just as it is often pedagogically, cognitively or communicatively helpful to do geometry with drawings on paper, chalkboards or computer screens, so it is often pedagogically, cognitively or communicatively helpful to do network analysis with visualizations on paper, chalkboards or computer screens (Munzner 2014). As Andy Clark (2002) has argued, human minds operate best when they are able to iteratively alternate between cognitive processes such as imagining, inferring, and evaluating and perceptual and agential processes such as seeing, feeling and manipulating. Shifting back and forth between an abstract network and its visual representation enables us to take advantages of both our cognitive powers and our pattern-recognizing visual acuity.

2. Workflow

In this section, we elaborate a recommended workflow for semantic network analysis (Figure 9.1). Much of the approach will be familiar to experimental philosophers, but there are some important differences.

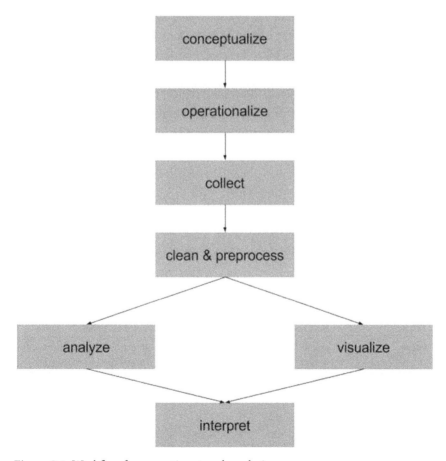

Figure 9.1 Workflow for semantic network analysis.

Conceptualization

As with any scientific research, we begin by conceptualizing the research question and all constructs involved. This step is essential to ensure face validity. Inadequate, imprecise and ambiguous conceptualization is arguably one of the root causes of the replication crisis currently racking psychology (Lurquin and Miyake 2017). It might at first blush seem that a bottom-up, big data approach that harnesses the power of visual analytics would make it unnecessary to worry about conceptualization issues, but recent developments have shown that perspective to be overly sanguine (Lynch 2016, Chapter 8). As Nobel economist Paul Krugman (2014) put it on his blog, you can't just 'let the data speak for itself – because it never does. [...] If you think the data are speaking for themselves, what you're really doing is implicit theorizing, which is a really bad idea (because you can't test your assumptions if you don't even know what you're assuming).' For our first use case – mining natural language for patterns of virtues, values and constituents of well-being (VVC) – the conceptualization step involves reflection on the philosophical literature on these constructs, as shown in Table 9.1.

Since the current chapter is illustrative rather than argumentative, we will not spend time here justifying our conceptualizations of virtues, values and constituents of well-being. Interested readers should refer to Alfano, Higgins and Levernier (2017) for details.

Operationalization

The second step in the workflow is to operationalize the constructs of interest. There is no way to directly observe and measure properties like virtues and

Table 9.1 Definitions of key terms.

virtue	admirable trait of personality or character (Zagzebski 2017)
value	stable attitude (or the object of such an attitude) that motivates behaviour and furnishes a standard for evaluation (Tiberius 2008)
well-being	pleasurable, desirable, valuable or otherwise good state of being (Parfit 1984)
cardinal VVC	VVC on which others turn; 'cardinal' derives from the Latin *cardo* meaning 'hinge' (Alfano 2012)
VVC conduit	high-traffic edge between VVC on measures like betweenness centrality (Van Norden 2008)

values. We must instead measure proxies for these phenomena. In the context of natural language processing and semantic network analysis, this means identifying accessible corpora that we expect to contain a large and diverse range of expressions that refer to virtues, values and constituents of well-being. Given the norm enjoining people to only speak well of the dead – especially in public announcements – we believe that obituaries are a good place to find proxies for virtues, values and constituents of well-being. This norm remains strong, as evidenced by the recent backlash to Randa Jarrar's tweet prompted by the death of former First Lady of the United States, Barbara Bush: 'Barbara Bush was a generous and smart and amazing racist who, along with her husband, raised a war criminal' (Wootson and Wong 2018). Other researchers interested in employing this workflow will need to think carefully about which corpora would best operationalize the constructs that interest them. For instance, research on lust and romantic love might use profiles from online dating apps, such as the OKCupid dataset (Kirkegaard and Bjerrekær 2014). By contrast, researchers interested in children's development and acquisition of concepts may prefer to consult the CHILDES database (MacWhinney 2000). Researchers interested in cross-cultural comparisons may wish to consult the Human Relations Area Files (HRAF 1967). And researchers interested in social epistemology may use the Twitter application programming interface.[2]

Data collection

This brings us to step three: collecting data. In more familiar experimental philosophy research, this step typically involves surveying live human participants with a questionnaire. Less commonly, live human participants may be invited to contribute other sorts of data, such as physiology, choice behaviours or brain scan data. In all cases, the resulting structure is a database in which the rows represent distinct participants and the columns record data (e.g. survey responses) provided by those participants. Our method diverges from the industry standard at this point. Instead of representing live participants, the rows in our database represent documents.

This approach affords several advantages over traditional questionnaire methods, though it also comes with drawbacks they lack. First, if one is analysing documents in the public domain, one needn't bother with IRB approval. Second,

---

[2] Available at https://developer.twitter.com/en/docs

documents in the public domain are free and, in many cases, plentiful, making it possible to have a very large sample size at zero financial cost (though researchers may need to invest considerable time into cleaning and organizing the data). Third, documents in the public domain do not suffer concerns about ecological validity that often plague laboratory and questionnaire studies. Fourth, it is often possible to ensure a diverse, intersectional (Crenshaw 1989) sample using public domain documents. Fifth, documents in the public domain are also available to other researchers, meaning that they automatically satisfy one of the desiderata of the open science: namely, open data. One significant drawback of relying on this methodology is that some data is hostage to the archivists. If there is no collection of suitable documents, the method cannot be employed. Another is that, especially when it comes to older corpora, the documents are often not machine-readable. This means that a significant amount of time and effort must be devoted to cleaning up regimenting the texts under study. A third drawback is that many corpora embody the interests and values of those who constructed them, leaving other interests and values tacit.

Cleaning and processing the data

The fourth step is to clean the collected data. Experimental philosophers will be familiar with the fact that raw data needs to be processed and regimented in various ways before it can be analysed. The same holds for natural language processing. The main difference here is that cleaning data is often at least as time-consuming and onerous as collecting it in the first place. The basic idea is to extract the constructs operationalized in step 2 from each of the documents, appending them as columns in the database. For instance, in our work on obituaries, we are interested in demographic variables such as date of birth, date of death, age at death, gender, religious affiliation, veteran status, killed-in-action status, education attained and marital status, among others. There are various natural language processing tools and software packages available for this kind of extraction, including ConText (Diesner 2014) and Python packages such as nltk (Bird, Loper and Klein 2009) and gensim (Radim and Sojka 2010).

In addition to demographic variables, we are interested in the virtues, values and constituents of well-being celebrated in people's obituaries. There are three main approaches available for extracting these variables: supervised, semi-supervised and unsupervised. In a supervised approach, the researchers establish in advance a codebook or set of principles for what to count as a textual expression of a virtue, value or constituent of well-being. This would be part of the operationalization step. The researchers then independently read through all of the obituaries and annotate them (putting a 1 in the column if the trait is represented, 0 otherwise) using the codebook. A measure such as Cohen's (1968) kappa can then be calculated to ensure adequate interrater agreement. Assuming an acceptable level of agreement is reached, it's possible to move to the double-barrelled next stage: analysis and visualization. However, if the number of obituaries (or other documents, in a different use case) is very large, hand-coding all of them may be daunting. This brings us to the other two approaches. In a semi-supervised approach, the researchers hand-code a subset (say, 10%) of the documents, then use machine learning to automatically code the remainder (Hastie et al. 2008). This can be done all at once, then spot-checked for validity. However, it's worth bearing in mind that machine learning suffers a recall/precision trade-off (Powers 2011). Essentially, the further out on a limb you force the machine learning algorithm to go, the more documents it will label, but the less reliable those labels will be. To handle this problem, it is often best to 'bootstrap' iteratively (Carlson et al. 2010). To illustrate, the researchers could hand-label 10% of the documents, then use machine learning to automatically label an additional 20% of them. The researchers then spot-check 10% of the automatically labelled documents for accuracy, correcting any biases they find. Next, the researchers again use machine learning on the 30% of labelled documents to label an additional 30% of them. After spot-checking again, they use machine learning one last time to label the remainder. The third approach throws out the codebook altogether, relying entirely on machine learning to unearth patterns that human coders might overlook or ignore because of bias (Hastie et al. 2008). While unsupervised learning is a fascinating methodology, we believe that researchers who have done a good job conceptualizing and operationalizing will enjoy more success using a codebook to support a supervised or semi-supervised approach. Even if unsupervised learning unearths patterns, it's often difficult or even impossible to give any meaning to those patterns.

The resulting database should now be ready to transform into a network format. The nodes of this network will be the column names, and the edges will be the frequencies of pairwise co-occurrence. To illustrate with the obituary use case, three of the traits often celebrated in obituaries are being a *friend,* being a *volunteer* and having a good *sense of humour*. In our data, people who are described as being a friend are more likely to also be described as being a volunteer than as having a good sense of humour. We encode each of these traits as a node and treat the frequency of co-occurrence as the weight (explained in more detail below) of the edge between them. There are multiple file formats for representing such a co-occurrence network. Perhaps the most common is an *edge list*, which is a comma separated values (.csv) format in which the first column represents origin nodes, the second column represents target nodes and the third column (if any) represents edge weights. Table 9.2 below is an edge list. A format we sometimes prefer to the edge list is called an *adjacency matrix*. You can generate an adjacency matrix from the row/column format described in this section by multiplying the database by its own transpose. To do this in R, first encode the data as described above and save it as 'data.csv', then run this script:

```
data <- read.csv("data.csv")
M <- data.matrix(data)
T <- t(M)
adjacency <- (T %*% M)
write.csv(adjacency, file="adjacency.csv")
```

We should also note that experimental philosophers may wish to reanalyse their existing datasets as networks. If your data is already formatted as a *correlation matrix*, you can use that just like an adjacency matrix, in which the variables are the nodes and the edge weights are the correlations between the variables. In any event, once you have either an edge list or an adjacency matrix, you are ready to proceed with analysis.

Analysis

For analytical purposes, it's helpful to associate various properties with the nodes and edges of a network. These measures are derived from the relational properties of the nodes. The basic relational property is a node's *degree*, or the number of edges it has (i.e. the number of other nodes it is connected to). This is a local property of nodes: a node with three connections could exist at the periphery of a network or function as a significant bottleneck in the network, depending on how the rest of the network is structured. To better understand the role played by different nodes, it's therefore helpful to employ more holistic properties. For example, the *betweenness* of a node is the sum of the number of geodesics that run through that node, where a *geodesic* is the shortest path through the network from one node to another. Nodes with high betweenness control the flow of information through the network. Another key property of a node is its *closeness*, defined as the average (or normalized average) of the lengths of all its geodesics to all other nodes in the network.

We have thus far been dealing solely with undirected, unweighted networks. Directedness and weight are in the first instance properties of edges. Whether edges are directed depends on whether the relation modelled by the network is symmetric or (at least potentially) asymmetric. For example, the 'is a sibling of' relation is symmetric, while the 'is a sister of' relation is (potentially) asymmetric. If x is a sibling of y, y is a sibling of x, but if x is a sister of y, y may or may not be a sister of x. When the relation is symmetric, we can rest content representing it as an undirected network. However, when the relation is asymmetric, it may be useful to represent it as a directed network. Doing so adds complexity to the representation and makes visualizations more onerous on the eye. Sometimes, though, it's worth the effort to represent networks in a directed way (and visualize them that way too). When we do so, the measures mentioned above need to be refined. Instead of simply *degree*, we can distinguish between *out-degree* (the sum of the number of edges directed from the node in question to another node) and *in-degree* (the sum of the number of edges directed from another node to a given node). In a directed network, a node's betweenness will also be affected because the number and length of geodesics through the network will typically decrease and increase, respectively. Likewise, a node's closeness will tend to increase as the length of paths increases.

A second way to complicate a network is to associate with each edge a *weight*. Whether edges are weighted depends on whether the relation modelled by the network is categorical or (at least potentially) scalar. For example, the 'cites n times' relation scales, while the 'cites' relation does not scale. If x cites y n times, then there is some whole number n characterizing the relationship. If x cites y (at all), there is no further question to ask. Likewise, the 'is correlated with strength n with' relation ranges from –1 to 1, whereas the 'is correlated with' relation is categorical. When the relation modelled in a network is not scalar, we can rest content representing it as an unweighted network. However, when the relation is scalar, it may be useful to represent it as a weighted network. As before, doing so adds complexity to the representation and makes visualizations more onerous on the eye, but may be worth the effort. When we do so, the measures mentioned above again need to be refined. Instead of simply *degree*, we can speak of the *strength* of a node (the sum of the weights of each of its edges). *Mutatis mutandis* for both betweenness and closeness.

It's often tempting to gussy up networks by making them both directed and weighted – and to try to represent both of these features visually – but researchers should always bear in mind the costs associated with additional complexity. As we argue in more detail below, without a commitment to careful pruning and asceticism about the number and variety of variables one aims to visualize, one is liable to produce visually illegible network visualizations. One may find it useful to consider weight and directedness when calculating network and node metrics while hiding these features in the visualization of the network.

That said, we here illustrate various network measures using a (small) visualized network that is both directed and weighted (Figure 9.2). Table 9.2 gives the abstract, mathematical characterization of the network.

While nothing in our analysis depends on what this network represents, we can imagine it as a business network where nodes are individual companies and edges are financial transactions. Those transactions are represented with varying weights, indicating the monetary value of the transactions, and with arrows, designating which company paid which.

In Figure 9.2, the nodes are represented by letters and the edges are represented by lines connecting those entities. Not all edges are equal. If, for example, we represented a friendship network, it would be useful to distinguish between close friends and acquaintances. To track the strength of friendship ties, we could give distinct edge weights to each (e.g. a weight of 2 for close friends and a weight of 1 for acquaintances). In Figure 9.2, edge weight is represented by the thickness of the edge, with wider lines representing strong ties and thin lines representing weak ties. If the edges were undirected, input and output would be interchangeable, but when working with a directed graph these matter.

As a quick visual inspection of Figure 9.2 will confirm, H paid J a lot more than J paid K, but it's quite difficult to see exactly how much more. Is it twice as much? Three times as much? Ten times as much? Table 9.2 indicates that the

Table 9.2 Network with 10 nodes and 17 edges. The first letter indicates the origin of the directed edge; the second letter indicates the target of the directed edge. For example, A–G indicates that there is an edge from node A to node G with a weight of 3.

Directed edges	Weights	Directed edges	Weights	Directed edges	Weights
A–E	1	C–D	4	F–E	2
A–F	1	D–E	1	F–G	1
A–G	3	D–G	2	G–C	3
B–A	2	D–H	3	H–J	5
B–C	4	E–F	4	J–K	1
B–G	2	E–H	2		

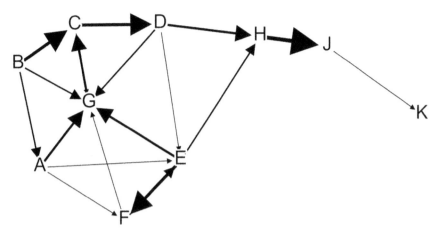

Figure 9.2 Visualized network with 10 nodes and 17 edges. Each letter represents a node. Each line represents an edge. Arrowheads indicate the direction of edges.

weight of H–J is five times the weight of J–K, but someone looking at Figure 9.2 would be hard-pressed to see this. In the next section, we expand on the difficulty of visually representing multiple features in the same network visualization and discuss best practices for informative visualizations.

As we explained above, the most significant properties of nodes are their relational properties. In our example, J has a degree of two because it is related to two other nodes. Degree is a limited measure because it only considers nodes in relation to their nearest neighbours and is insensitive to the significance of the connection. In an international trade network, for example, it would be important to know not just which countries trade with which, but also the quantity of goods traded. To track this information, we can sum up the weights of edges to assign strengths to nodes. J has a strength of six because J–K has a weight of one and H–J has a weight of five. This helps us to appreciate that, while A has more connections than C (i.e. a higher degree), C has more significant connections than A, giving it a weighted degree of 12 compared to A's weighted degree of 7, as seen in Table 9.3.

This information is still highly limited. In analysing a criminal or terrorist network, for example, we can learn something from the fact that A communicated with B, but we learn far more about A if we also know that B communicated with C, D and E, where these are high level figures in the illicit organization. To track such indirect connections, we can use measures of network centrality. Different algorithms define centrality in importantly distinct ways. Betweenness centrality, as mentioned above, measures how often a node occurs in the

Table 9.3 Properties of nodes in Table 9.1. In-D refers to in-degree, Out-D refers to out-degree, Weighted refers to weighted degree, Clustering refers to clustering coefficient, Eigenvect refers to Eigenvector centrality and Between refers to betweenness centrality.

Node	In-D	Out-D	Degree	Weighted	PageRank	Clustering	Eigenvect	Between
A	1	3	4	7	0.03035	0.41667	0.00599	3.5
B	0	3	3	8	0.02365	0.5	0	0
C	2	2	4	12	0.20708	0.33333	0.60526	11.5
D	1	3	4	10	0.11165	0.33333	0.35043	11.5
E	3	3	6	13	0.08819	0.25	0.39409	10.5
F	2	2	4	8	0.05725	0.5	0.26836	0
G	6	1	7	15	0.20790	0.26667	1	7
H	2	1	3	10	0.08029	0.16667	0.44308	14
J	1	1	2	6	0.09187	0	0.25800	8
K	1	0	1	1	0.10178	0	0.14887	0

geodesics between other nodes. Nodes with higher betweenness centrality are more likely to play an essential bridge or filter role in connecting two otherwise separate groups of nodes. As seen in Table 9.2, H has the highest betweenness because any connections between J or K and all the other nodes must pass through H. Similarly, in analysing the entities involved in criminal activity, bottlenecks (people who bridge the gap between two otherwise disconnected clusters) play an especially significant role (Diesner and Carley 2010).

Eigenvector centrality, by contrast, is a measure of the importance of a node in the network as measured by its connectedness to other nodes with high Eigenvector centrality (Newman 2018). This metric is similar to another measure of centrality, the *PageRank* metric for determining the relevance of websites in a search (Brin and Page 1998). The PageRank for website *W* is determined by considering the number of other websites with links to *W*, with greater weight given to linking websites that are themselves frequently linked. In a citations-based network, Eigenvector centrality and PageRank are measures of the relative centrality of an author to the discussion in their area of specialty. G has the highest Eigenvector centrality and PageRank because it has several connections with nodes that themselves have several connections. PageRank weights hubs and bottlenecks higher while Eigenvector centrality gives more weight to nodes at the geometric centre of the network. So, while PageRank ranks C as a close second (.207 compared to G's .208), by the standard of Eigenvector Centrality G is in a league of its own (1.0, compared to C's .6). A priori, there is no one correct or perfect centrality metric. Instead, researchers should choose the metric that best answers to their research questions, as articulated in the conceptualization stage.

Clustering coefficient, intuitively, is a measure of how insulated a node is from the wider network. More specifically, to measure the clustering coefficient for node N, we consider the connections between all of N's neighbours (the nodes directly connected to N), with more connections within the local neighbourhood generating higher scores (Watts and Strogatz 1998). Nodes that are mostly only connected to nodes which are connected to each other will have higher clustering coefficient scores. This is the inverse of a bottleneck. In the sample network above, B and F have high scores by this measure because their neighbours tend to also be neighbours of each other. In our earlier work studying the semantic networks in obituaries Alfano, Higgins and Levernier (2017), this measure was useful for identifying groups of traits that frequently co-occurred. For example, it allowed us to see a natural cluster involving the traits of being a traveller, camper, cyclist, backpacking and hunter.

A limitation of clustering coefficient is that it measures individual nodes only in relation to their immediate neighbourhood, and it's more useful in some cases to step back and see larger communities. *Modularity* serves this role by identifying natural groups of nodes on multiple scales of analysis. As a quick first pass, modularity parcels out the nodes into a network into different groups using an algorithm that aims to ensure that there are more intra-group connections than inter-group connections (Blondel et al. 2008). The threshold for group inclusion is the *resolution*, which needs to be fine-tuned for the particular network under analysis. Lower resolution will result in a higher threshold for group inclusion, and thus more communities, while a higher resolution will lead to fewer, larger groups. This measure is similar to the clustering coefficient in that it identifies groups of highly interconnected nodes but is better for seeing the larger picture. Going back to the example of obituaries, in Alfano, Higgins and Levernier (2017) we showed that being a traveller is directly related to being a hunter and a camper, but also indirectly associated with enjoying boating and mountaineering, as well as being in the military. As we show below, modularity is also useful in visualizing results because nodes can be marked to indicate group membership (e.g. with different colours), and this allows viewers to see connections that might have otherwise been obscured.

Thus far, we have been describing the sorts of metrics that can be extracted from networks and how to use them for network analysis. Experimental philosophers are likely to be wondering: what about significance tests? Suppose node X has a higher PageRank or betweenness than node Y. Is the difference significant? Familiar calculations in the null hypothesis significance testing paradigm cannot easily be employed for network metrics. In general, computer scientists seem uninterested in such metrics. That said, analytical tools that support inferential statistics on networks are just now being developed. Examples include the adaptive LASSO (Krämer et al. 2009), the IsingFit R package (Borkulo and Epskamp 2016), the bootstrapped difference test (Epskamp et al. 2016), the NetworkComparisonTest R package (Borkulo 2016) and network dynamics simulations (Costantini et al. 2014). We expect that further tools continue to be developed in the coming years.

Visualization

As we pointed out above, a network and its visualization are distinct. The former is an abstract structure, while the latter is an expression of that structure in visually perceivable form. In principle, everything that can be learned from a visualization could also be learned by analysing the bare mathematics of the network itself. However, given the relative strengths and weaknesses of human cognition and perception, it is sometimes (though by no means always) advantageous to employ visualizations. In his seminal work on the principles of visual analytics, Schneiderman (1996) argues that good visualizations typically work in the following way. First, they provide an overview of all the data. Second, they allow the user to zoom in on items of interest. Third, they allow the user to filter out uninteresting items. Fourth, they provide details on demand (rather than not providing them at all, or trying to provide all of them at once in the initial overview). All the while, they should enable users to view the relationships between items.

When building an interactive (preferably, online) tool, it is helpful to follow Schneiderman's prescriptions. However, when the research output is a journal article or book chapter that will be printed on dead trees, compromises need to be made. The goal then becomes providing an overview that also contains enough details and visual cues to be interesting without overloading viewers' perceptual capacities. Tamara Munzner's (2014) *Visualization Analysis and Design* is a one-stop-shop for best practices. Among the many things this book covers is a comprehensive ranking of so-called 'visual channels,' which carry information about either categorical or ordered (e.g. cardinal, ordinal) attributes (Munzner 2014, Chapter 5). Information about categorical attributes is communicated with decreasing effectiveness using spatial region > colour hue > motion > shape. In other words, if you want to convey that x and y belong to the same category as each other but a different category from z, the most effective way to do so is to put x and y in the same region, with z in a different region. Next most effective is using different colours of the rainbow. And so on. We should note that most red-green contrasts are discouraged because many people are red-green deficient. The prevalence of red-green deficiency in men ranges from approximately 3% (for African and Native American men) to 8% (for European men), while the prevalence in women is between.5% and 2% around the world (Birch 2012). Given that spatial region is a stronger signal than hue, that red-green deficiency affects such a large number of people, and that colour printing is often prohibitively expensive, we recommend using spatial location rather than or in addition to hue whenever possible to convey categorical information. If you do use hue, we recommend consulting www.colorbrewer2.org for a printer-friendly, colour-blind-safe palette. One set of three hues we endorse is, in HEX format, #e41a1c (a bright red), #377eb8 (a medium blue) and #4daf4a (a kelly green).[3]

For ordered attributes, Munzner (2014) recommends the following visual channels, again in decreasing order of effectiveness: position > length > angle > area > three-dimensional depth > colour luminance ≈ colour saturation. This ranking provides empirical backing for our observation above that it is difficult to tell exactly how much bigger the arrowhead representing the weight of H–J is than the arrowhead representing J–K. This is because the information is being conveyed through the fourth most effective visual channel: area (and, what's more, areas of arrowheads rather than circles). To take advantage of the more effective visual channels of position, length or angle, we could instead have represented edge weight using, respectively, a scatterplot, a histogram or a pie chart. However, doing so would have made it impossible to show the structure of the network in a way that provides an adequate overview (Scheiderman's first principle). These trade-offs constantly crop up in data visualization, so researchers should cleave to asceticism. Doing so will help them to avoid producing illegible network visualizations that resemble hairballs or Christmas trees.

This brings us to the topic of layouts: methods for visually organizing network graphs, some of which can be used together. The visual structure of the network can be determined manually or through the application of various algorithms.

---

[3] We should also point out that, in 2010, approximately 32,400,000 people were blind and 191,000,000 people suffered from moderate or severe visual impairment (Stevens et al. 2013). This points to an inherent drawback of visualizations and the need to develop additional tools that can be used effectively by people with visual disabilities.

Algorithms are often preferred for their sensitivity to the purely mathematical properties of the network, helping to avoid bias, and creating more elegant visual patterns that match the symmetries and asymmetries present in the data. Each algorithm has distinct (dis)advantages depending on the type and amount data, as well as which features researchers want to highlight. Most of our preferred algorithms are force directed, meaning the positions of nodes are determined by the iterated exertion of multiple forces. We'll explain this option in detail, but it's worth noting a few alternatives first. For ease of introduction for the uninitiated, the focus will be on layouts available in *Gephi* (gephi.org), a user-friendly software package for network analysis and visualization.

In very small networks (fewer than 20 nodes), circular layouts effectively communicate node connections, but edges in circular graphs are difficult to perceive in larger networks because of their high density, length and crisscrossing. In addition, the one-dimensional rim of the circle, where all the nodes reside, carries less visually salient information about the clustering of nodes into categorical groups. Arc diagrams suffer from the same shortcomings because all nodes exist along a single dimension. The visualization of a network with n nodes is the transformation of an n-dimensional space into a representational format with one, two or three dimensions. Where n is much greater than the dimensions of the layout, this distorts or loses potentially important information. As such, we discourage using one dimensional layouts for all but the smallest networks. Instead, we recommend using force directed and layered graph algorithms as starting points in visualizing networks.

The Sugiyama algorithm is a good example of a layered graph algorithm (Sugiyama et al. 1981). This is a worthy choice if your nodes form a hierarchy (e.g. business organizational structures or river systems). Sugiyama is most accurate with directed, acyclic graphs, though it will run with any edge list by forcing that structure onto your data and then reversing the alterations after computing positions. The result is a rigidly defined hierarchical structure. If your goal is not to create a purely hierarchical structure, force directed algorithms are more useful.

Force directed layout algorithms differ substantially from one another, but, for simplicity, we can think of them as operating with three basic forces: gravity, attraction and repulsion. Gravity draws all nodes closer to the centre of the graph, attraction pulls nodes together if and only if they are connected by an edge (with the strength of attraction based on edge weight), and, to prevent all nodes collapsing into the centre, repulsion pushes all nodes away from each other. For each layout, users can turn the dials on each force. The ideal force directed algorithm depends primarily on the size of the network. In order from largest (200,000+ edges) to smallest (fewer than 100), we recommend OpenOrd (Martin et al. 2011), Yifan Hu (Hu 2005), ForceAtlas2 (Jacomy et al. 2014), Force Atlas (Jacomy 2009) and Fruchterman-Reingold (Fruchterman and Reingold 1991).

Starting on the low end of the scale, Fruchterman-Reingold produces visually pleasing and accurate circular graphs for networks of up to 200 nodes. The only notable shortcoming, beyond the size restriction, is that the forced circular design can obscure the real distances between clusters. ForceAtlas, by contrast, produces less-elegant graphs that are truer to the distances between clusters in the network, as can be seen in Figure 9.3. ForceAtlas is more informative for graphs of this size, but processing speed limits its functionality with significantly larger networks.

Many algorithms that are well-suited to small and medium networks break down when applied to larger networks: taking several hours to compute, creating hairballs with little visual structure or converging on structures that do not accurately represent the patterns in the data. ForceAtlas2, Yifan Hu and OpenOrd overcome these problems with greater computational efficiency. This is done with parallel processing, only measuring local forces, combining forces and cutting long edges. ForceAtlas2 will give similar results to ForceAtlas, but with less accuracy for networks with fewer than 50,000 edges. In addition, ForceAtlas2 provides the option of dissuading hubs, pushing the most central nodes to the periphery. The Yifan Hu multilevel algorithm cuts even more corners, so if ForceAtlas2 is too slow this is the next step in trading accuracy for speed. With the largest graphs, OpenOrd is the best option. To illustrate with an example from our own research, in attempting to visualize a network of philosophy topics based on the categorization of books and articles by PhilPapers and PhilIndex (6.3 million original data points), we ran ForceAtlas2 for days without success, but OpenOrd finished in two hours (filtered results shown in Figure 9.4, left). However, OpenOrd cuts edges, so it should only be used if the more detail-oriented algorithms are inadequate.

Another challenge when representing especially large networks is finding ways to add semantic content to the syntactic structure of the representation. While OpenOrd created a clearly meaningful structure in the graph below, nothing in that image indicates what it is about, and adding hundreds of thousands of labels would not make the image more intelligible. One solution is to filter the edges or nodes by criteria such as edge weight, weighted degree or component membership. For example, to better represent the network of

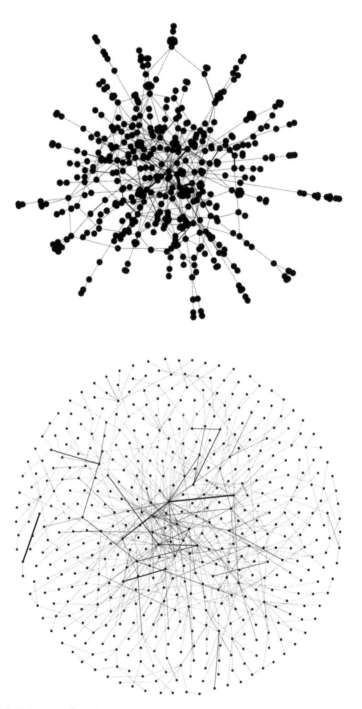

Figure 9.3 A network with 493 nodes and 612 undirected edges with the ForceAtlas layout (left) and Fruchterman-Reingold layout (right). The underlying mathematics is identical.

philosophy topics, we cut all edges with weight less than 20, along with any nodes no longer connected to the main component. Whereas ForceAtlas2 previously produced a meaningless hairball, when we applied it to the pared-down network it generated a meaningful structure where the remaining nodes could be given readable labels (if the image were significantly larger). To add content on the macro-scale, and labels that provide quick information about the network as a whole, groups were defined by modularity measure and described based on the general theme of the topics in each cluster. This can be seen to the right in Figure 9.4, where each specific topic is represented by a circle and the circles are coloured to reflect group membership. This provides more information than the first representation but obscures relations to an extent. This shortcoming can be addressed by treating a group of nodes as an individual and mapping the relations between groups rather than individuals, as was done at the bottom of Figure 9.4.

After using one of the algorithms above to generate the structure of the graph, one may wish to apply a few tweaks to the network to enhance its appearance. If the nodes are too spread out or too close together the graph can be expanded or contracted using the Expansion, Contraction or Noverlap layouts. If the nodes or edges are labelled, the Label Adjust layout will move nodes to avoid overlapping labels (since labels are horizontally aligned text, this can result in serious distortions). One can also manually adjust the positions of nodes, but this is discouraged since it potentially biases the results. All of these layouts are available in Gephi.

Once you are satisfied with the general layout, the next step is to work on the sizes and colours of the nodes. We typically use a centrality measure as the standard for node size, but one can base size on many other measures or determine size manually for each individual node. Colours for nodes, edges and labels can be based on group membership (modularity), various network measures (bearing in mind the caveats about colour mentioned above), or determined manually on a one-off basis.

In Gephi, the final adjustments are made in the Preview tab. Here, one should fine-tune the parameters determining sizes and colours. For our obituaries project, most graphs were developed by having 0% opacity on the nodes, showing labels with 12-point font (but with most nodes being larger due to their ranking-based size increases), and showing edges with between 0.3 and 1 thickness. The proportional size option is a way of making node labels larger based on their degree. We recommend turning this off and instead determining relative node sizes yourself in the Overview. Node labels can be easier to see if one applies an

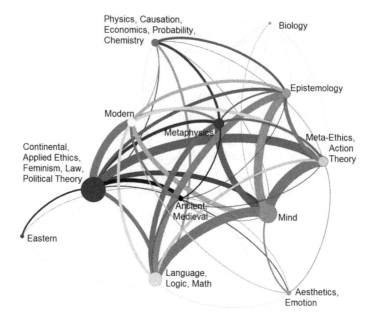

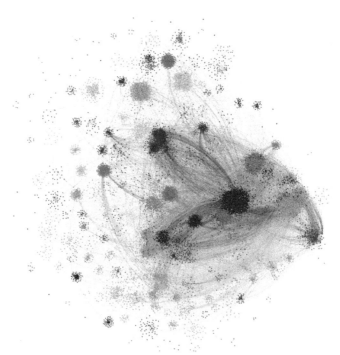

Figure 9.4 Publications in philosophy organized by OpenOrd without data editing (left), ForceAtlas2 with edges trimmed (right) and ForceAtlas2 with clusters grouped (bottom). The original data used to design these visualizations are identical. For zoomable versions of these visualisations see https://osf.io/kwuex/

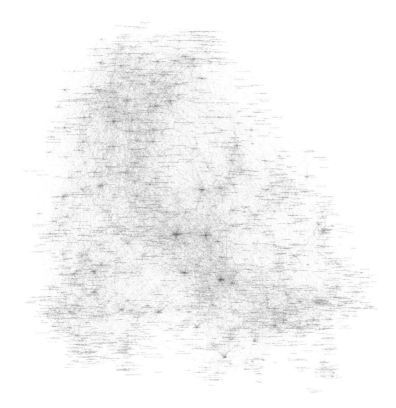

Figure 9.4 (*Continued*)

outline or box around the label. Edges can be either straight or curved, and their colour can be determined by a metric (e.g. weight), by their input/output node, or manually. Hit Refresh to visualize the network and select Export to save the image as a file.

Interpretation

Interpreting networks proceeds along two tracks. First, the researchers can make inferences from the mathematical properties of the network. Regarding nodes, the various centrality and clustering properties can be used to make inferences about both which nodes are more or less important (in particular ways). They can also be used to make inferences about subsets of the nodes in the network: to what extent they cleave into communities, as well as which nodes play important roles in connecting communities. To illustrate with our obituary case, we can use PageRank to identify cardinal virtues, vices and constituents of well-being. Alternatively, in a citation network, researchers could use betweenness to identify authors who connect communities or modularity to identify the communities themselves. Which measure one uses and how one interprets it depends on the research question, which should inform the conceptualization, operationalization and cleaning and processing steps.

Second, the researchers can make inferences from the visualizations of the network. As we mentioned above, in principle visualizations add nothing new. However, because they primarily engage perception rather than cognition, they may make it easier for human minds to notice and understand patterns in the data that would be difficult to comprehend with just the bare mathematics. If nodes have been sized based on their PageRank or betweenness, it may be obvious at first glance which ones are most important (assuming that PageRank or betweenness matter for the network in question). If modularity has been encoded categorically using hue, commonalities may pop out among neighbours in a community and differences between communities may seem obvious on inspection. We recommend allowing the conceptualization, operationalization and type of data to guide choices for visualization. In general, visualizations should be used for exploratory purposes, and researchers should rely on mathematical details for a more precise understanding of the results.

While visualizations do not add anything new, they do tend to make patterns more obvious. Indeed, one drawback of visualizations is that they can fool the eye into seeing patterns where none exist. For example, force directed layouts necessarily distort the geometry of networks by coercing them into a two-dimensional space. As such, some nodes end up adjacent to one another even though they are not connected in a meaningful way via the geometry of the network; and some nodes end up at the centre of the visualization even though they do not rank particularly high in Eigenvector Centrality, PageRank, Betweenness and so on. As we mentioned above, spatial region is the strongest visual cue of category-membership. Any layout will therefore inevitably end up suggesting spurious patterns and associations. We believe that this drawback is best handled by using visualizations primarily for *exploring* the data, whereas the bare mathematics of the network structure are better suited to (*dis*)*confirming* hypotheses suggested by exploration. Indeed, this is a well-recognized division of labour in exploratory graph analysis (Golino and Epskamp 2017).

3. Summary

In this chapter, we have provided a theoretical framework and workflow for semantic and social network analysis. Our intended audience has been experimental philosophers who are already familiar with more traditional methods such as surveys. We illustrated the approach using examples from our own published papers and works-in-progress. Our goal has not been to establish new results but to share methodological insights and tips that, we hope, other experimental philosophers will find interesting and promising for their own work. We believe that the approach outlined here will prove valuable in a variety of research contexts, and that adding network analytics to the toolbox will open up new prospects for collaboration with computer scientists and others who have yet to make much impact in experimental philosophy.

Suggested Readings

Alfano, M. (2015). Ramsifying virtue theory. In M. Alfano (ed.), *Current Controversies in Virtue Theory* (pp. 123–135). London: Routledge.

Alfano, M., Higgins, A., and Levernier, J. (2017). Identifying virtues and values through obituary data-mining. *Journal of Value Inquiry*, *52*(1), 59–79.

Brandt, M., Sibley, C., and Osborne, D. (2018). What is central to belief system networks? *PsyArXiv Preprints*. Doi: 10.17605/OSF.IO/TYR64. https://psyarxiv.com/tyr64.

Moretti, F. (2013). *Distant Reading*. London, New York: Verso.

Munzner, T. (2014). *Visualization Analysis and Design*. New York: AK Peters.

References

Alfano, M. (2012). The most agreeable of all vices: Nietzsche as virtue epistemologist. *British Journal for the History of Philosophy*, *21*(4), 767–790.

Alfano, M. (2015). Ramsifying virtue theory. In M. Alfano (eds.), *Current Controversies in Virtue Theory* (pp. 123–35). Routledge.

Alfano, M., Higgins, A., and Levernier, J. (2017). Identifying virtues and values through obituary data-mining. *Journal of Value Inquiry*, *52*(1), 59–79.

Alfano, V., and Stauffer, A. (2015). *Virtual Victorians: Networks, Connections, Technologies*. New York: Palgrave.

Betti, A., Gerrits, D., Speckmann, B., and van Den Berg, H. (2014). GlamMap: Visualizing library metadata. *VALA 2014* (17th Biennual Conference and Exhibition, Melbourne, Australia, 3–6 February 2014) (pp. 1–15).

Birch, J. (2012). Worldwide prevalence of red-green color deficiency. *Journal of the Optical Society of America A, 29*(3), 313–320.

Bird, S., Loper, E., and Klein, E. (2009). *Natural Language Processing with Python*. Sebastopol, CA: O'Reilly Media.

Blondel, V., Guillaume, J. L., Lambiotte, R., and Lefebvre, E. (2008). Fast unfolding of communities in large networks. *Journal of Statistical Mechanics: Theory and Experiment, 10*, P10008.

Borkulo, C. (2016). Statistical comparison of two networks based on three invariance measures. https://cran.rproject.org/web/packages/NetworkComparisonTest/NetworkComparisonTest.pdf. Accessed 3 October 2016.

Borkulo, C., and Epskamp, S. (2016), Fitting Ising models using the ELasso method. https://cran.r-project.org/web/packages/IsingFit/IsingFit.pdf. Accessed 3 October 2016.

Borsboom, D., and Cramer, A. (2013). Network analysis: An integrative approach to the structure of psychopathology. *Annual Review of Clinical Psychology, 9*, 91–121.

Brandt, M., Sibley, C., and Osborne, D. (2018). What is central to belief system networks? PsyArXiv Preprints. Doi: 10.17605/OSF.IO/TYR64. https://psyarxiv.com/tyr64

Brin, S., and Page, L. (1998). The anatomy of a large-scale hypertextual web search engine. *Computer Networks and ISDN Systems, 30*, 107–117.

Carlson, A., Betteridge, J., Kisiel, B., Settles, B, Hruschka, E., and Mitchell, T. (2010). Towards an architecture for never-ending language learning. *Proceedings of the Twenty-Fourth AAAI Conference on Artificial Intelligence* (AAAI-10), 1306–1313.

Chalmers, D. (2015). Why isn't there more progress in philosophy? *Philosophy, 90*(1): 3–31.

Chen, M., and Floridi, L. (2013). An analysis of information visualization. *Synthese, 190*(16), 3421–3438.

Clark, A. (2002). Towards a science of the bio–technological mind. *International Journal of Cognition and Technology, 1*(1), 21–33.

Cohen, J. (1968). Weighted kappa: Nominal scale agreement with provision for scaled disagreement or partial credit. *Psychological Bulletin, 70*(4), 213–220.

Costantini, G., Epskamp, S., Borsboom, D., Perugini, M., Mõttus, R., Waldorp, L., and Cramer, A. (2014). State of the aRt personality research: A tutorial on network analysis of personality data in R. *Journal of Research in Personality, 54*, 13–29.

Crenshaw, K. (1989). Demarginalizing the intersection of race and sex: A black feminist critique of antidiscrimination doctrine, feminist theory and antiracist politics. *University of Chicago Legal Forum*, (1), 139–67.

Diesner, J. (2014). ConText: Software for the integrated analysis of text data and network data, Social and Semantic Networks in Communication Research: Conference of International Communication Association, Seattle, WA, 2014.

Diesner, J., and Carley, K. (2010). Relational methods for analyzing and preventing crime (German: Relationale Methoden in der Erforschung, Ermittlung und Prävention von Kriminalität). In C. Stegbauer and R. Häußling (eds.), *Handbook of Network Research* (pp. 725–738). Wiesbaden: VS Verlag.

Epskamp, S., Borsboom, D., and Fried, E. (2016). *Estimating Psychological Networks and their Accuracy: A Tutorial Paper*. arXiv:1604.08462v3.

Epskamp, S., Cramer, A., Waldorp, L., Schmittmann, V., Borsboom, D. (2012). Qgraph: Network visualizations of relationships in psychometric data. *Journal of Statistical Software*, *48*(4), 1–18.

Fruchterman, T., and Reingold, E. (1991). Graph drawing by force-directed placement. *Software: Practice and Experience*, *21*(11), 1129–1164.

Golino, H., and Epskamp, S. (2017). Exploratory graph analysis: A new approach for estimating the number of dimensions in psychological research. *PLoS ONE*, *12*(6), e0174035.

Hastie, T., Tibshirani, R., and Friedman, J. (2008). *The Elements of Statistical Learning: Data Mining, Inference, and Prediction*, 2nd edition. Springer.

HRAF (1967). The HRAF quality control sample universe. *Cross-Cultural Research*, *2*(2), 81–88.

Hu, Y. (2005). Efficient, high–quality force–directed graph drawing. *The Mathematica Journal*, *10*: 37–71.

Jacomy, M. (2009). Force-atlas graph layout algorithm. http://gephi.org/2011/forceatlas2-the-new-version-of-our-home-brew-layout.

Jacomy, M., Venturini, T., Heymann, S., and Bastian, M. (2014). ForceAtlas2, a continuous graph layout algorithm for handy network visualization design for the Gephi software. *PLoS ONE*, *9*(6), e98679. https://doi.org/10.1371/journal.pone.0098679

Kirkegaard, E., and Bjerrekær, J. (2014). The OKCupid dataset: A very large public dataset of dating site users. *Open Differential Psychology*. https://openpsych.net/forum/showthread.php?tid=279.

Krämer, N., Schäfer, J., and Boulesteix, A.-L. (2009). Regularized estimation of large-scale gene association networks using graphical Gaussian models. *BMC Bioinformatics*, *10*, 38.

Krugman, P. (2014). Sergeant Friday was not a fox. *The New York Times*. https://krugman.blogs.nytimes.com/2014/03/18/sergeant-friday-was-not-a-fox/. Accessed 17 January 2018.

Lewis, D. (1966). An argument for the identity theory. *The Journal of Philosophy*, *61*(1), 17–25.

Lewis, D. (1970). How to define theoretical terms. *The Journal of Philosophy*, *67*(13), 427–46.

Lewis, D. (1972). Psychophysical and theoretical identifications. *Australasian Journal of Philosophy*, *50*, 249–258.

Lurquin, J., and Miyake, A. (2017). Challenges to ego-depletion research go beyond the replication crisis: A need for tackling the conceptual crisis. *Frontiers in Psychology*, *8*, 586.

Lynch, M. (2016). *The Internet of Us: Knowing More and Understanding Less in the Age of Big Data*. Liveright.

MacWhinney, B. (2000). *The CHILDES Project: Tools for Analyzing Talk*, 3rd edition. Lawrence Erlbaum.

Martin, S., Brown, M., Klavans, R., and Boyack, K. (2011). OpenOrd: An Open-Source Toolbox for Large Graph Layout. *SPIE Conference on Visualization and Data Analysis.*

Munzner, T. (2014). *Visualization Analysis and Design.* AK Peters.

Moretti, F. (2013). *Distant Reading.* Verso.

Newman, M. (2018). The mathematics of networks. In M. Newman. *Networks* (pp. 105–157).Oxford: Oxford University Press

Nichols, S., and Knobe, J. (2007). Moral responsibility and determinism: The cognitive science of folk intuitions. *Nous, 41*(4), 663–685.

Open Science Collaboration (2015). Estimating the reproducibility of psychology science. *Science, 349*(6251), 943.

Parfit, D. (1984). *Reasons and Persons.* Oxford: Oxford University Press.

Powers, D. (2011). Evaluation: From precision, recall and F-measure to ROC, informedness, markedness, and correlation. *Journal of Machine Learning Technologies, 2*(1), 37–63.

Radim, R., and Sojka, P. (2010). Software framework for topic modelling with large corpora, *Proceedings of the LREC 2010 Workshop on New Challenges for NLP Frameworks*, 45–50. ELRA.

Raichle, M. (2015). The brain's default mode network. *Annual Review of Neuroscience, 38*, 433–447.

Ramsey, F. (1931). Theories. In R. B. Braithwaite (ed.), *The Foundations of Mathematics* (pp. 212–236). London: Routledge and Kegan Paul.

Rathkopf, C. (2018). Network representation and complex systems. *Synthese, 195*(1), 55–78.

Schneiderman, B. (1996). The eyes have it: A task by data type taxonomy for information visualizations. *Proceedings of the 1996 IEE Symposium on Visual Analytics*, 336–343.

Schwitzgebel, E., and Cushman, F. (2012). Expertise in moral reasoning? Order effects on moral judgment in professional philosophers and non-philosophers. *Mind and Language, 27*(2), 135–153.

Stevens, G., White, R., Flaxman, S., Price, H., Jonas, J., Keeffe, J., Leasher, J., Naidoo, K., Pesudovs, K., Resnikoff, S., Taylor, H., Bourne, R., and Visual Loss Expert Group. (2013). Global prevalence of vision impairment and blindness: Magnitude and temporal trends, 1990–2010. *Ophthalmology, 120*(12), 2377–2384.

Sugiyama, K., Tagawa, S., and Toda, M. (1981). Methods for visual understanding of hierarchical system structures. *IEEE Transactions on Systems, Man, and Cybernetics, 11*(2), 109–125.

Tiberius, V. (2008). *The Reflective Life.* Oxford, New York: Oxford University Press.

Van Norden, B. (2008). *Mengzi: With Selections from Traditional Commentaries.* Indianapolis IN: Hackett.

Watts, D. J., and Strogatz, S. (1998). Collective dynamics of 'small-world' networks. *Nature, 393*(6684): 440–442.

Weinberg, J., Nichols, S., and Stich, S. (2001). Normativity and Epistemic Intuitions. *Philosophical Topics, 29*(1–2), 429–460.

Wootson, C. R., and Wong, H. (2018, April 19). After calling Barbara Bush an 'amazing racist,' a professor taunts critics: 'I will never be fired'. *The Washington Post*. https://www.washingtonpost.com/news/grade-point/wp/2018/04/18/after-calling-barbara-bush-an-amazing-racist-a-professor-taunts-critics-i-will-never-be-fired. Accessed 24 April 2018.

Zagzebski, L. (2017). *Exemplarist Moral Theory*. Oxford, New York: Oxford University Press.

10

History of Philosophy in Ones and Zeros

Arianna Betti, Hein van den Berg, Yvette Oortwijn and Caspar Treijtel

1. Introduction

This chapter presents the new field of 'data-driven history of philosophy' or 'computational history of philosophy' as a branch of the digital humanities. The field is at present rather small: up to now, philosophers have not been much involved in digital humanities projects (Bradley 2011; see also van den Berg et al. 2014; Betti, Reynaert, and van den Berg 2017). Exceptions include Overton (2013), who uses text mining techniques from natural language processing to discover patterns in the usage of the term 'explain', and Andow (2015), who provides a qualitative study of two corpora in order to examine instances of intuition talk. In addition, Fischer et al. (2015), who clarify the epistemological relevance of empirical findings about intuitions, and include

The authors' contribution is as follows: Arianna Betti (corresponding author) had the initial idea, did background qualitative research and collected existing hypotheses, set up and designed the research, carried out analysis of years 1888–1937, wrote the abstract, Sections 1 and 2 and co-wrote Sections 3 and 4; Hein van den Berg (Annotator 1) contributed to shaping the research, structured the paper, co-wrote Sections 3 and 4 and carried out analysis of years 1937–1959. Yvette Oortwijn (Annotator 2), contributed to data collection and data analysis for the background research, gathered secondary literature, set up tables, curated references, and carried out analysis of years 1888–1959; Caspar Treijtel collected the digital corpus including metadata, set up the search interface, and liaised with library staff responsible for licensing.

Acknowledgements
Thanks to Julia Turska for attracting our attention to the 1981 Quine quote in Section 2; to Jelke Bloem, Anna Bellomo, Lisa Dondorp, Silvan Hungerbühler, Maud van Lier, Thijs Ossenkoppele and Pauline van Wierst for discussion and valuable comments. Silvan also calculated for us Cohen's Kappa, Jelke ran the chi-squared test, and Lisa provided us with many valuable editing suggestions. Thanks also to Janneke Staaks, Lidie Koeneman and Marcel Ottenbros from the Library of the University of Amsterdam for help with licensing and data harvesting. The data that we have harvested was collected with the support and permission provided by the publishers of the investigated publications, Taylor & Francis (tandfonline. com), Ovid/Wolters Kluwer (ovid.com) and JSTOR (jstor.org).

a distributional semantics analysis of appearance words in Wikiwoods[1] to support reconstruction of arguments in historical texts. Alfano (forthcoming) uses digital humanities techniques in order to track Nietzsche's use of the terms 'drive', 'instinct' and 'virtue'.[2]

The relative lack of historians of philosophy involved in 'data-driven history of philosophy' is regrettable; valuable results can be obtained by applying even rather simple, well-known computational techniques to philosophically relevant texts (van Wierst et al. 2016). The present chapter is a further confirmation of the potential of using even more basic computational means to gather and set up access to data for philosophers to evaluate. We demonstrate the utility of a computational and data-driven approach in philosophy by providing a computational study of the spread of the notion of *'conceptual scheme'*. Investigating the history of a concept can contribute to a core technique of philosophy – *conceptual analysis*. Understanding the origins of a concept improves understanding of the notion itself and the identification of related ideas when the corresponding labels appear in a text (i.e. what the author intended by 'conceptual scheme'). Understanding the origins and development of a concept is key to correct interpretation of a text.

'Conceptual scheme' is one of the most intriguing notions in 20th century analytic philosophy and general philosophy of science. Roughly, the expression labels the ordinary notion of a perspective on the world. It appears in a celebrated 1973 article by Donald Davidson (Davidson 1973), is central to W. V. Quine's thought (see e.g. Quine 1960) and plays a key role in the work of Thomas Kuhn, in which it appears as both 'conceptual scheme' (Kuhn 1957) and – famously – 'paradigm' (Kuhn 1962), It is also of interest to other disciplines contributing to cognitive science, as it is related to *'schema'* in cognitive psychology (Rumelhart 1980) and *'frame'* in artificial intelligence (as introduced by Minsky 1974; see also Mey 1982: 106). Despite its importance, its origin, development and spread both inside and outside philosophy are still unclear.

Although integral to many of the arguments Quine puts forward, Quine nowhere defines or characterizes the expression 'conceptual scheme' as a technical term. The first occurrence in Quine, to our knowledge, is in Quine (1934), where the way he works with it suggests that he takes his audience to be thoroughly familiar with the phrase. If readers were familiar with the notion prior to reading Quine, its origin cannot be in Quine's work.

---

[1] http://moin.delph-in.net/WikiWoods
[2] See also this report on the first conference on data-driven history of philosophy (Torino, January 2017): https://dr2blog.hcommons.org/2017/06/23/dr2-conference-a-not-too-late-report/.

Traditional research in philosophy, psychology and sociology has put forward at least three hypotheses about the originator of 'conceptual scheme': William James (1843–1910) (Mey 1982, 25); Jean Piaget (1896–1980) (Preston 2008; Mey 1982, 106, 229) and Vilfredo Pareto (1848–1923) (Isaac 2012, 70). The Pareto hypothesis involves the mediation of Lawrence J. Henderson (1878–1942), a leading biochemist by training, and a key figure in the Harvard administration, who fostered both the general interdisciplinary climate in which Quine, a Harvard PhD, developed academically, and the specific reflection on the social sciences that took place in 1920s–30s Harvard. The three competing hypotheses have been independently formulated within (sub)fields in the social sciences and humanities (e.g. Mey 1982 does not mention Quine at all), and no attempt at a comprehensive study at an interdisciplinary history of the notion of conceptual scheme is to date available.

In this chapter we take a first step towards such a comprehensive study by focusing on one aspect of the Pareto hypothesis, namely the spread of a notion of 'conceptual scheme' we call '*Hendersonian*': we find it in about 42,000 articles from eight US psychology and sociology journals from 1888 to 1959. We assess the hypothesis by combining a quantitative procedure aided by basic computational techniques with qualitative elements informed by what Betti and van den Berg (2014) have previously called the 'model approach to the history of ideas'. We highlight practical and methodological issues arising from the use of this novel method, and formulate recommendations and plans for future work in this field.

2. Method and corpora

How do we find out where or from whom the notion of 'conceptual scheme' originated, how it has developed and the way it has, possibly, spread to disciplines other than the ones in which it originated? Traditional studies in the history of philosophy usually tackle questions such as this by qualitative analysis. The method typically adopted by these studies consists in careful, in-depth reading of a small number of primary sources (such as a selection of works by Quine, and/or James, Pareto, Henderson, and Piaget) combined with (sometimes less in-depth) reading of a certain number of secondary sources (e.g. works such as Mey 1982; Preston 2008 and Isaac 2012), which are used as interpretive guides to the primary sources. Textual passages that are deemed relevant for answering the research question at issue are then usually lifted from some of these sources, transcribed literally in the paper and offered as evidence for the historical findings presented. One such passage might be this:

> Let me clarify the status of the phrase. I inherited it some forty-five years ago
> through L. J. Henderson from Pareto, and I have meant it as ordinary language,
> serving no technical function. (Quine 1981, 41)

A selection of the sources actually used in the research are referred to in the
various sections of the paper, and their metadata is recorded in the bibliography
section.

In traditional studies of this kind, the concept of a properly delineated and
well-delimited corpus serving as the explicit textual basis for the research is rather
alien. Once the set of primary and secondary sources actually used (as well as
those to be *ideally* used) is specified, however, it becomes clear that the quantity
of source-material a researcher following the traditional method could possibly
process is rather small. Thus traditional research can deliver investigations only
on a small scale. This also means that wide-scope claims such as

> ... the conceptual scheme rapidly became such a pervasive idea in anglophone
> philosophy of science that it is hard to find a programmatic statement about
> science in the 1950s or 1960s without it. (Galison 1997, 788)

... which are common and generally accepted by peers, should not be taken
literally, because not every relevant source is taken into account. Tracing the
development of a concept across multiple disciplines through several decades
isn't a small-scale research goal, and cannot be tackled via analysis of the sort
that is adequate in those cases. What is more, the small-scale analysis intrinsic
in traditional research carries within itself a limitation of the research field that
tends to lead, inevitably, to an inward-looking, monodisciplinary attitude towards
the available spread of evidence. The evidence in question tends to be selectively
identified with the writings of a few known authors that are transmitted from
generation to generation of scholars as *the* authors to read, the presumption
being that these authors are the most influential in the relevant field.[3]

By contrast with the traditional method of investigation, in the present chapter
we rely on (1) a well-delimited, explicitly stated digital corpus of 41,462 articles
from a selection of US journals in psychology and sociology, to which we apply,
aided by a (2) very basic computational means, a novel approach consisting in
a mix of (3) qualitative and (4) quantitative analysis. We give details on each of
(1)–(4) here below.

---

[3] See on this point also Green et al. 2013; Sangiacomo 2018 [forthcoming]. See also https://
historyofphilosophy.net/rule-5-history-philosophy-take-minor-figures-seriously

(1) To construct the corpus that we use, we started by singling out *The Psychological Review* (founded in 1894) as a journal representative of the field of psychology in the United States. This was one of the most significant and prominent early American journals of psychology as a separate discipline (Green et al. 2013, 169; 2015, 16). A quick perusal and assessment of a simple search for 'conceptual scheme' in the years 1895–1960 of *The Psychological Review* (for details see (2) below), seemed to point to a focus of use in the field of *social psychology*: so we decided to expand the corpus to seven journals in the social sciences: *American Journal of Sociology, American Sociological Review, Journal of Social Psychology, American Catholic Sociological Review, Sociometry (Journal of), Social Forces* and *American Anthropologist*. One pragmatic reason to select these seven journals, alongside our wish to use as many journals in sociology and social psychology as possible, is that access to these was reasonably quick and easy (for details on data access see again (2) below). Other reasons include that *The American Journal of Sociology* and *American Sociological Review* were the top two journals in US sociology in the period: the former, founded in 1895, was the first sociology journal in the world and was dominated by the so-called Chicago school, while the latter was founded in 1936 as a reaction to the former (Lengermann 1979; Kinloch 1988, 190). The *Journal of Social Forces* (founded in 1922, later *Social Forces*) was the third major outlet in sociology. *Sociometry* (founded in 1937, later *Social Psychology Quarterly*), belongs to a subsequent cluster of more specialized sociology journals. The *American Anthropologist* (founded in 1888) expanded our coverage to anthropology, given the fluidity of the disciplinary borders in the social sciences in the period. The full table of our journals with the year of foundation, and the total count of articles published per journal follows:

Table 10.1 Articles per journal in the full corpus.

Title	Dates	No. of articles
American Journal of Sociology	1895–1959	12,361
American Anthropologist	1888–1959	8,614
American Sociological Review	1936–1959	6,257
Social Forces (Journal of Social Forces)	1959	5,724
The Psychological Review	1894–1959	3,396
American Catholic Sociological Review	1940–1959	2,784
The Journal of Social Psychology	1930–1959	1,324
Sociometry (Social Psychology Quarterly)	1937–1959	1,002

Note that although a total 41,462 journal articles might appear to be a massive corpus from the point of view of traditional research, the corpus is still very limited for our purposes – as we explain more in detail in Section 4.

(2) To collect the 41,462 journal article corpus and make it searchable and accessible in a user-friendly manner to our group, we approached the library of our university for aid with licensing and for technical support, which consisted in automatically downloading the articles, indexing them and setting up a simple interface based on the eXtensible Text Framework (XTF), an open source platform for access to digital content.[4] This kind of support is likely to be necessary for any other research group to download thousands of articles from more than one journal, plus the time-consuming process of seeking general agreement from several publishers and individual licences for thousands of articles. We used two different approaches to download the corpus. For *The Journal of Social Psychology* and *The Psychological Review* we downloaded the total number of journal articles (4,720). For the remaining 36,742 articles from the other six journals, since they are all hosted on JSTOR,[5] we decided to exploit JSTOR's 'data for research' facility.[6] This entailed downloading the full dataset of bigrams from the remaining six journals: for each of the 36,742 articles from *American Journal of Sociology, American Sociological Review, American Catholic Sociological Review, Sociometry, Social Forces* and *American Anthropologist*, we downloaded, courtesy of JSTOR, a.txt file containing all the possible bigrams, for example, two-word adjoining combinations present in the full text of the article. For example, for article 10.2307_277013, the list of bigrams starts as follows (the reader might quickly notice the spacing problem in the fourth and fifth entry, which is likely an OCR mistake caused by end-of-line splits; more on this in Section 4 below):

'social disorganization': 40
'group breakdown': 12
'we can': 11
'dis organization': 10
'disorgani zation': 8

Second, for each article containing the string 'conceptual scheme' anywhere in its full text (so occurrences of 'conceptual scheme' and 'conceptual schemes'

4   https://xtf.cdlib.org/
5   https://www.jstor.org/
6   https://www.jstor.org/dfr/?cid=dsp_j_dfr_01_2018&utm_source=jstor&utm_medium=display&utm_campaign=home_right_jstor_dfr

were included, but not of 'conceptual schemata'), we extracted the URLs of the corresponding.pdf, and downloaded the.pdf file manually from JSTOR. This gave us a harvest of 367 articles. We then uploaded, indexed and made accessible on the XTF interface those 367 articles, together with the 4,720 articles from *The Journal of Social Psychology* and *The Psychological Review* unfiltered for 'conceptual scheme', and of which 23 contained the bigram 'conceptual scheme' (note that this data mix between the filtered and the unfiltered sub-datasets is harmless, since for our analysis we decided to search only for the bigram 'conceptual scheme' and nothing else). The XTF interface offered a search function over the text layer of the pdf files, and faceted searches for journal titles and year of publication. No conversion of the pdfs to other formats was conducted.

(3) After gathering the corpus, we set up the qualitative part of the research, guided by the aim to have a way to interpret the data. Setting up the research in this way, so that the qualitative aspect guides the quantitative aspect of the research, conforms to a programmatic claim made in previous work (Betti and van den Berg 2016). The most important trait of the qualitative approach we adopt is its explicit nature, insofar as we employ an explicit – if rather rudimentary – model of (a particular conception) of conceptual scheme, which we use as a qualitative framework of interpretation for the quantitative research.[7] This is a model in the sense of Betti and van den Berg (2014).[8] In general, Betti and van den Berg's models represent ideas as complex relational structures with both stable (*continuity*) and variable (*discontinuity*) elements. Sophisticated shifts of meaning of ideas (concepts) through time can thus be measured as (dis)similarities between the stable and variable parts of such structures. This enables modelling of fine-grained discontinuities within a big picture of large-scale conceptual continuity enabling us to represent constantly shifting conceptual perspectives. The model approach is well suited to provide a proper theoretical foundation for tracing ideas computationally because it affords an ideal theoretical balance between fine-grained conceptual analysis and large-scale pattern finding.

In previous work, authors Betti and van den Berg have utilized a model of a traditional ideal of (axiomatic) science called the 'Classical Model of Science'

---

[7] For a similar approach see Sangiacomo 2018 [forthcoming], which in turns builds upon Betti and van den Berg 2014, 2016.

[8] In the following, we discuss the model approach as presented in Betti and van den Berg 2014 and Betti and van den Berg 2016. For a full account of the model approach, the reader is referred to the papers just mentioned.

(de Jong and Betti 2010). The Classical Model of Science models the idea of axiomatic science as a complex relational structure composed of seven conditions containing 10 sub-ideas or sub-concepts. This model is an interpretative tool on the basis of which continuities and discontinuities in the history of a particular idea of science can be studied.

For the present chapter, we have constructed a rudimentary model of a (particular conception of) the idea or concept of *conceptual scheme* that we have distilled – largely working by abduction – from Henderson (1932) – hence 'Hendersonian' – and which we identified with a minimal cluster of two conditions taken to be *necessary* to the notion (we say nothing about these conditions being also possibly *sufficient*):

(1) a (proper) science has a conceptual scheme;
(2) a conceptual scheme is a/the (or a part of the) theoretical framework of a (proper) science in terms of which empirical phenomena, facts or data are interpreted.

Both of these conditions can be said to also apply to the idea of conceptual scheme endorsed by Quine (e.g. Quine 1992 – given a broad enough reading of 'part of' and 'theoretical framework'). Note that the model in question is minimal, resulting in a broad conception fitting many variants of the notion of conceptual scheme, though it is not so broad as to fit, trivially, any conception whatsoever.[9] For example, naive realist theories holding that facts impose themselves on us and cannot thus be said to be 'interpreted' are excluded by the second condition, which includes the sub-concept of fact and/or data as *interpreted*. Note also that, as is the case in the Classical Model of Science, the model just sketched is highly abstract: the sub-concepts (e.g. *empirical phenomena, facts* or *data*) involved are determinables, and might be specified by different authors using different technical terms or sub-concepts. For instance, not everybody adhering to the model of conceptual scheme just sketched accepts facts (and the term 'fact' as a technical or semi-technical one), although everybody adhering to that model would accept that *some* items in their universe play at least some of the roles that facts play when they are accepted. Such items might be called 'data', 'phenomena', 'observations', etc. However, the relation between such items and a conceptual scheme seems most important: a conceptual scheme is that in terms of which the items in question are interpreted.

---

[9] This is in keeping with de Jong and Betti 2010, 191, 196.

(4) For the quantitative analysis: we manually constructed a derived dataset (in a spreadsheet) containing the following nine fields (mostly article metadata):

Journal title,
Article title,
Author name,
Year of publication,
Affiliation/location of Author at time of publication,
Discipline of affiliation of Author,
Qualitative notes, including academic career info, degrees, and other biographical notes
Context (passage of text) in which the Author uses the term 'conceptual scheme'
Annotation

The 'Annotation' field contains the annotator's *evaluation* of the Context (passage of text). We used two types of annotations: (A) the annotators indicated whether the Hendersonian notion appeared in the passage (Yes), either perhaps or vaguely appeared (Maybe) or was absent (No). That is, *Yes* means that the bigram 'conceptual scheme' is used in, or presupposes, the meaning fixed by our Hendersonian model. (B) The annotators indicated whether the passage provided philosophical or theoretical analysis, or additional information on the notion of *conceptual scheme* itself, instead of a mere use or application. Two expert annotators carried out both (A) and (B) independently for each passage. By adopting a model of the notion we set out to trace, we were able to interpret the substantial quantity of journal articles at our disposal more quickly. We could assess whether the article put forward the particular conception of conceptual scheme fixed by our model, and obtain a quantitative measure of it.

3. The bigram 'conceptual scheme' from 1888 to 1959 in psychology and sociology

The corpus shows a total of 367 articles including the bigram 'conceptual scheme' (about 0.9% of the total of articles published) in the eight journals we considered, authored by 287 different individuals at 133 different institutions. The *American Sociological Review* has by far the greatest number of articles using 'conceptual scheme' (about 30% of the total of 367), and Harvard University is the top affiliation of the authors who use the phrase. The 41 papers bearing a Harvard affiliation account for slightly more than 11% of the total. It is difficult to interpret the Harvard share adequately without normalizing for the affiliation of the authors of *all* of the papers published in the period. Additionally, we would need to know how many sociology departments there were in US at the time, their staff count, and how many sociologists in general were active in the period. Nevertheless, we might observe the following: qualitative research states uncontroversially that the dominant US sociology department from the 1890s until World War II was at Chicago (Harvey 1987, Chapter 7; Abbott 2009, 18.1). If we wished to take this qualitative information into account, then our data is significant insofar as it shows that the number of Harvard papers using 'conceptual scheme' is twice that of Chicago, which is our second-ranked institution. Note that the situation varies per decade, as the gap between Harvard and the other institutions thins out: prior to the 1930s, only three papers used the bigram, and none of their authors was affiliated to Harvard. In the 1930s, 12 out of the 24 articles were published by authors associated with Harvard (50%), whereas just two (less than 10%) were published by authors affiliated to that decade's second-ranked institution (Chicago again). In the 1940s Harvard ranked as the top affiliation again, with 18 papers out of 114 (about 16%), whereas the second-ranked, this time Fordham, counts half that number (9 out of 114 papers, about 8%). In the 1950s Harvard was overtaken by Chicago, which tops the list with 14 papers out of 226 (about 6%), while Harvard ranks second with 11 papers out of 226 (about 5%). Furthermore, our annotation revealed that every article in which the phrase 'conceptual scheme' occurs (every article in our tables), uses it (or appears to use it) in the sense that we have called 'Hendersonian': to restate, *every* article in our tables expresses, or possibly expresses, the Hendersonian notion; we comment further on this in Section 4 below. One possible interpretation of this data is that the Hendersonian notion of conceptual scheme was a hallmark of Harvard sociology in the 1930s, and that it spread to other institutions in the course of three decades.

This hypothesis of the spread of conceptual scheme as a notion needs further refinement. For one thing, our data include a significant number of authors who had no formal Harvard affiliation at the time of publication, but had a Harvard history (below we call them 'invisible Harvardians'). Among the top 10 authors we find Talcott Parsons (Harvard), a major figure in the history of sociology, ranking first with 14 papers, and 6 other (in)visible Harvardians

Table 10.2 Articles in which bigram 'conceptual scheme' appears by journal.

Journal (total 8)	Number of articles using 'conceptual scheme' (total 367)
American Sociological Review	108
Social Forces	65
American Journal of Sociology	64
American Anthropologist	51
American Catholic Sociological Review	41
Psychological Review	17
Sociometry	15
Journal of Social Psychology	6

Table 10.3 Number of articles in which the bigram 'conceptual scheme' features by affiliation (threshold ≥ 5) of author.

Affiliation (total 133) (threshold ≥5)	Number of authored articles using 'conceptual scheme'
Harvard	41
University of Chicago	16
Fordham	14
University of Michigan	10
Columbia	8
Ohio State	8
University of Wisconsin	8
Michigan state	8
Princeton	7
Indiana University	7
Northwestern University	6
University of California	6
University of Washington	5
Smith College	5
Cornell University	5
Catholic University of America	5
Rutgers	5
University of Colorado	5
University of Minnesota	5

(Timasheff, Kluckhohn, Davis, Moore, Merton, Devereux).[10] Thus, *70%* of the top 10 authors are Harvardians or invisible Harvardians. Taking the invisible Harvardians into account arguably leaves the 'spread' part of hypothesis untouched, while strengthening the 'Harvard hallmark' part. We come back to this point below.

In the remainder of this section, we break down our findings for the period prior to the 1930s, for the 1930s, the 1940s and the 1950s. We should signal that the decision to break the period up in this way has been conventional, but

Table 10.4 Articles featuring use of 'conceptual scheme' by author.

Authors (total 287) (threshold ≥3)	Number of authored articles containing 'conceptual scheme'
Parsons, Talcott	14
Timasheff, Nicholas S.	10
Kluckhohn, Clyde	6
Davis, Arthur K.	5
Morris, Rudolph E.	4
Moore, Wilbert E.	4
Bain, Read	4
Merton, Robert K.	4
Devereux, George	3
Hartnett, Robert C.	3
Whyte, William F.	3
Strauss, Anselm L.	3
Riley, Matilda W.	3
Jurczak, Chester A.	3
Pellegrin, Roland J.	3
Seeman, Melvin	3
Loomis, Charles P	3
Brewer, Earl D.C.	3
Moreno, Jacob L.	3
Hallowell, A. Irving	3

---

[10] Note that our check for invisible Harvardians was done manually for each author, and depended on readily available information on the internet. The biographical details of some authors are not easily accessible.

recognize that this decision might influence the analysis. We come back to this point briefly in Section 4 below.

Prior to the 1930s

From 1888 to 1929, we identified three journal articles containing the bigram 'conceptual scheme' (about 0.026% of the total):

The affiliations of the authors who use the phrase are collected in the table below, followed by a table with an overview of the authors who use the phrase.

Table 10.5 Journals featuring bigram 'conceptual scheme' before 1930.

Journal (total 8)	Number of articles containing 'conceptual scheme' (total: 367 | total for the period: 3)
Psychological Review	2
Social Forces	1

Table 10.6 Articles containing 'conceptual scheme' by author affiliation (Prior to 1930).

Affiliation (total: 133)	Number of authored articles containing 'conceptual scheme' (total: 367 | total period: 3)
Bristol	1
Columbia	1
Lucknow	1

Table 10.7 Articles using 'conceptual scheme' by author, prior to 1930.

Author (total: 287)	Number of authored articles containing 'conceptual scheme' (total: 367 | total period: 3)
Morgan, Conwy L.	1
Haeberlin, Herman K.	1
Mukerjee, Radhakamal	1

The authors who use the phrase are, as we can see, Conwy Lloyd Morgan, a famous British ethologist and psychologist,[11] Herman K. Haeberlin, an anthropologist who studied under Franz Boas (Boas 1919), and Radhakamal Mukerjee, a leading Indian thinker and social scientist.[12] The earliest occurrences of the bigram within our corpus thus occur within the disciplines of psychology, anthropology and sociology.

Although these mentions of the phrase 'conceptual scheme' occur before 1932, the year Henderson developed the account of conceptual scheme on which we have based our model, we have annotated the occurrences in order to see whether they conform to the Hendersonian model. The reason for doing this is to see whether the model in question was anticipated prior to its formulation by Henderson, which might suggest that Henderson developed his account of conceptual scheme earlier, or under the influence of someone else or both. Both annotators scored one article (Haeberlin) as conforming to the Hendersonian model of conceptual scheme (Yes), and two articles (Morgan and Mukerjee) as maybe conforming to the Hendersonian model of conceptual scheme (Maybe). No article includes a philosophical elaboration of the notion. The interrater reliability (Cohen's kappa) of both the question of whether the Hendersonian notion was contained in the articles and the question of whether the article included additional philosophical information was kappa $= 1$. Thus there is some scant evidence that the notion of conceptual scheme we are considering was already adopted prior to Henderson, but given the small dataset we are working with we cannot base any strong conclusions on this evidence.

If we consider the passages from a qualitative perspective, we may note that Morgan's and Mukerjee's are difficult to interpret, since they simply use the phrase without much elaboration or explanation. However, in Haeberlin's case the suggestion seems to be that a conceptual scheme must be interpreted as similar to (a part of) a (scientific) theory. This interpretation of the phrase 'conceptual scheme', which may be taken to conform to the Hendersonian model of conceptual scheme and which we also encounter in the 1940s and 1950s, is taken from the following passage:

> Wundt has devised a remarkable foundation of concepts upon which to build up a new science of the folk-soul. His concepts of the higher synthesis, the social mind, the reality of folk-psychological actuality, etc., are all seemingly firmly

---

[11] https://en.wikipedia.org/wiki/C._Lloyd_Morgan
[12] https://en.wikipedia.org/wiki/Radhakamal_Mukerjee

anchored in a monumental philosophical system; but Wundt's conceptual scheme breaks down when applied. (Haeberlin 1916, 301 [Y|Y|N|N][13])

Here, the conceptual scheme of Wundt is interpreted as his philosophical system or as the structure of concepts upon which Wundt builds up a science of the folk-soul.

1930s

In the 1930s, our corpus shows 24 articles that contain the bigram 'conceptual scheme' (about 0.3% of the total) – about 11 times the number of papers with respect to the previous period (the increase is normalized per number of papers published). The journals in which we found the bigram are:

As we can see, while the phrase 'conceptual scheme' figures in the sociological literature of the 1930s, and we also find a mention in the *American Anthropologist*, there are *no* occurrences within the psychology journals of our corpus in this period.

The affiliations of the authors who use the bigram is presented in Table 10.9, followed by an overview of the authors who use the bigram. Note that each author of a multi-author paper gets one article to their name.

Table 10.8 Journals in which 'conceptual scheme' appears, 1930s.

Journal (total: 8)	Number of articles containing 'conceptual scheme' (total: 367| total period: 24)
American Sociological Review	11
American Journal of Sociology	9
American Anthropologist	2
Sociometry	1
Social Forces	1

[13] We append to each passage that we quote in our qualitative analysis a four-letter code for the score of Annotator 1 and 2 as to A and B. For instance: M|Y|Y|N stands for the circumstance that Annotator 1 scored the passage as maybe (M) containing the Hendersonian notion of conceptual scheme, and Annotator 2 scored it as containing it (first occurrence of 'Y'); moreover, Annotator 1 scored the passage as containing additional information (second occurrence of 'Y') and Annotator 2 scored the passage as not containing additional information (N).

Table 10.9 Articles using 'conceptual scheme' by author affiliation, 1930s.

Affiliation (total: 133)	Number of articles containing 'conceptual scheme'
Harvard	12
University of Chicago	2
SSRC	1
Beacon Hill	1
Columbia	1
New York University	1
Smith College	1
University of Texas	1
Miami	1
Hawaii	1
Iowa	1
Ohio State	1
Yale	1
Chicago	1
University of Pennsylvania	1

Table 10.10 Articles using 'conceptual scheme' by author, 1930s.

Author	Number of articles containing 'conceptual scheme'
Parsons, Talcott	5
Merton, Robert K.	3
Kluckhohn, Clyde	1
Young, Donald	1
Davis, Kingsley	1
Brinton, Crane	1
Moreno, Jacob L.	1
Timasheff, Nicholas S.	1
Bierstedt, Robert	1
Cressey, Paul G.	1
DeNood, Neal B.	1
Gettys, Warner E.	1
Burgess, Ernest W.	1
Winch, Robert F.	1

Table 10.10 (*Continued*)

Author	Number of articles containing 'conceptual scheme'
Lind, Andrew W.	1
Bain, Read	1
Reuter, Edward B.	1
Leighton, J[oseph] A.	1
Dollard, John	1
Dunham, Warren H.	1
Hallowell, A. Irving	1

Harvard is by far the top affiliation for authors who employ the bigram. This finding is consistent with the idea that the phrase was introduced in the sociological literature by (or via) Henderson at Harvard in the 1930s. Among the authors who use the bigram most frequently are the famous Harvard-affiliated sociologists Talcott Parsons (who uses it in five published articles in our journal corpus) and Robert K. Merton (who uses the bigram in three). If we now also add to the count the 'invisible Harvardians', namely authors with a different affiliation but with a Harvard degree or a Harvard history, we see that of the 21 names listed, a third are (invisible) Harvardians: Parsons, Merton, Timasheff, Kluckhohn, Davis, Brinton and DeNood (Page 1985, 119).

Our annotation of the corpus gave the following results: out of the 24 articles, Annotator 1 scored 18 as articulating the Hendersonian notion of conceptual scheme (Yes), five articles as maybe articulating it (Maybe) and zero articles as not containing it (No). One passage was left without a score because it contained only a reference to another article with the bigram 'conceptual scheme' in the title. Moreover, four articles were identified as containing additional philosophical information on the notion, whereas 19 articles applied it without providing additional explanation. Annotator 2 scored 17 articles as articulating the Hendersonian notion of conceptual scheme (Yes) and six articles as maybe articulating it (Maybe) and zero articles as not containing it (No). Again, one article was left unscored. In addition, three articles were marked as containing additional information on the notion of conceptual scheme, and the other 20 articles as not containing additional information. The interrater reliability (Cohen's kappa) on whether the articles contained the Hendersonian notion of conceptual scheme was kappa = 0.89958159. The interrater reliability on whether the articles contained additional information on the notion of conceptual scheme was kappa = 0.868852459. These results indicate that a large portion – either 75% (annotator 1) or 70.8% (annotator 2) – of the authors who use the bigram 'conceptual scheme' adopted the Hendersonian notion – 'Yes' to (A): furthermore, whenever the phrase was used, it was (possibly) to express the Hendersonian notion ('Yes' or 'Maybe'). No single use of the phrase wasn't at least *possibly* Hendersonian ('No').

Having provided a quantitative analysis of the use of the phrase 'conceptual scheme' in our corpus, we may now turn to a qualitative analysis. It is striking that the use of the phrase in our corpus in the 1930s strictly follows the Hendersonian model. For example, as we have seen, condition (2) of the Hendersonian model fixes that a conceptual scheme is a/the (or a part of the) theoretical framework of a science in terms of which facts, data, or phenomena are interpreted. This conception is clearly articulated by Parsons:

> The relation between the relative and the non-relative aspects of scientific knowledge, in Weber's view, can best be illuminated by a somewhat further development of the subject. In the first place, the actual object or phenomenon which the scientist studies is not a 'fully concrete' reality but is a 'construction' which brings together in a coherent descriptive whole those aspects of concrete reality which are significant to the investigator. Such a construction Weber calls a 'historical individual.' The essence of his view is contained in the common current formulas that observation takes place 'in terms of a conceptual scheme,' 'within a frame of reference.' (Parsons 1936: 677 [Y|Y|Y|Y])

The sociologist Nicholas Timasheff, at the time also affiliated with Harvard, expresses the same idea while quoting Henderson: 'The sociology of law is a science based on observation. Observation means the description of facts in the terms of a conceptual scheme' (Timasheff 1938, 219 [Y|Y|Y|Y]). In a later paper, again in line with condition (2) of the Hendersonian model of conceptual scheme, Parsons licenses a conception of *data* and (scientific) *facts* as something which is stated in terms of a conceptual scheme. Parsons again writes:

> Furthermore, it is a fairly well-recognized methodological principle of general application that data are not simply 'facts' but that, like all the facts of science, they are stated 'in terms of a conceptual scheme'. (Parsons 1937, 479 [Y|Y|Y|N])

Our qualitative analysis of the 1930s suggests that the Hendersonian notion of conceptual scheme was accepted in the sociological articles we have considered.

1940s

From the 1940s, we have identified 114 journal articles that contain the bigram 'conceptual scheme' (about 1% of the total) – this is, again, an increase of almost four times the number of articles we had found for the 1930s (normalized per number of papers published). The breakdown per journal:

The bigram 'conceptual scheme' was, again, used most often in the *American Sociological Review* and the *American Journal of Sociology*, which together published roughly half of the papers in our corpus containing the bigram (this is rather uninformative, since these two journals also published roughly half of the total number of papers for the period).

Table 10.11 Articles in which 'conceptual scheme' features by journal, 1940s.

Journal (total: 8)	Number of articles containing 'conceptual scheme' (total: 367 | for the period: 114)
American Sociological Review	30
American Journal of Sociology	25
Social Forces	17
American Anthropologist	16
American Catholic Sociological Review	14
Psychological Review	6
Sociometry	5
Journal of Social Psychology	1

Table 10.12 Articles in which 'conceptual scheme' occurs, by author affiliation (threshold ≥ 3), 1940s.

Affiliation (threshold ≥3)	Occurrences
Harvard	18
Fordham	9
Princeton	4
Northwestern University	4
University of North Carolina	3
University of California	3
Columbia	3
Cornell University	3

Table 10.12 (*Continued*)

Affiliation (threshold ≥3)	Occurrences
Catholic University of America	3
University of Oklahoma	3
Union College	3

Table 10.13 Articles featuring 'conceptual scheme' by author (threshold ≥ 2), 1940s.

Authors (threshold (threshold ≥2)	Occurrences
Parsons, Talcott	7
Timasheff, Nicholas S.	6
Kluckhohn, Clyde	5
Davis, Arthur K.	5
Devereux, George	3
Hartnett, Robert C.	3
Whyte, William F.	3
Bain, Read	2
Moore, Wilbert E.	2
Hallowell, A. Irving	2
Kimball, Solon T.	2
Homans, George C.	2
Demerath, Nicholas J.	2

A survey of the affiliations of the authors who published articles using 'conceptual scheme' (table below) shows that, as with the 1930s, Harvard is the most common. Eighteen articles were published by authors working at Harvard, double the number published by authors working at Fordham University (nine). Other universities appear less frequently.

To unveil the true import of the Harvard link with the notion, however, we need to dig a bit deeper.

We find authors with a Harvard formal affiliation at the time of writing. We see, for example, Parsons, who again tops the 1940s chart with seven articles, and Clyde Kluckhohn,[14] third equal with five articles, as well as George C. Homans with two articles. The link between Harvard and the bigram 'conceptual scheme',

---

[14] Kluckhohn received his PhD in anthropology at Harvard in 1936 and later became to professor in Social Anthropology at the same institution, https://en.wikipedia.org/wiki/Clyde_Kluckhohn.

however, is far stronger when one considers, again, the invisible Harvardians in the list: Timasheff (second in the list, six papers), who came to the United States in 1936 as a visiting lecturer at Harvard University, and later became a Fordham University Professor, authored six of the nine Fordham articles[15]; Arthur K. Davis (tied third with five articles), PhD at Harvard under Parsons (1941), later moved to Union College (Nock 2002), and wrote, next to two Harvard papers, all three of the Union papers; George Devereux (fourth equal with three articles), was part of the so-called 'Parsonage', Parson's Sociological Group (Gerhardt 2016, 39, fn 150, counts Devereux among 'Parson's students'; Camic 1991, xliii fn 120 does the same; see also Hamilton 1992, 209). He authored one paper in 1940 under a Harvard affiliation, alongside two subsequent Wyoming papers. William Foote Whyte (seventh on the list), who was affiliated with the University of Oklahoma and wrote all three Oklahoma papers, had been a junior fellow at Harvard. Solon Kimball, (Harvard PhD 1936), authored two papers while affiliated to Alabama and Michigan State which, being under the threshold, do not show in the table. Nicholas J. Demerath, also a Harvard graduate, wrote two of the three North Carolina papers. Hadley Cantril and Wilbert E. Moore had Harvard degrees, and, taken together, were involved in three of the four Princeton papers. This means that of the 13 top individuals in the list, 10 were Harvardians (77%). Other Harvard graduates not shown on this table (having one paper each) are Robert K. Merton (Columbia) who was at Harvard up to 1938 and wrote one of the Columbia papers, Kingsley Davis (Penn State), Conrad M. Arensberg (Brooklyn College), Albert K. Cohen (Indiana), Charles P. Loomis (Michigan State), Muzafer Sherif (Princeton) and George H. Hildebrand (Princeton; Ehrenberg et al. 2007). From this we can safely conclude that the bigram was popular among Harvard staff and students. This is consistent with the idea that the phrase was a hallmark of Harvard sociology.

Of our 114 journal articles from the 1940s, Annotator 1 scored 72 passages as articulating the Hendersonian notion of conceptual scheme (Yes), 30 passages as maybe containing it (Maybe) and 0 passage as not containing the Hendersonian notion (No). Twelve passages were left without a score because it was not possible to score them because they contained only a reference to an article with the bigram 'conceptual scheme' in the title. Annotator 1 also noted that of the 114 journal articles, 14 contained additional philosophical information on or analysis of conceptual scheme, whereas 88 articles merely

---

[15] https://www.nytimes.com/1970/03/10/archives/dr-nicholas-timasheff-dies-sociologist-on-fordham-faculty.html.

used the notion without providing additional information or analysis. Again, 12 articles were not scored in this manner because providing a score was not possible. Annotator 2 scored 66 passages as articulating the Hendersonian notion of conceptual scheme (Yes), 36 passages as maybe containing it (Maybe) and 0 passage as not containing the Hendersonian notion (No). For the same reasons as Annotator 1, 12 passages were left unscored. In addition, Annotator 2 thought that 14 out of the 114 articles contained additional information on the notion of conceptual scheme, 88 contained no such information and 12 were left unscored. The interrater reliability (Cohen's kappa) on whether the articles contained the Hendersonian notion was kappa = 0.9025641026. The interrater reliability on whether the articles contained additional philosophical information on the notion of conceptual scheme was kappa = 0.9535830619. These results again indicate that whenever the bigram 'conceptual scheme' was used, it was to express the Hendersonian notion (or maybe so), as no single use of the phrase was not *possibly* Hendersonian. In addition, Annotator 1 scored 58 out of 72 articles using the Hendersonian notion of conceptual scheme as not containing any additional information on this notion (81%). Annotator 2 scored 52 out of 66 articles in this way (79%). Given this high score on the lack of additional information, we may presume that the notion was (at least considered as) well understood and perhaps accepted as a matter of course. In short: that 'conceptual scheme' belonged to a set of standard terms in at least some quarters of US sociology in the 1940s.

Let's again turn to a tentative qualitative analysis of the notion of conceptual scheme as it was used in 1940s sociology. The analysis provides results that are comparable to our analysis of the 1930s. First, we may note that, in line with condition (2) of our model, conceptual schemes were conceived of as 'systems' of concepts that are 'integrated' as one would expect a framework to be, and on the basis of which we analyse empirical phenomena. As Talcott Parsons writes:

> A generalized social system is a conceptual scheme, not an empirical phenomenon. It is a logically integrated system of generalized concepts of empirical reference in terms of which an indefinite number of concretely differing empirical systems can be described and analyzed (see L.J. Henderson, *Pareto's General Sociology,* Cambridge: Harvard University Press, 1935, chap. iv and n.3). (Parsons 1940, 844 [Y|Y|Y|Y])

In a quite similar fashion, the anthropologist Conrad Arensberg describes conceptual schemes as frameworks that allow one to organize *observations*

(Arensberg 1941 [Y|Y|N|N]); similarly, Timasheff notes that although science is based on observation, observation conducive to scientific knowledge requires more than perceptions: in addition to sense perceptions, 'there must be a conceptual scheme in terms of which the observations are formulated' (Timasheff 1947, 201 [Y|Y|Y|Y]).

Timasheff further notes that conceptual schemes often consist of (systems of) *definitions* (Timasheff 1947, 202), which serve the following function: they are tools of scientific analysis by allowing identification of objects and they are tools for the communication and preservation of knowledge (Timasheff 1947, 201). Mature sciences, such as economics, have, according to Timasheff, a conceptual scheme consisting of systems of definitions in which basic concepts, such as *price, supply, demand*, etc., are 'almost as well systematized as the concepts of the natural sciences' (Timasheff 1947, 202). Hence, a conceptual framework of definitions in which terms are defined in terms of a set of basic concepts seems to be a good example of a conceptual scheme.

The idea that conceptual schemes are frameworks of concepts or definitions used for interpreting *observations* and *phenomena* fits condition (2) of our Hendersonian model of the idea of conceptual scheme. Furthermore, in line with this condition, some authors in the 1940s also clearly take *facts* to be empirical statements in terms of a conceptual scheme. Thus, for example, the sociologist Gwynne Nettler, in an article in the *American Sociological Review*, quotes Henderson when he says that a fact is 'an empirically verifiable statement about phenomena in terms of a conceptual scheme' (Nettler 1946, 180 [Y|Y|Y|Y]).

Finally, condition (1) of our Hendersonian model of the idea of a conceptual scheme is that sciences *must have* a conceptual scheme. In other words, having a conceptual scheme is normatively prescribed to sciences in order to be a proper science. As an example of this view in the 1940s, one might consider this passage from Warner E. Gettys, who, writing about the science of human ecology, notes that human ecology must be based on an ecological conceptual scheme if it is to be counted as a science:

> If 'human' ecology [...] will concentrate on the study of the distributive aspects of human beings and their institutions by methods suited to such material and then analyze and interpret its data in terms of its ecological conceptual scheme, it may well make a place for itself among the social sciences and beyond the criticism of its friends and the mockery of its detractors. (Gettys 1940, 474 [Y|Y|N|N])

1950s

In the 1950s, 226 journal articles from our corpus included the bigram 'conceptual scheme' (slightly less than 2% of the total). This represents a normalized increase of about one and half times the number of occurrences in the 1940s, which shows that the bigram increased in popularity. The occurrences of the bigram within the journals of our corpus are displayed in Table 10.14.

The affiliations of authors who published articles containing the bigram above a certain threshold shows a larger spread than in the 1940s. From 11 affiliations in the 1940s (with a threshold of ≥ three articles) the total for the 1950s, adopting the same threshold, increased to 23 affiliations. Moreover, the highest-ranked institution by number of published papers using 'conceptual scheme' is now Chicago. We take these data to indicate a spread of the notion of conceptual scheme. A caveat, however: the increase in the range of affiliations could be due to an increase in the number of sociology departments from the 1940s to the 1950s, while the change in the top ranking institution could be due to a change in departmental situations (increase or decrease in hiring, mergers, etc). Should hard data on these factors become available, our numbers should be normalized against them.[16] Table 10.15 shows the spread of affiliations in our corpus.

Table 10.14 Articles featuring 'conceptual scheme' by journal, 1950s.

Journal (total: 8)	Number of articles containing 'conceptual scheme' (total: 367 | total for the period: 226)
American Sociological Review	67
Social Forces	46
American Anthropologist	33
American Journal of Sociology	30
American Catholic Sociological Review	27
Psychological Review	9
Sociometry	9
Journal of Social Psychology	5

[16] We do know there is an increase in papers published, namely 12,224 in the 1950s vs 9,728 in the 1940s, and though we do not have data on the affiliation of the authors of all these papers, we calculated that the number of unique, non-Harvard affiliations present in our corpus increases when normalized against the total numbers of papers published, and decreases (though far less than in the

Table 10.15 Articles in which 'conceptual scheme' is used by author affiliation (threshold ≥ 3), 1950s.

Affiliation (threshold ≥ 3)	Occurrences
University of Chicago	14
Harvard	11
University of Michigan	9
University of Wisconsin	7
Michigan state	6
Ohio State	6
Fordham	5
Indiana University	5
University of Colorado	5
Rutgers	5
University of Minnesota	5
University of Washington	4
Marquette University	4
Princeton	3
University of California	3
Columbia	3
Pennsylvania State College	3
Smith College	3
Wayne University	3
Brown University	3
Oberlin College	3
Louisiana State University	3
Berkeley	3

We take as further evidence for the spread of the *notion* of conceptual scheme the fact that the number of scientists who use the *bigram* 'conceptual scheme' in the 1950s in two or more articles (30 authors) is more than double the corresponding number in the 1940s (13 authors). We should also consider that the number of Harvard-affiliated papers in the 1950s decreases with respect to

1950s than before, and has a sharp downward trend, thus showing growing variety) if normalized against the number of 'conceptual scheme' papers. The latter finding is consistent with the fact that some institutions other than Harvard start publishing a higher number of papers per institution (e.g. Chicago in the Fifties): and indeed, the *ratio* of the number of non-Harvard authored papers to unique non-Harvard institutions grows through the decades.

the 1940s, while the number of papers containing 'conceptual scheme' published in the 1950s is almost twice the number of papers published in the 1940s (normalized growth: more than one and half times).

As per Table 10.16, the author who used the bigram 'conceptual scheme' most in the 1950s is the social scientist Rudolph E. Morris, (Marquette University), who uses it in four published articles within our 1950s corpus. Timasheff follows as the second highest ranked author for the second consecutive decade, with three papers. Anselm L. Strauss, of the University of Chicago, Matilda White Riley, a research assistant at Harvard in the 1930s and later professor at Rutgers, Chester A. Jurczak, a social scientist affiliated with Duquesne University, and Roland J. Pellegrin, a social scientist affiliated with Louisiana State university, all use the bigram in three papers, viz:

Table 10.16 Articles using bigram 'conceptual scheme' by author (threshold ≥ 2), 1950s.

Author (threshold ≥ 2)	Occurrences
Morris, Rudolph E.	4
Timasheff, Nicholas S.	3
Strauss, Anselm L.	3
Riley, Matilda W.	3
Jurczak, Chester A.	3
Pellegrin, Roland J.	3
Parsons, Talcott	2
Seeman, Melvin	2
Moore, Wilbert E.	2
Loomis, Charles P.	2
Brewer, Earl D. C.	2
Moreno, Jacob L.	2
Lindesmith, Alfred R.	2
Pfautz, Harold W.	2
Duncan, Otis D.	2
Jonassen, Christen T.	2
Hart, Charles W. M.	2
Simpson, George E.	2
Abel, Theodore	2
Fallers, Lloyd A.	2

Table 10.16 (*Continued*)

Author (threshold ≥ 2)	Occurrences
Theodorson, George A.	2
Alpert, Harry	2
Wilkening, Eugene A.	2
Form, William H.	2
Schuyler, Joseph B.	2
Wax, Murray	2
Schnore, Leo F.	2
Etzioni, Amitai	2
Spuhler, J. N.	2
Taylor, Walter W.	2

As before, we looked at the so-called 'invisible Harvardians'. In the 1950s, the invisible Harvardians include Timasheff, who authored three of the Fordham papers, Matilda White Riley, who authored three of the Rutgers papers, Charles P. Loomis, who published two papers containing the notion while at Michigan State, Wilbert E. Moore (two articles while at Princeton) Bernhard Barber with one article at Smith College (not on table), Muzafer Sherif, one article at the University of Oklahoma (not on table), and Hiroshi Daifuku, one article at the University of Wisconsin (not on table). The only author with an active Harvard affiliation on our list is Talcott Parsons. Thus, in the 1950s, of the 30 individuals on our list, 5 are Harvardians or invisible Harvardians – about 16.5%. In the 1940s, they were 77%. Hence, although there are invisible Harvardians in the 1950s, these are far lower in (normalized) percentage than in the 1940s, and our talk of a spread of the notion of conceptual scheme from Harvard to other universities seems justified – although note that our use of a threshold and possible lack of data on the total number of invisible Harvardians in the period at issue might affect this reasoning.[17]

We once again investigated the extent to which the Hendersonian notion of conceptual scheme was employed. Out of our 226 journal articles for the

---

[17] So the number of invisible Harvardians did not grow from the 1940s to the 1950s (at least if we consider the authors above a certain threshold): what if it had grown? Would this datum be in conflict with (at least one take on) the hypothesis that the notion 'spread' to other Universities'? For it might be contentious to say that a notion 'spread to other Universities', if all that is going on is that ex-Harvardians moved to other Universities and took a Harvard hallmark expression with them. We say that it is legitimate to talk of *spread* in this case (while we might talk of 'influence' in case that authors with no demonstrable Harvard history would use the notion more and more).

1950s, Annotator 1 scored 146 passages as containing the Hendersonian notion of conceptual scheme (Yes), 69 passages as maybe containing it (Maybe) and 0 passage as not containing it (No). Eleven passages could not be scored. In addition, Annotator 1 thought that 9 out of the 226 articles contained additional theoretical or philosophical information on the notion of conceptual scheme, whereas 206 articles used the term without providing additional theoretical information. Again, 11 articles could not be scored. Annotator 2 scored 133 passages as containing the Hendersonian notion of conceptual scheme (Yes), 82 passages as maybe containing it (Maybe) and 0 passage as not containing it (No). Eleven passages could not be scored. Annotator 2 noted that 8 out of the 226 articles offered additional philosophical information on the notion of conceptual scheme, 207 did not provide additional information and 11 were unscored. The interrater reliability (Cohen's kappa) on whether the articles contained the Hendersonian notion of conceptual scheme was kappa = 0.8864716566. The interrater reliability on whether the articles contained extra info was kappa = 0.9725761437. Annotator 1 scored 137 out of 146 articles using the Hendersonian notion of conceptual scheme as not containing any additional information or explanation on this notion. Annotator 2 scored 125 out of 133 articles containing the Hendersonian notion of conceptual scheme in the same way. The fact that about 94% (Annotator 1 and Annotator 2) of the authors who used the Hendersonian notion of conceptual scheme simply applied this notion without providing additional explanation suggests that in the 1950s, the idea is well-entrenched in (at least part of) US sociology. Again, in all cases in which the bigram was used, it was (possibly at least) used in the Hendersonian sense.

If we look at the occurrences of the phrase 'conceptual scheme' in our corpus from a qualitative perspective, we find much that is familiar from earlier decades. Thus, in line with condition (2) of our Hendersonian model, a conceptual scheme is often understood as part of a theoretical framework of a science in terms of which facts, data or phenomena are interpreted. For example, Alvin Boskoff writes:

> Scientific theory, then, is composed of generalized conceptual schemes which are ultimately confronted for verification by facts. But while sound theory must fit the facts, it does not follow that 'the facts' determine what the theory is to be. 'Fact' is always guided by a theoretical scheme, even if it is implicit, which formulates what we know and tells us what we want to know. (Boskoff 1950, 393 [Y|Y|Y|Y])

A new topic that was much discussed in the 1950's was Robert Merton's critique of Talcott Parsons' paper *'The Position of Sociological Theory'* (1948). In his comments to this paper, as well as in other works, Merton criticized sociological works that were derived from a single, all-inclusive master conceptual scheme. Rather than finding a single conceptual scheme for the whole of sociology from which uniformities of social behaviour can be derived, Merton argued that sociology should articulate multiple special theories for limited data. As he puts it himself:

> I believe that our major task *today* is to develop special theories applicable to limited ranges of data—theories, for example, of class dynamics, of conflicting group pressures, of the flow of power and interpersonal influence in communities—rather than to seek here and now the 'single' conceptual structure adequate to derive all these and other theories. (Merton 1948, 166)

(This passage was not scored because it did not contain the phrase 'conceptual scheme'. Parson's paper, on which Merton commented, was scored [Y|Y|Y|Y].)

Merton's critique of the idea of an all-inclusive conceptual scheme for sociology was often discussed in the 1950s (e.g. see Boskoff 1950 [Y|Y|Y|Y]; Kilzer 1950 [Y|Y|Y|Y]; Bain 1950 [Y|Y|N|N]; Vance 1952 [Y|Y|N|N]; Nett 1952 [Y|Y|N|N]). Note the use of 'derive' in 'derive all these [...] theories' which suggests a specification of the Hendersonian model similar to that of Timasheff (1947) as seen above.

Apart from Merton's critique of grand theorizing in sociology, the phrase 'conceptual scheme' is used in the 1950s in a manner that suggests that conceptual schemes are similar to *theories*. Thus, for example, theories developed by Marx and Engels are described as conceptual schemes (Jonassen 1951 [Y|Y|N|N]). The core meaning in these uses of conceptual scheme seems to be simply that of a theoretical framework of a science (see e.g. Luchins 1951 [Y|Y|N|N]).

4. The spread of the notion of conceptual scheme: preliminary findings and strengths and limitations of the method

On the basis of our analysis of the use of the phrase 'conceptual scheme' from 1888 to 1959 in our journal corpus, we conclude the following.

First, all cases of use of the bigram, even those from 1888 to 1929, are at least *possibly* Hendersonian, that is, they conform or possibly conform to a conception of conceptual scheme according to which (1) a (proper) science has a conceptual scheme; (2) a conceptual scheme is a/the (or a part of the) theoretical framework of a (proper) science in terms of which empirical phenomena, facts or data are interpreted.[18] We witness a sensible growth in the use of the phrase through time: the articles containing the bigram found in 1888–1929 count for 0.03% of the total number of papers published in that period by the eight journals in our corpus, and for 1.85% of the total in 1950–1959. In 1930–1939, Harvard emerges from the data as the locus of the (Hendersonian) notion of conceptual scheme, as in this period half of the papers containing the bigram are authored by Harvard-affiliated individuals; afterwards, this percentage decreases to about 16% (1940–1949) and 5% (1950–1959): in the final period Chicago takes over as the top-ranked institution for papers containing the bigram 'conceptual scheme'.[19] Thus, we submit, the notion spread from Harvard to other institutions: the number of (non-unique) affiliations other than Harvard indicated in authored papers containing the bigram increases in each decade (by about 21% in the 1940s with respect to the 1930s, and a further 12% in the 1950s with respect to the 1940s). In particular, in the 1950s the number of unique authors who use the bigram 'conceptual scheme' in two or more papers (30) is more than double that of the 1940s (13 authors), while the number of papers published in the 1950s is less than double the number published in the 1940s. Meanwhile, the Harvard share decreases sharply. We venture the hypothesis that the Hendersonian notion of conceptual scheme was a hallmark of Harvard sociology in the 1930s and that it spread to other institutions over the course of three decades.

Second, given the strict association between the bigram and its Hendersonian meaning our research so far seems to lend some plausibility to the hypothesis that Quine had the notion of conceptual scheme from Henderson. This seems to be the case at least as far as the expression 'conceptual scheme' is concerned, and under a rather liberal interpretation of what can be considered 'part of (a) framework' in the Hendersonian model. This would mean that at least part of the 1981 quote from Quine mentioned in Section 2 is trustworthy ('I inherited it some forty-five years ago through L. J. Henderson'). Considering the last part ('from Pareto'), whether there were sources other than Henderson, or whether Henderson himself inherited the notion, together with the bigram, from Pareto

---

[18] For the breakdown: *Does the phrase 'conceptual scheme' express the Hendersonian notion as fixed by the model?* < 1930s: Annotator 1 Yes 33% Maybe 67% | Annotator 2 Yes 33% Maybe 67%; 1930s: A1 Yes 75% Maybe 29% | A2 Yes 71% Maybe 25%; 1940s: A1 Y 63% M 26%| A2 Y 58% M 32%; 1950s: A1 Y 65% M 31% | A2 Y 59% M 36%. *Does the article contain additional information on the notion of conceptual scheme?* < 1930s, A1 Y 0% N 100% | A2 Y 0% N 100%; 1930s: A1 Y 17% N 79% | A2 Y 13% N 83%; 1940s: A1 Y 12% N 77% | A2 Y 12% N 77%; 1950s: A1 Y 4% N 91% |A2 Y 4% N 92%.

[19] If we take into account the 'invisible Harvardians', the percentages are higher, but we still see a significant decrease: 54% in the 1930s; 39% in the 1940s; 8% in the 1950s. Note, however, that as we say in footnote 11, biographical details might be difficult to ascertain for every author, so these percentages might not be fully correct.

or from someone else is a matter for further research. It is, however, interesting to note in this connection that we took the main *locus* of the phrase and source of our model to be Henderson 1932, but we have found uses of the bigram in the Hendersonian sense as early as 1905 (A1: M; A2: M) or at least 1916 (A1: Y; A2: Y). Moreover, Henderson himself is curiously absent from our list of authors, although other authors do refer to his work in connection with the bigram.

This brings us directly to strengths and limitations of our method. Traditional research on the origin, development and spread of ideas or concepts, such as the idea of a conceptual scheme, relies on a small, selected number of sources. In this chapter, we have presented research that relies on a large corpus (large compared to traditional studies in the history of philosophy) which serves as a broad quantitative basis on which hypotheses concerning the origin, development and spread of an idea such as conceptual scheme can be properly supported.

By constructing a large dataset of journal articles and interpreting these on the basis of our Hendersonian model of conceptual scheme, we were able to perform a quantitative analysis of the way in which the bigram 'conceptual scheme' was used in our corpus. The quantitative analysis in question allowed us, among other things, to establish the frequency with which the bigram was used, to identify the journals with the most occurrences of the bigram, to determine the authors who used the bigram the most, and to determine their affiliations. This information is important to study the disciplinary origin of the bigram and the dissemination of the bigram within different disciplines.

Having described the strengths of our method we may now turn to its limitations. The digital corpus we use in this chapter, though unusually large if compared with the number of sources used in traditional research in the history of philosophy, is still quite limited. Ideally, the corpus to answer any research question on the history of any idea in any period would be *universal*: it would consist of the high-quality, digital version of every single text produced in the period at issue – or at least of every single *published* text. A universal ideal corpus such as this is far beyond the practical reality of any corpus that can be obtained in a sensible amount of time by any known means. The biggest challenge, and the most time-consuming part of any research in the computational history of philosophy is building a suitable digital corpus (Betti, Reynaert, and van den Berg 2017). In short, building a universal digital corpus of the size needed to answer the question at issue properly is, at present, a quixotic endeavour. For this reason, we initially designed the research based on a restricted corpus composed by a combination of *two* datasets: articles from journals arguably representative for the disciplines of sociology and psychology in US in the period at issue (*Dataset 1*, the one we use in this pilot) and the complete *oeuvres* of each of the individual authors known to have contributed to the traditional hypotheses (James, Henderson, Pareto, Piaget, etc.) (*Dataset 2*, which we are yet to build). From a scholarly point of view, such a combined corpus strikes us as a good methodological compromise to tackle our research question given the practical constraints. The construction of even this restricted corpus, however, poses hard challenges. We plan to extend Dataset 1 with additional psychology and philosophy journals. Then we will construct Dataset 2 and run a study on its basis. The ultimate aim will be to combine the two datasets.

As a first next step we plan to enlarge Dataset 1 with the *American Journal of Psychology, Open Court, Pedagogical Seminary* and two major philosophy journals, *Mind* and *The Journal of Philosophy*.

Another limitation of the method we employ is that the corpus we have built is made up of pdf files. This means that we cannot perform sophisticated computational analyses on the corpus, but have had to restrict ourselves to very basic searches, restricted by the quality of the textual layer OCR provided by the publisher. This means, for example, that although the visual appearance of the pdf might have contained the string 'conceptual scheme' three times for a human reader, our search results might have yielded only one (or even zero), because the OCR of the textual layer might have 'misread' the remaining two occurrences as, say, 'conceptual sche me'. Such OCR errors can render occurrences of the target phrase invisible to a computer search (see the case of 'dis organization 10/ disorgani zation 8' that we mention in Section 2).

For more sophisticated analyses we would need to have the corpus transformed into a more reliably machine-processed format (plain text or .txt, or, ideally, TEI).[20] A high-quality corpus in .txt (if not in basic TEI) is a necessary requisite for sophisticated computational analysis in the history of philosophical ideas, where we need access to an (in principle) indefinitely extendable number of books and journals that contain a concept of scholarly interest (Betti, Reynaert, and van den Berg 2017). At the moment however, turning a 42,000-pdf corpus into a high-quality corpus in .txt format would be a laborious effort exceeding our purpose. Right now, we are more interested in quick adaptation and/or growth of the corpus: as said, our corpus of more than 42,000 articles is still very limited for the kind of research question we want to tackle. However, as we have

---

[20] For a nice example of effectively using TEI as a research tool on a philosophical text, see http:// digitallockeproject.nl/. Note however that the TEI was created/transcribed manually.

shown, when restricted to simple searches (such as that for the string 'conceptual scheme') plus a qualitative-quantitative manual analysis of the kind examined in this chapter, working with pdfs on a simple textual interface might be the best short-term strategy. Soon however, we would want to search for the *concept* of conceptual scheme, even when the bigram is not used. For this, we would need a high-quality textual corpus in a more machine-friendly format.

A further limitation of the method is that it does not allow for in-depth qualitative analysis as traditionally performed in the history of philosophy. To conduct such an analysis on a corpus as large as that used constructed for the research in this chapter, would mean researching in depth the views of all of the authors identified in the dataset – hardly feasible. We address this by adopting the *model approach* to the history of ideas, which allows for qualitative analysis of large bodies of text.

The choice of non-homogenous periodization (sometimes a decade, sometimes a longer period) presents a further limitation, because of possible distortions of the analyses arising from the particular periodization chosen.

Note, however, that our quantitative analysis proceeded on the basis of a qualitative interpretative framework, that is, the Hendersonian model of the idea of conceptual scheme. Hence, our analysis of the data is not theory-free but rather theory-laden. It is our belief that scholars in the history of philosophy, and more broadly the humanities, can only adequately analyse large quantities of data on the basis of such qualitative interpretative frameworks or models (see Betti and van den Berg 2016, 2014). Our experience of the research for this chapter confirmed this fact. Initially, we collected our dataset of journal articles and tried to make sense of the many occurrences of the bigram 'conceptual scheme' without a theoretical model. However, we soon discovered that we could not make sense of the large quantity of data bottom-up. Only after we introduced the Hendersonian model of a conceptual scheme could we adequately interpret the volume of journal articles that we needed to study. The Hendersonian model made the interpretation of a large corpus of texts possible. One would have to acknowledge, however, that our model was rudimentary and that for a more sophisticated analysis a better, more articulated model is needed, one that would allow us to be more discriminating as to what fits and what does not fit, and afforded fine-grained analyses and distinctions. The current working model of conceptual scheme should be considered a prototype. We plan to improve on it in the future.

That said, we have found no convincing evidence for the existence of a competing notion of conceptual scheme: that is neither annotator found even a single passage in which the bigram is not possibly (at least) expressing the Hendersonian notion. The fact that the we have found no discussion of alternative conceptions (i.e. we have no combination 'there is no Hendersonian notion' and 'there is further information on the notion'), can be taken to be evidence that, at least in the corpus we considered, no such alternative conception appears. We cannot exclude that the model we built is too general – that it captures too much. If this is the case, then the annotators' *Maybes* might be reconsidered: some of them might turn into 'Yes' or 'No' should an annotator be asked to base evaluation on different variants of the Hendersonian model, or to choose between two competing models. Nevertheless, there is nothing necessarily puzzling in the results we describe.

The insight that models are needed to interpret texts is not new, although there is still little awareness of it in traditional research in the history of ideas. It has long been established by cognitive psychologists who have developed the so-called schema theory of knowledge (Betti and van den Berg 2014; Betti 2014) According to these psychologists (see e.g. Anderson 1977; Rumelhart 1980), readers need interpretative schemas to interpret texts, and without schemata understanding is impossible (see also Chapter 2 in this volume). As Anderson remarks in a discussion of reading comprehension: 'text is gobbledygook unless the reader possesses an interpretative framework to breathe meaning into it' (Anderson 1977, 423). This view was once again confirmed when we did the research for this article.

5. Conclusion

Traditional research has provided conflicting qualitative hypotheses on the origin of the notion of conceptual scheme. These hypotheses are based on selective reading of small corpora. In this chapter we provided a new, mixed method for studying the spread of this notion, in principle generalizable to any notion, concept or idea whatsoever. Using our *model approach* to the history of ideas, we provided a qualitatively informed quantitative analysis of the spread of the use of the bigram 'conceptual scheme' in about 42,000 journal articles in the social sciences and psychology from 1888 to 1959. We argued that having such a corpus, large relative to traditional qualitative studies in history of philosophy, is necessary to provide adequate evidence for hypotheses concerning the origin, development and spread of an idea. Our analysis provides some alternative support for the hypothesis that Quine adopted the notion of conceptual scheme following Lawrence J. Henderson (1878–1942), although at this stage we cannot exclude multiple influences or a common ancestor for its origin.

Suggested readings

Betti, A., and van Den Berg, H. (2014). Modelling the history of ideas. *British Journal for the History of Philosophy*, *22*(4), 812–835.

Betti, A., Reynaert, M., and van Den Berg, H. (2017). @PhilosTEI: Building corpora for philosophers. In J. Odijk, and A. van Hessen (eds.), *CLARIN in the Low Countries* (pp. 371–384). London: Ubiquity Press.

Green, C. D., Feinerer, I., and Burman, J. T. (2015). Searching for the structure of early american psychology: Networking psychological review, 1909–1923. *History of Psychology*, *18*(2), 196–204.

Sangiacomo, A. (2018). Modelling the history of early modern natural philosophy: The fate of the art-nature distinction in the Dutch universities. *British Journal for the History of Philosophy*. https://doi.org/10.1080/09608788.2018.1506313

References

Abbott, A. (2009). Organizations and the Chicago School. In P. S. Adler (ed.), *The Oxford Handbook of Sociology and Organization Studies*. Oxford: Oxford University Press.

Alfano, M. (forthcoming). Digital humanities for history of philosophy: A case study on Nietzsche. In L. Levenberg, and T. Neilson (eds.), *Handbook of Methods in the Digital Humanities*. Lanham MD: Rowman & Littlefield.

Anderson, R. C. (1977). The notion of schemata and the educational enterprise: General discussion of the conference. In R. C. Anderson, R. J. Spiro, and W. E. Montague (eds.), *Schooling and the Acquisition of Knowledge* (pp. 415–431). Hillsdale, NJ: Erlbaum.

Andow, J. (2015). How distinctive is philosophers' intuition talk? *Metaphilosophy*, *46*(4–5), 515–538.

Arensberg, C. M. (1941). Review: English villagers of the thirteenth century by George C. Homans. *American Journal of Sociology*, *47*(3), 491–493.

Bain, R. (1950). Review: Social theory and social structure. Towards the codification of theory and research by Robert K. Merton. *Social Forces*, *28*(4), 443–44.

van Den Berg, H., Parra, G., Jentzsch, A., Drakos, A., and Duval, E. (2014). Studying the history of philosophical ideas: Support tools for research discovery, navigation, and awareness, in I-Know 2014. *Proceedings of the 14th International Conference on Knowledge Technologies and Data-driven Business*, Article no. 12. Graz, Austria: ACM.

Betti, A. (2014). On Haslanger's focal analysis of race and gender in resisting reality as an interpretive model. *Krisis, 1,* 13–18.

Betti, A. and van Den Berg, H. (2014). Modelling the history of ideas. *British Journal for the History of Philosophy, 22*(4), 812–835.

Betti, A. and van Den Berg, H. (2016). Towards a computational history of ideas. In L. Wieneke, C. Jones, M. Düring, F. Armaselu, and R. Leboutte (eds.), *Proceedings of the Third Conference on Digital Humanities in Luxembourg with a Special Focus on Reading Historical Sources in the Digital Age.* CEUR Workshop Proceedings, CEUR-WS.Org, Aachen. http://ceur-ws.org/Vol-1681/Betti_van_den_Berg_computational_history_of_ideas.pdf.

Betti, A., Reynaert, M., and van Den Berg, H. (2017). @PhilosTEI: Building corpora for philosophers. In J. Odijk, and A. van Hessen (eds.), *CLARIN in the Low Countries* (pp. 371–384). London: Ubiquity Press. http://dx.doi.org/10.5334/bbi.32.

Boas, F. (1919). In memoriam: Herman Karl Haeberlin. *American Anthropologist, 21*(1), 71–74.

Boskoff, A. (1950). The systematic sociology of Talcott Parsons. *Social Forces, 28*(4), 393–400.

Bradley, P. (2011). Where are the philosophers? - Thoughts from THATCamp pedagogy. *Journal of Digital Humanities, 1*(1), online.

Camic, C. (1991). Introduction: Talcott Parsons before 'The Structure of Social Action'. In T. Parsons (eds.), *The Early Essays* (pp. ix–lxix). Chicago, IL: University of Chicago Press.

Davidson, D. (1973). On the very idea of a conceptual scheme. *Proceedings and Addresses of the American Philosophical Association, 47,* 5.

Ehrenberg, R. G., Jakubson, G., Groen, J., and Price, J. (2007). Inside the black box of doctoral education: What program characteristics influence doctoral students' attrition and graduation probabilities? *Educational Evaluation and Policy Analysis, 29*(2), 134–150.

Fischer, E., Engelhardt, P. E., and Herbelot, A. (2015). Intuitions and illusions: From explanation and experiment to assessment. In E. Fischer, and J. Collins (eds.), *Experimental Philosophy, Rationalism, and Naturalism: Rethinking Philosophical Method* (pp. 259–292). Abingdon: Routledge.

Galison, P. (1997). *Image and Logic: A Material Culture of Microphysics.* Chicago, IL: University of Chicago Press.

Gerhardt, U. (2016). *The Social Thought of Talcott Parsons: Methodology and American Ethos.* Abingdon: Routledge.

Gettys, W. E. (1940). Human ecology and social theory. *Social Forces, 18*(4), 469–76.

Green, C. D., Feinerer, I., and Burman, J. T. (2013). Beyond the schools of psychology 1: A digital analysis of psychological review, 1894–1903: Beyond the schools. *Journal of the History of the Behavioral Sciences, 49*(2), 167–89.

Green, C. D., Feinerer, I., and Burman, J. T. (2015). Searching for the structure of early American psychology: Networking psychological review, 1909–1923. *History of Psychology, 18*(2), 196–204.

Haeberlin, H. K. (1916). The theoretical foundations of Wundt's folk-psychology. *Psychological Review*, *23*(4), 279–302.

Hamilton, P. (1992). *Talcott Parsons: Critical Assessments*. Abingdon: Taylor and Francis.

Harvey, L. (1987). *Myths of the Chicago School of Sociology*. Abingdon: Ashgate Publishing Limited.

Henderson, L. J. (1932). *An Approximate Definition of Fact*. Cambridge, MA: Johnson Reprint Corporation.

Isaac, J. (2012). *Working Knowledge*. Cambridge, MA: Harvard University Press.

Jonassen, C. T. (1951). Some historical and theoretical bases of racism in northwestern Europe. *Social Forces*, *30*(2), 155–161.

de Jong, W. R. and Betti, A. (2010). The classical model of science: A millennia-old model of scientific rationality. *Synthese*, *174*(2), 185–203.

Kilzer, E. (1950). Review: Social theory and social structure toward the codification of theory and research by Robert K. Merton. *The American Catholic Sociological Review*, *11*(1), 39–40.

Kinloch, G. C. (1988). American sociology's changing interests as reflected in two leading journals. *The American Sociologist*, *19*(2), 181–194.

Kuhn, T. S. (1957). *The Copernican Revolution: Planetary Astronomy in the Development of Western Thought*. Cambridge, MA: Harvard University Press.

Kuhn, T. S. (1962). *The Structure of Scientific Revolutions*. Chicago, IL: University of Chicago Press. https://archive.org/details/structureofscien00kuhn.

Lengermann, P. M. (1979). The founding of the American Sociological Review: The anatomy of a rebellion. *American Sociological Review*, *44*(2), 185–198.

Luchins, A. S. (1951). An evaluation of some current criticisms of gestalt psychological work on perception. *Psychological Review*, *58*(2), 69–95.

Merton, R. K. (1948). Discussion of the position of sociological theory, by Talcott Parsons. *American Sociological Review*, *13*(2), 164–68.

de Mey, M. (1982). *The Cognitive Paradigm: Cognitive Science, a Newly Explored Approach to the Study of Cognition Applied in an Analysis of Science and Scientific Knowledge*, Dordrecht, Holland: D. Reidel Pub. Co.

Minsky, M. (1974). *A Framework for Representing Knowledge*. Cambridge, MA: Massachusetts Institute of Technology.

Nett, R. (1952). System building in sociology: A methodological analysis. *Social Forces*, *31*(1), 25–30.

Nettler, G. (1946). The relationship between attitude and information concerning the Japanese in America. *American Sociological Review*, *11*(2), 177–191.

Nock, D. A. (2002). Prophetic versus priestly sociology: The salient case study of Arthur K. Davis. *The American Sociologist*, *33*(2): 57–85.

Overton, J. A. (2013). 'Explain' in scientific discourse. *Synthese*, *190*(8), 1383–1405.

Page, C. H. (1985). *Fifty Years in the Sociological Enterprise: A Lucky Journey*. Amherst MA: University of Massachusetts Press.

Parsons, T. (1936). Review: Max Webers Wissenschafslehre, by Alexander von Schelting. *American Sociological Review, 1*(4), 675–681.

Parsons, T. (1937). Review: Economics and sociology, by Adolf Lowe. *American Journal of Sociology, 43*(3), 477–481.

Parsons, T. (1940). An analytical approach to the theory of social stratification. *American Journal of Sociology, 45*(6), 841–862.

Preston, J. (2008). *Kuhn's 'The Structure of Scientific Revolutions' A Reader's Guide.* London/New York: Continuum.

Quine, W. V. O. (1934). Lectures on Carnap. In R. Creath (ed.). *Dear Carnap, Dear Van: The Quine-Carnap Correspondence and Related Work* (pp. 45–103). Berkeley: University of California Press.

Quine, W. V. O. (1960). *Word and Object,* 3rd edition. Cambridge, MA: MIT Press.

Quine, W. V. O. (1981). On the very idea of a third dogma. In *Theories and Things* (pp. 38–42). Cambridge, MA: Harvard University Press.

Quine, W. V. O. (1992). Interview with W. V. O. Quine by Yasuhiko Tomida: Quine, Rorty, Locke: Essays and discussions on naturalism hildesheim. Zurich and New York: Georg Olms (2007), pp. 127–164.

Rumelhart, D., and Everett E. (1980). Schemata: The building blocks of cognition. In R. J. Spiro (eds.), *Theoretical Issues in Reading Comprehension.* Hillsdale, NJ: Lawrence Erlbaum.

Sangiacomo, A. (2018 [forthcoming]). Modelling the history of early modern natural philosophy: The fate of the art-nature distinction in the dutch universities. *British Journal for the History of Philosophy.*

Timasheff, N. S. (1938). The sociological place of law. *American Journal of Sociology, 44*(2), 206–221.

Timasheff, N. S. (1947). Definitions in the social sciences. *American Journal of Sociology, 53*(3), 201–209.

Vance, R. B. (1952). Is Theory for Demographers? *Social Forces, 31*(1), 9–13.

van Wierst, P., Vrijenhoek, S., Schlobach, S., and Betti, A. (2016). Phil@Scale: Computational methods within philosophy. In L. Wieneke, C. Jones, M. Düring, F. Armaselu, and R. Leboutte (eds.), *Proceedings of the Third Conference on Digital Humanities in Luxembourg with a Special Focus on Reading Historical Sources in the Digital Age,* CEUR Workshop Proceedings, CEUR-WS.Org, Vol. 1681. Aachen.

Index

Ingram Content Group UK Ltd.
Milton Keynes UK
UKHW020216100323
418354UK00005B/108